Applied and Industrial Mathematics

Mathematics and Its Applications

Volume 56

Applied and Industrial Mathematics

Venice – 1, 1989

edited by

Renato Spigler

Department of Mathematics,
University of Padova,
Padova, Italy

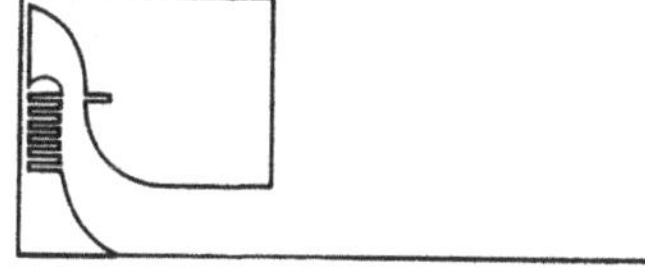

KLUWER ACADEMIC PUBLISHERS
DORDRECHT / BOSTON / LONDON

Library of Congress Cataloging-in-Publication Data

Applied and industrial mathematics : Venice-1, 1989 / edited by Renato Spigler.
p. cm. -- (Mathematics and its applications ; v. 56)
"Papers presented at the 'Venice-1/Symposium on Applied and Industrial Mathematics' held in Venice, Italy, between 2 and 6 October 1989"--Pref.
Includes index.
ISBN 0-7923-0521-3 (HB : acid free paper)
1. Engineering mathematics--Industrial applications--Congresses. 2. Manufacturing processes--Mathematical models--Congresses. I. Spigler, Renato, 1947- . II. Venice-1/Symposium on Applied and Industrial Mathematics (1989 : Venice, Italy) III. Series: Mathematics and its applications (Kluwer Academic Publishers) ; v. 56.
TA331.A63 1990
510--dc20 90-48304

ISBN 0-7923-0521-3

Published by Kluwer Academic Publishers,
P.O. Box 17, 3300 AA Dordrecht, The Netherlands.

Kluwer Academic Publishers incorporates
the publishing programmes of
D. Reidel, Martinus Nijhoff, Dr W. Junk and MTP Press.

Sold and distributed in the U.S.A. and Canada
by Kluwer Academic Publishers,
101 Philip Drive, Norwell, MA 02061, U.S.A.

In all other countries, sold and distributed
by Kluwer Academic Publishers Group,
P.O. Box 322, 3300 AH Dordrecht, The Netherlands.

Printed on acid-free paper

Printed in the Netherlands

SERIES EDITOR'S PREFACE

'Et moi, ..., si j'avait su comment en revenir, je n'y serais point allé.'

Jules Verne

The series is divergent; therefore we may be able to do something with it.

O. Heaviside

One service mathematics has rendered the human race. It has put common sense back where it belongs, on the topmost shelf next to the dusty canister labelled 'discarded non-sense'.

Eric T. Bell

Mathematics is a tool for thought. A highly necessary tool in a world where both feedback and non-linearities abound. Similarly, all kinds of parts of mathematics serve as tools for other parts and for other sciences.

Applying a simple rewriting rule to the quote on the right above one finds such statements as: 'One service topology has rendered mathematical physics ...'; 'One service logic has rendered computer science ...'; 'One service category theory has rendered mathematics ...'. All arguably true. And all statements obtainable this way form part of the raison d'être of this series.

This series, *Mathematics and Its Applications*, started in 1977. Now that over one hundred volumes have appeared it seems opportune to reexamine its scope. At the time I wrote

> "Growing specialization and diversification have brought a host of monographs and textbooks on increasingly specialized topics. However, the 'tree' of knowledge of mathematics and related fields does not grow only by putting forth new branches. It also happens, quite often in fact, that branches which were thought to be completely disparate are suddenly seen to be related. Further, the kind and level of sophistication of mathematics applied in various sciences has changed drastically in recent years: measure theory is used (non-trivially) in regional and theoretical economics; algebraic geometry interacts with physics; the Minkowsky lemma, coding theory and the structure of water meet one another in packing and covering theory; quantum fields, crystal defects and mathematical programming profit from homotopy theory; Lie algebras are relevant to filtering; and prediction and electrical engineering can use Stein spaces. And in addition to this there are such new emerging subdisciplines as 'experimental mathematics', 'CFD', 'completely integrable systems', 'chaos, synergetics and large-scale order', which are almost impossible to fit into the existing classification schemes. They draw upon widely different sections of mathematics."

By and large, all this still applies today. It is still true that at first sight mathematics seems rather fragmented and that to find, see, and exploit the deeper underlying interrelations more effort is needed and so are books that can help mathematicians and scientists do so. Accordingly MIA will continue to try to make such books available.

If anything, the description I gave in 1977 is now an understatement. To the examples of interaction areas one should add string theory where Riemann surfaces, algebraic geometry, modular functions, knots, quantum field theory, Kac-Moody algebras, monstrous moonshine (and more) all come together. And to the examples of things which can be usefully applied let me add the topic 'finite geometry'; a combination of words which sounds like it might not even exist, let alone be applicable. And yet it is being applied: to statistics via designs, to radar/sonar detection arrays (via finite projective planes), and to bus connections of VLSI chips (via difference sets). There seems to be no part of (so-called pure) mathematics that is not in immediate danger of being applied. And, accordingly, the applied mathematician needs to be aware of much more. Besides analysis and numerics, the traditional workhorses, he may need all kinds of combinatorics, algebra, probability, and so on.

In addition, the applied scientist needs to cope increasingly with the nonlinear world and the extra mathematical sophistication that this requires. For that is where the rewards are. Linear models are honest and a bit sad and depressing: proportional efforts and results. It is in the non-linear world that infinitesimal inputs may result in macroscopic outputs (or vice versa). To appreciate what I am hinting at: if electronics were linear we would have no fun with transistors and computers; we would have no TV; in fact you would not be reading these lines.

There is also no safety in ignoring such outlandish things as nonstandard analysis, superspace and anticommuting integration, p-adic and ultrametric space. All three have applications in both electrical engineering and physics. Once, complex numbers were equally outlandish, but they frequently proved the shortest path between 'real' results. Similarly, the first two topics named have already provided a number of 'wormhole' paths. There is no telling where all this is leading - fortunately.

Thus the original scope of the series, which for various (sound) reasons now comprises five sub-series: white (Japan), yellow (China), red (USSR), blue (Eastern Europe), and green (everything else), still applies. It has been enlarged a bit to include books treating of the tools from one subdiscipline which are used in others. Thus the series still aims at books dealing with:

- a central concept which plays an important role in several different mathematical and/or scientific specialization areas;
- new applications of the results and ideas from one area of scientific endeavour into another;
- influences which the results, problems and concepts of one field of enquiry have, and have had, on the development of another.

The present volume contains mainly the initial papers plus a selection of contributed ones of the first large scale conference on industrial applied mathematics in Italy. Together they give the prospective reader quite a good global picture of what is going on in the field at the moment. And that is important, for after a long period of neglect applied and industrial mathematics is definitely enjoying an upwards trend in popularity. Rightly so though there is currently a concurrent danger of neglect of fundamental (pure) mathematics which should yield the 'raw material' in terms of insights, theorems and techniques for a next future wave of applications.

There is a sharp distinction between technical research papers and invited survey and state-of-the-art papers which give one a good feeling for, and an idea of, what is going on and possible in a given field. Both kinds are important but the latter take precedence. Indeed if one has not somehow acquired the relevant global where-things-are-at information, for instance by perusing this book, of what is available and possible in the way of applied industrial mathematics relevant to some concrete problem, one might never find the technical research papers needed to do one's job well. Accordingly, the bulk of this book is given over to the invited papers and it is a pleasure to welcome this authoritative collection in this series.

The shortest path between two truths in the real domain passes through the complex domain.

J. Hadamard

La physique ne nous donne pas seulement l'occasion de résoudre des problèmes ... elle nous fait pressentir la solution.

H. Poincaré

Never lend books, for no one ever returns them; the only books I have in my library are books that other folk have lent me.

Anatole France

The function of an expert is not to be more right than other people, but to be wrong for more sophisticated reasons.

David Butler

Bussum, October 1990

Michiel Hazewinkel

PREFACE

In this book are collected the main (invited) papers along with a strict selection of contributed papers presented at the "Venice-1/Symposium on Applied and Industrial Mathematics" held in Venice, Italy, between 2 and 6 October 1989. The main papers are presented in alphabetical order, while the contributed papers are grouped into small sections according to subject. Needless to say, the latter subdivision is somewhat artificial, since a paper on Applied or Industrial Mathematics often bridges two or more topics. The process itself of selecting such papers was extremely difficult, since so many good papers were delivered at the Symposium. A selection, however, had to be made in order to contain the size of the volume. This task was accomplished with the help of the Scientific Committee. During this process, an effort was made to represent a wide variety of subjects so as to give an idea of the topics that were "hot" around the time when the Symposium was being organized and held. Most papers are largely original.

An assessment of the impact and success of the Symposium is now in order. The meeting was intended as the first general Symposium, on a large scale, on modern Applied and Industrial Mathematics, to be organized and held in Italy. The aim was not only to show the state-of-the-art in the field, but also to promote research in Italy in these areas and to attempt at bridging the still existing large gap between the academic and industrial worlds. Whether and to what extent these goals have been achieved will be seen only in the near future.

About 170 people from a score of countries attended the Symposium and almost 80 papers were presented in 23 sessions, besides the 15 one-hour keynote addresses. The sessions were devoted to Inverse and Ill-Posed

Problems, Nonlinear Fluid-Mechanics, Mathematical Problems in Image Analysis, Numerical Modelling in Meteorology, Wave Propagation in Random Media, Control and Optimization, Dynamical Systems, Probabilistic Methods in Applied Mathematics, Composite Materials and Random Media, Data Assimilation for Geophysical Models, Linear and Nonlinear Waves, Numerical Analysis for Fluid-Flow, Approximation Methods for Partial Differential Equations, Mathematics of Semiconductors, Computational Fluid-Dynamics, Mathematical Modelling in Biology, Chemical Kinetics and Transport Phenomena, and Information and Data Processing.

The Symposium had 37 sponsors and/or patrons who are listed below. Among these, I wish to thank the Italian C.N.R., whose financial support was decisive, and SIAM for encouragement. Finally, I wish to thank KLUWER Academic Publishers for including this book in its series on Mathematics and Its Applications.

Renato Spigler
(Chairman of the Organizing Committee)

Padua and Venice, July 1990

SPONSORS AND PATRONS

- High Patronage of the President of the Italian Republic
- Ministero degli Affari Esteri
- Ministero dell'Industria
- Regione del Veneto
- Provincia di Venezia
- Assessorato alla Cultura di Venezia
- Azienda di Promozione Turistica Venezia (APT)
- Camera di Commercio, Industria, Artigianato ed Agricoltura di Venezia
- Camera di Commercio, Industria, Artigianato ed Agricoltura di Padova
- Cassa di Risparmio di Padova e Rovigo
- Istituto Federale delle Casse di Risparmio delle Venezie
- Istituto di Credito Fondiario delle Venezie
- Istituto Bancario San Paolo di Torino
- Consiglio Nazionale delle Ricerche (CNR)
- Gruppo Nazionale per la Fisica Matematica del CNR
- Society for Industrial and Applied Mathematics (SIAM)
- Università degli Studi di Padova
- Dipartimento di Metodi e Modelli Matematici per le Scienze Applicate, Università di Padova
- Centro Interdipartimentale di Analisi dei Sistemi, Università di Padova
- Consorzio Padova Ricerche
- International Centre for Theoretical Biology (ICTB)
- Istituto Universitario di Architettura di Venezia (IUAV)
- Istituto Veneto di Scienze, Lettere ed Arti
- Accademia Nazionale dei Lincei

- UNESCO
- European Consortium for Mathematics in Industry (ECMI)
- Associazione degli Industriali della Provincia di Venezia
- Associazione degli Industriali della Provincia di Padova
- NECSY Network Control Systems, S.p.A.
- CISE, S.p.A.
- FIDIA, S.p.A.
- Associazione Italiana Prove non Distruttive (AIPnD)
- KLUWER Academic Publishers Group
- ALITALIA
- Palazzo Pisani Moretta
- and with the collaboration of SIP (Società Italiana per l'Esercizio delle Telecomunicazioni)
- and of Ministero delle Poste e Telecomunicazioni

TABLE OF CONTENTS

Invited Papers

PHYSICAL PROBLEMS AND RIGOROUS RESULTS IN KINETIC THEORY

C. CERCIGNANI
Department of Mathematics
Politecnico di Milano
Piazza Leonardo Da Vinci, 32
20133 Milano
Italy

ABSTRACT. After a brief introduction to the applications of kinetic theory (with particular concern for aerospace problems) the kinetic equations introduced by Boltzmann and Enskog are briefly discussed. Then a survey of recent rigorous results in the mathematical theory of these equations is given.

1. Introduction

The idea that gases are formed of elastic molecules rushing hither and thither at large speeds, colliding and rebounding according to the laws of elementary mechanics is, of course, an old idea. The physical development of the kinetic theory of gases was, however, accomplished in the nineteenth century [1,2].

Although the rules generating the dynamics of these systems are easy to describe, the phenomena associated with this dynamics are not so simple, especially when the region where the molecules move is nonconvex or the number of molecules is large. Today the matter is relatively well understood and a rigorous kinetic theory is emerging, as this talk is trying to illustrate. The importance of this matter stems from the need of a rigorous foundation of such a basic physical theory not only for its own sake, but also as a prototype of a mathematical construct central to the theory of non-equilibrium phenomena in large systems.

The typical evolution equation in kinetic theory is the Boltzmann equation which quickly became an important tool for investigating the properties of dilute gases. We cannot deal in detail with the history of the subject, but we cannot refrain from mentioning that in 1912 David Hilbert [3] indicated how to obtain approximate solutions of the Boltzmann equation by a series expansion in a parameter, inversely proportional to the gas density. The paper is also reproduced as Chapter XXII of his treatise entitled *Gründzüge einer allgemeinen Theorie der linearen Integralgleichungen* [4]. The reasons for this are clearly stated in the preface of the book ("Neu hinzugefügt habe ich zum Schluss ein Kapitel über kinetische Gastheorie. [...] erblicke ich in der Gastheorie die gläzendste Anwendung der die Auflösung der Integralgleichungen

R. Spigler (ed.), Applied and Industrial Mathematics, 3–14.

betreffenden Theoreme"). Almost unnoticed, however, the rigorous theory of the Boltzmann equation started much later (1933) with a paper by Carleman [5], who proved a theorem of global existence and uniqueness for a gas of hard spheres in the so-called space homogeneous case.

The Boltzmann equation became a practical tool for the aerospace engineers, when they started to remark that flight in the upper atmosphere must face the problem of a decrease in the ambient density with increasing height. This density reduction would alleviate the aerodynamic forces and heat fluxes that a flying vehicle would have to withstand. However, for virtually all missions, the increase of altitude is accompanied by an increase in speed; thus it is not uncommon for spacecraft to experience its peak heating at considerable altitudes, such as, *e.g.*, 70 km. When the density of a gas decreases, there is, of course, a reduction of the number of molecules in a given volume and, what is more important, an increase in the distance between two subsequent collisions of a given molecule, till one may well question the validity of the Euler and Navier-Stokes equations, which are usually introduced on the basis of a continuum model which does not take into account the molecular nature of a gas. It is to be remarked that the use of those equations can also be based on the kinetic theory of gases , which justifies them as asymptotically useful models when the mean free path is negligible. As remarked above, however, according to kinetic theory the basic description of the evolution of a not-too-dense gas is in terms of a function of position $\underline{x}$, velocity $\underline{v}$, time t, $f=f(\underline{x},\underline{v},t)$, which gives the probability density of finding a molecule at position $\underline{x}$, with velocity $\underline{v}$, at time t. The usual quantities such as density ρ, bulk velocity $\underline{v}$, stresses p_{ij} (including pressure), temperature T, heat flux $\underline{q}$, are obtained as moments of the basic unknown f, through simple formulas that one can find in the textbooks [6,7].

The Boltzmann equation reads as follows:

$$\frac{\partial f}{\partial t} + \underline{v}\cdot\frac{\partial f}{\partial \underline{x}} + \underline{X}\cdot\frac{\partial f}{\partial \underline{v}} = Q(f,f) \qquad (1.1)$$

where

$$Q(f,f)= \iint (f'f'_* - ff_*)B(\underline{v},\underline{v}_*,\underline{u})d\underline{v}_*d\underline{u} \qquad (1.2)$$

Here $B(\underline{v},\underline{v}_*,\underline{u})$ is a kernel associated with the details of the molecular interaction, f'_*, f_* are the same thing as f, except for the fact that the argument $\underline{v}$ is replaced by $\underline{v}'$, $\underline{v}'_*$, $\underline{v}_*$. The latter is an integration variable having the meaning of the velocity of a molecule colliding with the molecule of velocity $\underline{v}$, whose evolution we are following, while $\underline{v}'$ and $\underline{v}'_*$ are the velocities of two molecules entering a collision which will bring them into a pair of molecules with velocities $\underline{v}$ and $\underline{v}_*$. $\underline{u}$ is a unit vector defining the direction of approach of two colliding molecules.

How does one handle the rather complicated Boltzmann equation which is usually accompanied by similarly complicated boundary conditions? Attempts began in the late 1950's and early 1960's. One of the first fields to be explored was that of the "simple flows", such as Couette and Poiseuille flows in tubes and between plates; here it turned out that the equation to be solved is still formidable and various approximation methods where proposed. Some of these were perturbation methods: for large or small Knudsen numbers or about an equilibrium solution (Maxwellian). The first two approaches gave useful results in the limiting regimes, while the third method led to studying the so called Linearized Boltzmann Equation [6,7], which produced predictions which are in a spectacular agreement with experiment and have shed considerable light on the basic structure of rarefied flows, whenever nonlinear effects can be neglected .

This gave confidence to further use of the Boltzmann equation for practical problems and the development of important codes for numerically solving that equation . Other problems which were treated with the linearized equation were the half space problems which are basic in order to understand the structure of kinetic boundary layers.

Another particularly interesting problem is related to the structure of a shock wave; this is not a discontinuity surface as in the theory of compressible Euler equation but a thin layer (having, usually, a thickness of the order of a mean free path).

Another interesting equation appearing in the kinetic theory of gases is the Enskog equation [8], which takes into account the finite volume of the molecules (assumed to be hard spheres of diameter σ). Accordingly the collision operator is written as follows

$$Q(f,f) = \int_{\mathcal{S}_+ \times \mathbb{R}^3} (f' f'_- \kappa_- - f f_+ \kappa_+) B(\underline{v}, \underline{v}_*, \underline{u}) d\underline{v}_* d\underline{u} \qquad (1.3)$$

where $\underline{u}$ varies on the hemisphere $\mathcal{S}_+$ ($\underline{u}$: $|\underline{u}|=1$, $(\underline{v}-\underline{v}_*)\cdot\underline{u}\geq 0$ and

$$B(\underline{v},\underline{v}_*,u) = \sigma^2 \max((v-v_*,u),0) \qquad (1.4)$$

The arguments of f', f'_-, f, f_+ are $(\underline{x},\underline{v}')$, $(\underline{x}-\sigma\underline{u},\underline{v}'_*)$, $(\underline{x},\underline{v})$, $(\underline{x}+\sigma\underline{u},\underline{v}_*)$, where

$$\underline{v}' = \underline{v} - \underline{u}\ (\underline{v}-\underline{v}_*,\underline{u}), \quad \underline{v}'_* = \underline{v}_* + \underline{u}\ (\underline{v}-\underline{v}_*,\underline{u}) \ . \qquad (1.5)$$

The factors $\kappa_\pm$ take into account the fact that the collision frequency is different when the molecules have a finite volume because of the presence of recollisions.

The Enskog equation, although much less studied than the Boltzmann equation is important to describe the phenomena occurring at high densities from the standpoint of kinetic theory and also those occurring when acoustic or electromagnetic waves with wavelength comparable with the size of a molecule travel through a dense gas.

2. The mathematical theory of the Boltzmann equation.

As mentioned in the introduction, the purely mathematical aspects of the Boltzmann equation began to be investigated in the thirties by the Swedish mathematician T. Carleman [5], who provided an existence proof for the purely initial value problem with homogeneous data (*i. e.* data independent of $\underline{x}$). The same problem was revisited by Arkeryd [9] in 1972; he provided solutions in a (weighted) L^1 space, rather than in a (weighted) L^∞ space. Solutions depending on the space variables are much more difficult to handle, if we do not restrict our attention to solutions existing only locally in time but look for solutions existing for an arbitrarily long time interval; the first results were obtained by several Japanese authors [10-12] and referred to solutions close to a (homogeneous) Maxwellian distribution. Then Illner and Shinbrot [13] provided an existence proof for a global solution close to vacuum; their assumptions were later relaxed by Bellomo and Toscani [14], while Toscani [15] has recently considered solutions close to a nonhomogeneous Maxwellian (which, however, must be a solution of the Boltzmann equation).

I proved [16] existence for a very particular case with data arbitrarily far from equilibrium, the so-called affine homoenergetic flows. Arkeryd, Esposito and Pulvirenti [17] proved existence for solutions close to a homogeneous solution (different from a Maxwellian).

One should also mention the important paper of Arkeryd [18] who proved an existence theorem in the context of non-standard analysis (actually the Loeb L^1 solutions are equivalent to standard Young measure solutions).

Quite recently, R. DiPerna and J.P. Lions [19] provided an existence theorem (without uniqueness) for the general case of inhomogeneous data; their proof is quite clever and makes use of a compactness lemma by Perthame, Golse, Sentis and Lions [20] to overcome the difficulties met by other authors. Before DiPerna and Lions [19] provided their ingenious proof of existence, a general existence theorem was available only in the Loeb L^1 frame of nonstandard analysis [18].

It should be realized that any equation similar to the Boltzmann equation but having a little more of compactness in the dependence upon the space variables is rather easy to deal with; this was shown by Morgenstern [21] in the 1950' and Povzner [22] in the 1960'; they introduced mollifying kernels in the collision term of the Boltzmann equation, producing 8-fold and 6-fold integrations, respectively, in place of the original 5-fold integration. In the 1970's W. Greenberg, P. Zweifel and myself [23] indicated that a theorem of existence and uniqueness can be proved if the particles can sit only at discrete positions on a lattice.

3. The mathematical theory of the Enskog equation.

The possibility of using compactness properties is in a slightly better shape for the Enskog equation, especially in the case of data depending on one or two space variables [24,25]. After the DiPerna and Lions result, however, one faced the paradoxical situation that the theory of the Boltzmann equation suddenly gained the upper hand. In order to re-establish a sort of equilibrium Arkeryd and myself have proved existence theorems for the Enskog equation with general L^1 data: thus one can prove wellposedness and regularity for small data globally [26], as well as for large data globally [27] in R^3. Arkeryd's proof [27] requires the initial data to possess finite moments of any order r in the velocity variable, but also delivers uniqueness. In another paper [28], an existence theorem was proved under less stringent requirements on the initial data f_0 but with the (unphysical) assumption, first introduced in Ref. [25], that the collision kernel is symmetrized; in other words, the integral over the direction of the line joining the centers of two colliding particles is extended to the entire unit sphere rather than to a hemisphere. The proof utilized the same basic ingredients as Ref. [19], but the detailed analysis of the collision operator was substituted by the use of an (equivalent) iterated integral form of the equation, giving a considerably shorter proof. The domain was assumed to be a box. At about the same time, J. Polewczak [29] sketched a proof for the unsymmetrized Enskog equation in R^3; he introduced, however, an assumption on the high density factor κ in the Enskog equation, which essentially amounts to having a collision term dominated by a linear operator.

In a recent paper [30] it was shown how to remove the restrictions of Refs. [28-29] and prove a global existence theorem in L^1 for the Enskog equation in a periodic box for the unsymmetrized case with a constant κ. The proof goes through for initial values f_0 with $(1+v^2+|\log f_0|)f_0$ in L^1_+ and can be easily extended to the case of R^3 (with the additional assumption that $x^2f_0\in L^1$).

A second result of the paper concerns the question of the asymptotic equivalence of the Enskog and Boltzmann equation when the diameter σ of the particles tends to zero. This matter was previously investigated in the case of smooth data with small norm [31] for various forms of the Enskog equation and for the case of general data with finite mass, energy and entropy and a symmetrized collision term [28].

In Ref. [30] Eq. (1.3) was considered in a periodic box Λ, which, after rescaling, can be taken to be R^3/Z^3, with initial data $f_0=f(0)$ such that:

$$f_0(1+|\underline{v}|^2+\log f_0) \in L^1_+(\Lambda x R^3) \tag{3.1}$$

The extension to R^3 only requires to add the assumption $|\underline{x}|^2f_0\in L^1$.

The key point of the proof was the use of a modified form of the H-theorem valid for the Enskog gas; the first example of theorem of this kind appears to the best of my knowledge, in a paper by Resibois [32]. In fact the trick was that of introducing a modified H-functional, inspired by Resibois' paper [32, 33]; I repeatedly used this functional in private conversations in the last few years and used a variant of it in the discussion of the validity of the Boltzmann equation for soft spheres [34]. The functional turns out to be the same as a functional used by Polewczak in his paper [29] (in the particular case of a constant κ). It is, however, the further rearrangement that I shall presently explain that makes this functional useful in the existence proof.

Let us start now from the collision operator

$$Q(f) = \sigma^2\kappa\int_{\mathcal{S}_+ x\mathbb{R}^3} (f'f'_- - ff_+)(\underline{v}-\underline{v}_*)\cdot\underline{u}d\underline{v}_*d\underline{u} \tag{3.2}$$

and the definitions of particle density and flow:

$$\rho = \int fd\underline{v} \;\; ; \;\; \underline{j} = \int \underline{v}fd\underline{v} \;. \tag{3.3}$$

Using the elementary inequality $g(\log g-\log h)\ge g-h$, we obtain, for a sufficiently regular f:

$$\int_{\mathbb{R}^3\times\Lambda} Q(f)\log f\,d\underline{v}d\underline{x} \le$$

$$= \frac{1}{2}\sigma^2\kappa\int_{\mathcal{S}_+\times\mathbb{R}^3\times\mathbb{R}^3\times\Lambda} (ff_- - ff_+)(\underline{v}-\underline{v}_*)\cdot\underline{u}\,d\underline{u}d\underline{v}_*d\underline{v}d\underline{x} =$$

$$\frac{1}{2}\sigma^2\kappa\int_{\mathcal{S}_+\times\mathbb{R}^3\times\mathbb{R}^3\times\Lambda} ff_-(\underline{v}-\underline{v}_*)\cdot\underline{u}\,d\underline{u}d\underline{v}_*d\underline{v}d\underline{x} = \frac{1}{2}\kappa\sigma^2\int_{\mathcal{S}\times\Lambda}(\underline{j}\rho_- - \underline{j}_-\rho)\cdot\underline{u}\,d\underline{u}d\underline{x} =$$

$$\frac{1}{2}\kappa\sigma^2\int_{\mathcal{S}\times\Lambda}(\underline{j}_+\rho - \underline{j}_-\rho)\cdot\underline{u}\,d\underline{u}d\underline{x} = \kappa\sigma^2\int_{\mathcal{S}\times\Lambda}\rho\underline{j}_+\cdot\underline{u}\,d\underline{u}d\underline{x} =$$

$$\kappa\sigma^2\int_{\mathcal{B}_\sigma\times\Lambda}\mathrm{div}\hat{\underline{j}}\rho\,d\underline{y}d\underline{x} = -\kappa\sigma^2\int_{\mathcal{B}_\sigma\times\Lambda}\frac{\partial\hat{\rho}}{\partial t}\rho\,d\underline{y}d\underline{x} = -\frac{1}{2}\kappa\sigma^2\frac{d}{dt}\int_{\mathcal{B}_\sigma\times\Lambda}\hat{\rho}\rho\,d\underline{y}d\underline{x}. \quad (3.4)$$

Here $\mathcal{B}_\sigma$ is the ball $|\underline{x}-\underline{y}|\le\sigma$ and a caret (^) denotes that the argument is $\underline{y}$ rather than $\underline{x}$.

Hence $H=H_B+\frac{1}{2}\kappa\sigma^2\int_{B_\sigma\times\Lambda}\hat{\rho}\rho\,d\underline{y}d\underline{x}$, if H_B is the usual H functional, is a decreasing quantity. The additional term is not larger than $\frac{1}{2}\kappa\sigma^2||f||^2_{L^1}$ and non-negative, hence *a priori* bounded from above and below. This implies that H_B is *a priori* bounded in terms of the initial data at any time, if f is regular enough and Eq. (3.1) holds.

We further remark that

$$\frac{dH}{dt} = -\frac{\kappa}{2}\int_{\mathcal{S}_+\times\mathbb{R}^3\times\mathbb{R}^3\times\Lambda} f'f'_-\,\ell\left(\frac{f'f'_-}{ff_-}\right)B\,dudv_*dvdx \le 0 \quad (3.5)$$

where

$$\ell(x) = \log(x) + \frac{1-x}{x} \ge 0. \quad (3.6)$$

Please note that $\ell(x)$ is decreasing for x<1 and increasing for x>1. The form (3.5) of the entropy relation does not only provide information on the decrease of H and hence an *a priori* inequality on flogf, but also an inequality on the right hand side integrated with respect to the time variable. This allows to control the gain part of the collision term in terms of the loss part. This type of estimate was first used by Arkeryd for the symmetrized Enskog equation [25] and made popular by DiPerna and Lions [19], who used it in their celebrated existence proof for the Boltzmann equation. In both cases the required inequality is available since 1872 [2], but only recently I discovered Eq. (3.5).

In fact f satisfies, according to the previous discussion :

$$0 \le \int_0^\infty ds \int_M f'f'_- \,\ell\left(\frac{(f f_-)'}{f f_-}\right) B \, d\mu < C < \infty, \qquad (3.7)$$

where $M = \Lambda \times R^3 \times R^3 \times \mathcal{S}$, $d\mu = d\underline{x} d\underline{v} d\underline{v}_* d\underline{u}$. and the function $\ell(x)$ was defined in Eq. (3.6). The above-mentioned comparison of the gain and loss terms reads as follows:

$$\int_{O\times\mathcal{S}_+} f'f'_- \, B \, d\mu ds \le j \int_{O\times\mathcal{S}_+\cap\Omega_{jn}} f f_- \, B \, d\mu ds +$$

$$+ \frac{1}{\ell(j)} \int_{O\times\mathcal{S}_+\cap\Omega_{jn}} f'f'_- \,\ell\left(\frac{f'f'_-}{f f_-}\right) B \, d\mu ds$$

$$\le j \int_{O\times\mathcal{S}_+\cap\Omega_{jn}} f f_- B d\mu ds + \frac{C}{\ell(j)} \qquad (3.8)$$

where O is a measurable set in $\Lambda \times R^3 \times R^3 \times R_+$, $j>1$, and

$$\Omega_{jn} = ((x,v,v_*,u,s):\ f^{n'} f_+^{n'} \le j f^n f_+^n\ ) \qquad (3.9)$$

$$\Omega'_{jn} = ((x,v,v_*,u,s):\ f^{n'} f_+^{n'} \ge j f^n f_+^n\ ) \qquad (3.10)$$

In order to construct the desired solutions of Eq. (1.3) a well adapted approximation scheme is needed, which retains the essential structure of the H-theorem for the EE which we have just discussed. This inequality plays the same role as the usual H-theorem in DiPerna and Lions' proof [19], but, because of the fact that a part of the H-functional arises from the right hand side one has to be particularly careful when constructing approximating sequences,

In addition one has to use various forms of solution: the renormalized, mild and exponential multiplier forms as defined in, e.g., Ref. [19] together with the following iterated integral form introduced in Ref. [25]:

DEFINITION: f satisfies the EE in <u>iterated integral form</u> , if

$$Q^{\pm}(f)^{\#}(x,v,.) \in L^1_{loc}(R_+) \quad \text{for a. a. } (x,v) \in \Lambda \times R^3 \quad (3.7)$$

and

$$\int_{\Lambda x R^3} f^{\#}(t)\psi(t)dxdv = \int_{\Lambda x R^3} f_0\psi(0)\, dxdv + \int_0^t \int_{\Lambda x R^3} f^{\#}(s)\, \partial_s\psi\, dxdvds +$$

$$+ \int_{\Lambda x R^3} \left(\int_0^t \psi(s)Q(f)^{\#}(s)ds \right) dxdv, \quad t>0, \ \psi \in CL. \qquad (3.8)$$

Here CL is the linear space of all functions ψ in $C^1(R_+;L^{\infty}(\Lambda x R^3))$ with bounded support and with $\psi(x,v,.) \in C^1(R_+)$ for a. a. $(x,v) \in \Lambda x R^3$. The last integral is an iterated integral. It is not required that $|\psi Q(f)^{\#}| \in L^1(\Lambda x R^3 x(0,t))$, only that $\int_0^t \psi(s)Q(f)^{\#}(s)ds \in L^1(\Lambda x R^3)$. Finally

$$Q^{+}(f) = \int_{\mathcal{S}_+ x\mathbb{R}^3} f'f'_{-}\kappa B(v,v_*,u)dv_*du \qquad (3.9)$$

$$Q^{-}(f) = \int_{S_+ x\mathbb{R}^3} ff_{+}\kappa B(v,v_*,u)dv_*du \qquad (3.10)$$

One can prove [28] the following

LEMMA 1. The above four solution forms are equivalent for the EE, if

$$\frac{Q^{\pm}(f)}{1+f}, \quad \int_{S_+ x\mathbb{R}^3} f_+ Bdv_*du \in L^1_{loc}(\Lambda x R^3 x R_+). \qquad (3.11)$$

The first step is to study the well-posedness of the initial value problem for an equation approximating Eq. (1.3) in $\Lambda x\mathbb{R}^3$ by means of estimates of the type introduced in Ref. [27]. Then we can now proceed to a discussion of the proof of an existence theorem for the EE (1.3) by the type of argument given in Ref. [28] for the symmetrized Enskog equation. The actual proof becomes more technical in the present case due to two consecutive approximations with different and more involved entropy estimates.

Given f_0 satisfying Eq. (3.1) we obtain an approximating sequence (3.1) with an initial value vanishing at large speeds. This sequence $(f^n)_{n\in N}$ satisfies mass, energy and entropy inequalities. Applying these and other estimates to the mild form of (1.3) gives that $(f^n)_{n\in N}$ is uniformly equicontinuous from $[0,T]$ to $L^1(\Lambda x R^3)$. Hence there is a subsequence $(f^{n'})$ converging weakly in $L^1(\Lambda x R^3 x[0,T])$ as well as in $L^1(\Lambda x R^3)$ for $0\le t\le T$ to a function $f\in C([0,T],L^1_+(\Lambda x R^3))$. Various subsequences of (f^n) will also be denoted by $(f^{n'})$ in the sequel without any further comment.

The averaging technique of Golse *et al.* [20] is used in the form stated in Ref. [19]:

LEMMA 2 . Let (E,μ) be an arbitrary measure space, and let $\psi \in L^{\infty}(\Lambda x R^3 x[0,T]; L^1(E))$.

i) If g^n and G^n belong to a weakly compact set in $L^1(K)$ for any compact set K in $\Lambda x R^3 x(0,T)$, and $(\partial_t + v\partial_x)g^n = G^n$ in distribution sense, $n\in N$, then $\int_{R^3} g^n\psi dv$ belongs to a compact set in $L^1(\Lambda x(0,T)xE)$ for any compact set K in $\Lambda x R^3 x(0,T)$, provided that supp $\psi \subset KxE$.

ii) If, in addition g^n belongs to a weakly compact set in $L^1(\Lambda x R^3 x[0,T])$, $n\in N$, then $\int_{R^3} g^n\psi dv$ belongs to a compact set in $L^1(\Lambda x[0,T]xE)$.

We shall apply this lemma to the renormalized, approximated Enskog equation

$$(\partial_t + v\partial_x)g_\delta = Q^n_\delta(g) \qquad (3.12)$$

with

$$g_\delta = \delta^{-1}\log(1+\delta g), \quad Q^n_\delta(g) = Q^n(g)/(1+\delta g) \qquad (3.13)$$

Eq. (3.12) is satisfied in the distribution sense by $g = f^n$ and, in order to apply the lemma, f^n_δ and Q^n_δ should satisfy the conditions on g^n and G^n in Lemma 2. But f^n_δ, $n\in N$, belongs to a weakly compact set in $L^1(\Lambda x\mathbf{R}^3 x[0,T])$ since $0\le f^n_\delta \le f^n$, and, as discussed above, $\{f^n\}_{n\in\mathbf{N}}$ is weakly pre-compact in L^1 . In addition

$$\{f^n f^n_+ \; \chi(x,v)|v-v_*|/(1+\delta f^n)\}_{n\in\mathbf{N}}$$

is weakly precompact in $L^1(Mx[0,T])$ for any characteristic function χ of a measurable set of bounded support in $\Lambda x R^3$.

It follows that for $\delta>0$, $\{Q^{n-}_\delta(f^n)\}_{n\in N}$ is a weakly precompact subset of $L^1(K)$ for any compact subset K of $\Lambda x R^3 x[0,T]$. This, together with an estimate following from Eq. (3.5) implies the same weak L^1-precompactness for $\{Q^{n+}_\delta(f^n)\}_{n\in N}$, when $\delta>0$. So we have the relevant compactness properties of $\{f^n_\delta\}_{n\in N}$ and $\{Q^n_\delta(f^n)\}_{n\in N}$ for an application of Lemma 2.

With rather lengthy proofs one can arrive at the following

THEOREM 1: Let f_0 satisfy (2.5). Then there exists a function $f\in C([0,\infty), L^1(\Lambda x R^3))$, satisfying the EE (1.3) (in any of the four equivalent forms of Lemma 1) with initial value f_0. For t>0 mass and energy are bounded by their initial values and the entropy satisfies

$$\int_{\Lambda x\mathbb{R}^3} f(t)\log f(t)dxdv \le \int_{\Lambda x\mathbb{R}^3} f_0(\log f_0 + v^2)dxdv + \frac{\kappa\sigma^2}{2}\left(\int_{\Lambda x\mathbb{R}} f_0 dxdv\right)^2 \qquad (3.14)$$

The type of arguments used to prove the existence theorem imply that the solutions of the EE provided by Theorem 1 converge to solutions of the BE when the diameter σ tends to zero and we assume that, with a suitable rescaling, κ and B remain constant in the process. Following again the line of argument of Ref. [28] we can prove the following

THEOREM 2. For any sequence $\{\sigma_j\}_{j\in N}$ of reals with $\lim_{j\to\infty} \sigma_j = 0$, and any corresponding sequence of solutions $\{f_{\sigma_j}\}_{j\in N}$ of Theorem 1, there is a subsequence $\{\sigma_{j'}\}$ for which $\{f_{\sigma_{j'}}\}$ converges weakly in L^1 to a solution f of the BE for elastic spheres.

4. **Boundary value problems.**

While the initial value problem for the Boltzmann equation has received a great deal of attention, comparatively less work has been done for the steady problems, which, after all, are those of paramount interest for the space engineer. These problems, in a linearized form, were satisfactorily dealt with in the 1960's, with the exception of the important half-space problems that were completely treated only in the last few years. An early result for the nonlinear problem in a slab with data close to equilibrium was obtained by Pao [35] in 1967, by a suitable use of a previous result of mine [36] dealing with the corresponding linearized problem. Much later Ukai and Asano [37] were able to treat the small Mach number flow past a solid body.

The case of data arbitrarily removed from equilibrium was not considered till R. Illner, the late M. Shinbrot and myself [38] wrote a paper on the slab problem; for technical reasons we took the molecular velocities to be discrete and obtained existence for arbitrarily large data and domains. The treatment has been extended to the case of a rectangle [39] (but only for a very particular discrete velocity model) and to a half space (in collaboration with M. Pulvirenti [40]).

From these short remarks it will be clear that there is still a lot of work to be done on the boundary value problems in kinetic theory. In particular there is nothing (not even a proper formulation) on the boundary value problem for the Enskog equation.

REFERENCES

[1] Maxwell, J. C. (1867) "On the Dynamical Theory of Gases", *Philosophical Transactions of the Royal Society of London* 157, 49-88.

[2] Boltzmann, L. (1872) "Weitere Studien über das Wärmegleichgewicht unter Gasmolekülen", *Sitzungsberichte Akad. Wiss.,* Vienna, part II, 66, 275-370

[3] Hilbert D. (1912) "Begründung der kinetischen Gastheorie", *Mathematische Annalen* 72, 562-577.

[4] Hilbert D. (1912) Chapter XXII of *Gründzüge einer allgemeinen Theorie der linearen Integralgleichungen*, Leipzig.

[5] Carleman, T. (1933) "Sur la théorie de l'équation intégrodifferentielle de Boltzmann", *Acta Math.*, 60, 91-146.

[6] Cercignani, C. (1988) *The Boltzmann equation and its applications,*

Springer, New York.
[7] Cercignani, C. (1969) *Mathematical Methods in Kinetic Theory,* Plenum Press, New York.
[8] Enskog D. (1922) "Kinetische Theorie der Wärmeleitun, Reibung und Selbstdiffusion in gewissen verdichteten Gasen und Flûssigkeiten", *Kungl. Svenska Vetenskaps Akademiens Handl.* **63,** No 4.
[9] Arkeryd L. (1972) "On the Boltzmann equation. Part I: Existence", *Arch. Rat. Mech. Anal.* **45,** 1-16 and "On the Boltzmann equation. Part II: The full initial value problem", *Arch. Rat. Mech. Anal.* **45,** 17-34.
[10] Ukai, S. (1974) "On the existence of global solutions of mixed problem for non-linear Boltzmann equation", *Proceedings of the Japan Academy* 50, 179-184.
[11] Nishida, T. and Imai, K. (1977) "Global solutions to the initial value problem for the nonlinear Boltzmann equation", *Publications of the Research Institute for Mathematical Sciences, Kyoto University* 12, 229-239.
[12] Shizuta, Y. and Asano, K. (1974) "Global solutions of the Boltzmann equation in a bounded convex domain", *Proceedings of the Japan Academy* 53, 3-5.
[13] Illner, R. and Shinbrot, M. (1984) "The Boltzmann equation: Global existence for rare gas in an infinite vacuum" *Comm. Math. Phys.*, 95, 217-226.
[14] Bellomo, N. and Toscani, G. (1985) "On the Cauchy problem for the nonlinear Boltzmann equation: global existence, uniqueness and asymptotic stability", *J. Math. Phys.*, **26**, 334 .
[15] Toscani, G. (1989) "Global solution of the initial value problem for the Boltzmann equation near a local Maxwellian" *Arch. Rat. Mech. Anal.* (to appear).
[16] Cercignani, C. (1989) "Existence of Homoenergetic Affine Flows for the Boltzmann Equation", *Arch. Rat. Mech. Anal.* 105, 377-387.
[17] Arkeryd, L., Esposito, R. and Pulvirenti, M. (1988) "The Boltzmann equation for weakly inhomogeneous data" *Commun. Math. Phys.*, 111,393.
[18] Arkeryd, L. (1984) "Loeb solutions of the Boltzmann equation", *Arch. Rat. Mech. Anal.* **86**, 85-97 (1984).
[19] DiPerna, R. and Lions, P. L. (1989) "On the Cauchy problem for Boltzmann equations", *Ann. of Math.*, to appear.
[20] Golse,F. , Lions, P. L., Perthame, B., Sentis, R. (1988) "Regularity of the moments of the solution of a transport equation " *J. Funct. Anal.* **76**, 110-125 .
[21] Morgenstern, D. (1955) "Analytical studies related to Maxwell-Boltzmann equation", *J. Rational Mech. Anal.*, **4**, 533-565.
[22] Povzner, A. Ya. (1962) "The Boltzmann equation in the kinetic theory of gases", *Mat. Sbornik*, **58**, 65-86.
[23] Cercignani, C., Greenberg, W. and Zweifel, P. (1979) "Global solutions of the Boltzmann equation on a lattice", *J. Stat. Phys.*, **20**, 449 (1979).
[24] Cercignani, C. (1987) "Existence of Global Solutions for the Space Inhomogeneous Enskog Equation", *Transport Theory Stat. Phys.* 16, 213-221 .
[25] Arkeryd, L. (1986) "On the Enskog Equation in two space variables", *Transport Theory and Stat. Phys.* 15, 673-691.
[26] Cercignani, C. (1988) "Small Data Existence for the Enskog Equation in L^1", *J. Stat. Phys.* **51**, 291-297.

[27] Arkeryd, L. (1989) "On the Enskog Equation with large initial data", to appear in *SIAM Journ. on Math. Anal.*.

[28] Arkeryd, L. and Cercignani, C. (1989) "On the Convergence of Solutions of the Enskog Equation to Solutions of the Boltzmann Equation", *Comm. in PDE* **14**, 1071-1090.

[29] Polewczak, J. (1988) "Global existence in L^1 for the Modified Nonlinear Enskog Equation in R^3", Jour. Stat. Phys. **56**, 159 .

[30] Arkeryd, L. and Cercignani, C. (1989) "Global existence in L^1 for the Enskog equation and convergence of the solutions to solutions of the Boltzmann equation" submitted to *J. Stat. Phys.*

[31] Bellomo, N. and Lachowitz, M. (1988) "On the Asymptotic Equivalence between the Enskog and the Boltzmann Equations", *J. Stat. Phys.* **51**, 233-247.

[32] Résibois, P. (1978) "H-theorem for the (Modified) Nonlinear Enskog Equation", *J. Stat. Phys.* **19**, 593-609.

[33] Cercignani, C. and Lampis, M. (1988) "On the Kinetic Theory of a Dense Gas of Rough Spheres", *J. Stat. Phys.* **53**, 655-672.

[34] Cercignani, C. (1983) "The Grad Limit for a System of Soft Spheres", *Comm. Pure Appl. Math.* **36**, 479-494.

[35] Pao, Y. P. (1968) "Boundary-value problems for the linearized and weakly nonlinear Boltzmann equation", *J. Math. Phys.* **9**, 1893-1898 .

[36] Cercignani, C. (1967) "Existence and uniqueness for nonlinear boundary value problems in kinetic theory" *J. Math. Phys.* **8**, 1653-1656 (1967).

[37] Ukai, S. and Asano, K. (1983) "Steady solutions of the Boltzmann equation", *Arch. Rational Mech. Anal.* **84**, 249.

[38] Cercignani, C., Illner, R. and Shinbrot, M. (1987) "A boundary value problem for discrete-velocity models", *Duke Math. Journal* **55**, 889-900 .

[39] Cercignani, C., Illner, R. and Shinbrot, M. (1988) "A boundary-value problem for the two-dimensional Broadwell model", *Comm. Math. Phys.* , **114**, 687 - 698.

[40] Cercignani, C., Illner, R., Pulvirenti, M. and Shinbrot, M. (1988) "On nonlinear stationary half-space problems in discrete kinetic theory", *J. Stat. Phys.*, **52**, 885-896.

STATISTICAL MECHANICS OF VORTEX FILAMENTS

A. CHORIN
University of California, Berkeley

Turbulent flow can be idealized as a collection of filiform vortex filaments. In an appropriate range of scales, these filaments can be viewed as being in thermal equilibrium. That equilibrium can be analyzed by a variety of tools, including renormalization, simulation on a lattice, pivot-like algorithms, and Monte-Carlo methods. The analysis reveals the mechanics by which integral constraints (for example, conservation of helicity and of energy) create correlations and scaling behavior in physical space, and provides reasonable estimates of characteristic exponents and of the values of the parameters that describe the higher statistics of the flow.

The simplest of the analyses consists of creating on a computer a collection of self-avoiding, energy bounded lattice random walks and calculating the correlations directly. The walks can be created by a pivot algorithm modified by a Metropolis exclusion rule. The resulting calculation is similar in several respects to calculations relating to the statistics of polymers.

R. Spigler (ed.), Applied and Industrial Mathematics, 15.

THE HAMILTONIAN WAY FOR COMPUTING HAMILTONIAN DYNAMICS

FENG KANG
ACADEMIA SINICA COMPUTING CENTER
P.O.BOX 2719
BEIJING 100080
CHINA

ABSTRACT. We present a survey of a recent comprehensive study on the numerical methods for Hamiltonian systems based on symplectic geometry, together the motivations for the research, justification for the symplectic approach adopted, some of the main results, their ramifications and their implications [1-9].

1. BACKGROUNDS AND MOTIVATIONS

Hamilton originated a mathematical formalism for the foundation of geometrical optics in 1820's and several years later he applied it to a quite different field, i.e. classical mechanics, as an alternative and mathematically equivalent formalism to the then well-founded Newtonian and Lagrangian ones. Hamilton had moderate expectations about his own contributions for still wider applications and for a possible formation of a new discipline [10]. However, the acceptance of Hamiltonian formalism had been slow and sceptical; it was considered generally as "beautiful but useless" [11]. This attitude was best summarized by Klein's recognition of the mathematical elegance of the theory but with an additional remark of caution: "Physicists can make use of these theories only very little, and engineers nothing at all" [12].

This "non-Kleinian" remark of Klein was refuted definitively, at least so far as physicists are concerned, by the founding of quantum mechanics, whose formalism is Hamiltonian. It is by now almost beyond dispute that, all real physical processes with negligible dissipation can always be cast in suitable Hamiltonian form; and, as Schroedinger has put it, "Hamiltonian principle has become the cornerstone of modern physics,... If you wish to apply the modern theory to any particular problem, you must start with putting in Hamiltonian form" [13].

At present, Hamiltonian formalism lay at the basis of many diverse fields of physical and engineering sciences such as classical mechanics, satellite orbits, rigid bodies and robotic motions, geometrical optics, plasma confinement, accelerator design, optimal control, WKB and ray asymptotics, fluid dynamics, elasticity, electrodynamics, non-linear waves, solitons, quantum mechanics, relativity, etc. So, a systematic research and development of numerical methodology for Hamiltonian systems is well motivated. If it turns out to be successful, it would imply wide-ranging applications and give additional evidences for the refutation of Klein's remark by showing that Hamiltonian formalism is not only beautiful but also useful, and, not only useful for theoretical analysis but also useful for practical computations.

After a search into the cumulated wealth of existing literature on numerical methods for differential equations, we were surprised by the puzzling fact that, the pertinent works related specifically to Hamiltonian differential equations are virtually, if not absolutely, void. This gives us added incentive to make an inquiry into this grossly neglected but fertile and promising field.

2. THE SYMPLECTIC APPROACH

After the initiation of Hamilton, developed further by Jacobi, the most significant progress of the formalism was its geometrization, started since Poincare, resulted in so-called *symplectic geometry*, a new discipline as anticipated by Hamilton, which has become the natural language for Hamiltonian dynamics. Symplectic geometry has the merit to make the deep-lying and intricate properties of dynamics explicit, transparent and easy to grasp. Hamiltonian systems are simply canonical systems on symplectic manifold, i.e. phase space endowed with symplectic structures, the dynamical

R. Spigler (ed.), Applied and Industrial Mathematics, 17–35.

evolutions are simply symplectic transformations. see e.g., [14]. The desirable algorithms should preserve as much as possible the relevant symplectic properties of the original system. The best way to achieve this aim is to work out the analog within the same framework of the original. So we conceive, construct, analyse and assess numerical algorithms exclusively from within the framework of symplectic geometry. This judiciously chosen approach, inspired by the author's previous experience with the development of the finite element method for solving elliptic equations, is quite natural and justified in view of the innate relationships between Hamiltonian dynamics and symplectic geometry. The approach turns out to be fruitful and successful, and leads to the effective construction as well as the theoretical understanding of an abundance of what we call *symplectic difference schemes*, or symplectic algorithms, or simply *Hamiltonian algorithms*, since they present the *proper* way, i.e., *the Hamiltonian way* for computing *Hamiltonian dynamics*.

Symplectic geometry was developed, due to historical circumstance, mostly for the purposes of theoretical analysis; it lacks the computational component; for example, there is no such theory as that of symplectic approximations for symplectic operators. So, our program might be considered also as an attempt to fill this blank. Generating functions play a key role in the classical transformation theory in symplectic geometry, but they are rather of limited scope in the present setting. So we worked out an extended theory of generating functions in the context of linear Darboux transformations [2,4,6]. This theory leads to our general methodology for the construction and analysis of symplectic schemes. It seems to have its own independent interest in symplectic geometry too.

3. Some Preliminaries

In real linear space R^{2n}, the vectors will be represented by column matrices $z = (z_1, ..., z_{2n})'$, the prime $'$ denotes matrix transpose. The *standard symplectic form* in R^{2n} is given by the anti-symmetric, non-degenerate bilinear form

$$\begin{gathered} \omega(x,y) = x'Jy, \quad \forall x,y \in R^{2n}, \\ J = J_{2n} = \begin{pmatrix} 0 & -I \\ I & 0 \end{pmatrix}, \quad J' = -J = J^{-1}. \end{gathered} \tag{1}$$

The symplectic group $Sp(2n)$ consists of the real *symplectic* matrices S preserving ω, i.e. satisfying $S'JS = J$. The corresponding Lie algebra $sp(2n)$ consists of real *infinitesimally symplectic* matrices L satisfying $L'J + JL = 0$, or equivalently $L = JA$, for some symmetric matrix A. The *symplectic product* for two symmetric matrices A, B is defined as $\{A, B\} = AJB - BJA$. Then the space of real *symmetric* matrices $sm(2n)$, under the symplectic product, forms a Lie algebra which is isomorphic to $sp(2n)$ under the usual commutator bracket.

A diffeomorphism g of R^{2n} onto itself is called a *symplectic* operator (or,in classical language, *canonical* transrformation) if its Jacobian matrix g_z is everywhere symplectic. A differential map of R^{2n} into itself is called an *infinitesimally symplectic* operator (or, transformation) if its Jacobian matrix is everywhere infinitesimally symplectic. A differential map f of R^{2n} into itself is called a *symmetric* operator (or, transformation) if its Jacobian matrix is everywhere symmetric; in this case, there exists, at least locally, a real function ϕ of R^{2n}, unique up to an additive constant, such that $f = \nabla\phi$, where $\nabla = (\partial/\partial z_1, ..., \partial/\partial z_{2n})'$. The space of smooth real functions forms a Lie algebra (of infinite dimensions) $C^\infty(R^{2n})$ under the *Poisson bracket* defined by

$$\{\phi, \psi\} = (\nabla\phi)'J(\nabla\psi). \tag{2}$$

An operator S is symplectic iff it preserves the Poisson bracket, i.e., $\{\phi \circ S, \psi \circ S\} = \{\phi, \psi\} \circ S$. The totality of quadratic forms $\phi_A(z) = (\frac{1}{2})z'Az, A' = A$ forms a Lie subalgebra of $C^\infty(R^{2n})$ under Poisson bracket, which is isomorphic to the Lie algebra $sm(2n)$ under the symplectic product.

Choose a smooth function $H \in C^\infty(R^{2n})$, it defines a *Hamiltonian canonical system*

$$\frac{dz}{dt} = J\nabla H(z) \tag{3}$$

on the symplectic space, i.e. phase space R^{2n}, endowed with the symplectic form $\omega(x,y) = x'Jy$. H is called the *Hamiltonian* or *energy* function defining the system. It is customary to split the $z \in R^{2n}$ into two parts $z = (p,q)$, $p_k = z_k$, $q_k = z_{n+k}$, $k = 1,\cdots,n$. In classical language p, q are called position and momentum vectors conjugate to each other. Then $H(z) = H(p,q)$, and equation (3) can be written as

$$\begin{aligned} \frac{dp}{dt} &= -H_q(p,q) \\ \frac{dq}{dt} &= H_p(p,q) \end{aligned} \tag{4}$$

The *fundamental theorem on Hamiltonian formalism* says that the solution $z(t)$ of the canonical system (3) of energy H with initial value $z(0)$ can be generated by a *one-parameter group* $g^t = g_H^t$, called the *phase flow* of H, of symplectic operators of R^{2n} (locally in t) such that

$$z(t) = g_H^t z(0)$$

The system (3) can be written as

$$\frac{d}{dt}(g_H^t z) = J(\nabla H) \circ g_H^t z, \quad \forall z \in R^{2n}$$

The sympleticity of the phase flow implies the class of *conservation laws of phase area* of even dimensions $2m$, $m = 1,\cdots,n$) for the Hamiltonian system (3), the case $m = n$ is the Liouville's conservation law.

Moreover, the Hamiltonian system possesses another class of *conservation laws* related to the *energy* $H(z)$. Consider any smooth function $F(z)$ in R^{2n}, then the composite function $F \circ g_H^t(z)$ satisfies the equation

$$\frac{d}{dt} F \circ (g_H^t z) = \{F, H\} \circ g_H^t z$$

$F(z)$ is called an *invariant* function under the system (3) with Hamiltonian H if $F = F \circ g_H^t$ for all t, this is equivalent to the condition $\{F, H\} = 0$. The energy H itself is always an invariant function.

4. Simple symplectic difference schemes

Consider now the difference schemes for Hamiltonian system (3), restricted mainly to the case of single step (i.e. 2-level) schemes. Time t is discretized into $t = 0, \pm\tau, \pm 2\tau, \cdots$; τ is the step size, $z(k\tau) \approx z^k$. Each 2-level-scheme is characterized by a *transition* operator relating the old and new states by

$$\hat{z} = G^\tau z, \quad z = z^k, \quad \hat{z} = z^{k+1} \tag{5}$$

$G^\tau = G_H^\tau$ depends on τ, H and the mode of discretization. From the standard-point of symplectic geometry, it is natural and even mandatory to require G_H^τ to be symplectic. A difference scheme for Hamiltonian system (3) is called *symplectic* (or, *canonical*) if its transition operator G_H^τ is symplectic for all Hamiltonian functions H and step sizes τ. We shall also consider difference schemes non-symplectic in this general sense, but are symplectic for special classes of Hamiltonians. The true solution relating the old and new states is

$$z((k+1)\tau) = g_H^t z(k\tau),$$

So we are confronted with the problem of symplectic approximation to the symplectic phase flow g_H^τ. For the latter there is a Lie expansion in τ

$$(g_H^\tau z)_i = z_i + \tau\{z_i, H\} + \frac{\tau^2}{2!}\{\{z_i, H\},\} + \cdots, \quad i = 1,\cdots,2n. \tag{6}$$

Although the truncations give "legitimate" approximations, but they are non-symplectic in general, so undesirable.

Consider the simplest one-legged weighted Euler scheme for (3)

$$\hat{z} = z + \tau J(\nabla H)(c\hat{z} + (1-c)z) \tag{7}$$

where c is a real constant. It is easy to show that it is symplectic iff $c = \frac{1}{2}$. The only symplectic case corresponds to the *time-centered Euler scheme*

$$\hat{z} = z + \tau J(\nabla H)(\frac{\hat{z}+z}{2}) \tag{8}$$

This simple proposition illustrate a general situation: apart from some very rare exceptions, the vast majority of conventional schemes are non-symplectic. However, if we allow c in (7) to be a *real matrix of order* $2n$, we get a far-reaching generalization: (7) *is symplectic* iff

$$c = \frac{1}{2}(I_{2n} + J_{2n}B), \quad B' = B \tag{9}$$

The symplectic case corresponds to

$$\hat{z} = z + \tau J(\nabla H)(\frac{1}{2}(I+JB)\hat{z} + \frac{1}{2}(I-JB)z), \quad B' = B, \tag{10}$$

They are all implicit, (10) is solvable for new $\hat{z}$ in terms of old z whenever $|\tau|$ is small enough. (10) enjoy many properties in common with (3), they may be considered as *Hamiltonian difference equations.*

Let the transition in (10) to be

$$z^{k+1} = G^{\tau}_{H,B} z^k, \tag{11}$$

for $B = 0$, the accuracy order is 2, For $B \neq 0$, the order is 1; however, the alternating composite schemes given by

$$z^{k+1} = G^{\tau}_{H,\pm B} z^k, \tag{12}$$

where

$$\begin{aligned} G^{\tau}_{H,\pm B} &= G^{\frac{\tau}{2}}_{H,B} \circ G^{\frac{\tau}{2}}_{H,-B} \\ G^{\tau}_{H,\mp B} &= G^{\frac{\tau}{2}}_{H,-B} \circ G^{\frac{\tau}{2}}_{H,B} \end{aligned} \tag{13}$$

have accuracy order 2,

So we get a great variety of simple symplectic schemes of order 1 or 2, classified according to type matrices $B \in sm(2n)$, which is a linear space of dimension $2n^2 + n$.

The conservation properties of the above symplectic schemes are well-understood. In fact we have proved:

Let $\phi_A(z) = \frac{1}{2}z'Az$, $A' = A$ be a quadratic form and A commutes symplectically with the type matrix B, i.e. $AJB = BJA$. Then ϕ_A is invariant under the system with Hamiltonian H iff ϕ_A is invariant under symplectic difference schemes (10) of types $G^{\tau}_{H,B}$, $G^{\tau}_{H,-B}$.

We list some of the most important types B together with the corresponding form of symplectic matrices A of the conserved quadratic invariants ϕ_A:

$B = 0$, $\quad A$ arbitrary.

$B = \pm\begin{pmatrix} 0 & I_n \\ I_n & 0 \end{pmatrix} = \pm E_{2n}$, $\quad A = \begin{pmatrix} 0 & b \\ b' & 0 \end{pmatrix}$, $\quad b$ arbitrary; angular momentum type.

$B = \pm I_{2n}$, $\quad A = \begin{pmatrix} a & b \\ -b & a \end{pmatrix}$, $\quad a' = a$, $b' = -b$; Hermitian type.

$B = \pm\begin{pmatrix} I_n & 0 \\ 0 & -I_n \end{pmatrix} = \pm I_{n,n}$, $\quad A = \begin{pmatrix} a & b \\ -b & -a \end{pmatrix}$, $\quad a' = a$, $b' = -b$.

5. Explicit symplectic schemes

The symplecticity of Hamiltonian dynamical evolutions implies some kind of symmetry between the past and the future. So symplectic schemes are implicit by nature. However, for special classes of Hamiltonians of practical importance, and for certain type matrices B (see 4), the corresponding simple symplectic schemes will be practically explicit, leading to simple and fast algorithms. For example, consider the " separable " Hamiltonians of the form

$$H(p,q) = \phi(p) + \psi(q), \tag{14}$$

where ϕ, ψ are functions of n variables, then

$$H_p(p,q) = \phi_p(p), \quad H_q(p,q) = \psi_q(q).$$

Most of the energy functions in classical mechanics are separable in the above sense. Apply symplectic schemes (10) of type E_{2n}, and we get explicit schemes

$$\begin{array}{lll} \text{Type} & E_{2n}; & \hat{p} = p - \tau\psi_q(q), \quad \hat{q} = q + \tau\phi_p(\hat{p}). \\ \text{Type} & -E_{2n}; & \hat{p} = p - \tau\psi_q(\hat{q}), \quad \hat{q} = q + \tau\phi_p(p). \end{array} \tag{15}$$

The accuracy of this scheme can be raised without additional labor by putting p (or q) at integer times and q (or p) at half-integer times, we get *time-staggered symplectic* scheme of order 2, explicit for separable Hamiltons:

$$p^{k+1} = p^k - \tau\psi_q(q^{k+\frac{1}{2}}), \quad q^{k+1+\frac{1}{2}} = q^{k+\frac{1}{2}} + \tau\phi_p(p^{k+1}). \tag{16}$$

This explicit symplectic scheme was used and explained in detail by Feynman in his Lectures for computing the motions of an oscillating spring and of a planet around the sun [11]. The staggering (in time and/or in space) is a useful numerical technique for improving accuracy and symmetry.

The oldest and simplest among all difference schemes is the explicit Euler method

$$\hat{z} = z + \tau J\nabla H(z) = (I + \tau J\nabla H)(z) = G_H^\tau(z). \tag{17}$$

It is interest to ask: under what conditions on H the operator G_H^τ will be symplectic. We have proved: If, in R^{2n}

$$H(p,q) = \phi(Ap + Bq), \qquad AB' = BA' \tag{18}$$

where ϕ is a function of n variable, A, B are matrices of order n; then G_H^τ is symplectic and exact, i.e. $G_H^\tau = g_H^\tau$.

If, moreover

$$H(p,q) = \sum_{i=1}^{m} H_i(p,q), \quad H_i(p,q) = \phi_i(A_i p + B_i q), \quad A_i B_i' = B_i A_i' \tag{19}$$

Where ϕ_i are functions of n variables, A_i, B_i are matrices of order n. Let $[s_1, s_2, \cdots, s_m]$ be an arbitrary permutation of $[1, 2, \cdots, m]$; then the following scheme is symplectic and of order 1:

$$z \to \hat{z} = (G^\tau[s_1, s_2, \cdots, s_m])z, \quad G^\tau[s_1, s_2, \cdots, s_m] = (G^\tau[s_1]) \circ (G^\tau[s_2]) \circ \cdots \circ (G^\tau[s_m]) \tag{20}$$

where $G^\tau[i] = G_{H_i}^\tau$. Moreover,, the following alternating composite is symplectic and of order 2:

$$z \to \hat{z} = (G^{\frac{\tau}{2}}([s_1, s_2, \cdots, s_m]) \circ (G^{\frac{\tau}{2}}[s_m, \cdots, s_2, s_1])z \tag{21}$$

Note that the explicit symplectic schemes of order 1 and 2 of type $\pm E_{2n}$ for separable Hamiltonian in the sense of (14) are special cases of the above situation. We have computed with success using (19-21) the systems with Hamiltonians [18,19]

$$H_k(p,q) = \sum_{j=1}^{k} \cos(p\cos\frac{2\pi j}{k} + q\sin\frac{2\pi j}{k}) \tag{22}$$

with k-fold rotational symmetry in phase plane. They are not separable in the sense of (14) when $k \neq 1, 2, 4$.

For the general characterization for the symplecticity of explicit Euler schemes, we have proved: $G_H^\tau = I + \tau J\nabla H$ in R^{2n} is symplectic iff

$$H_{zz} J H_{zz} = 0, \quad \forall z \in R^{2n} \tag{23}$$

where H_{zz} is the Hessian matrix of H.

Furthermore we have an as yet incomplete but interesting characterization of (23) as follows:

A necessary and sufficient condition for $H_{zz} J H_{zz} \equiv 0$ in R^{2n} is that

$H(z) = H(p,q)$ can be expressed, under suitable *linear symplectic and orthogonal transformation* $z = Sw$, $w = (u,v)$, as

$$H(S(u,v)) = \phi(u) + \psi(v), \quad \psi \text{ is linear}, \quad S \in Sp(2n) \cap O(2n) \tag{24}$$

i.e. separable in sense of (14) with additional linearity in ψ.

The "sufficiency" part follows easily from the propositions stated above. However, the "necessity" part is not yet fully established. We have proved for the case that H is a quadratic form. Moreover for $n = 1$, (23) is equivalent to the Monge-Ampere equation $|H_{zz}| = 0$, so H defines a global surface of zero Gaussian curvature in R^3, the assertion is a consequence of a global theorem on developable surface in R^3 [16,17]. Since the global strong degeneracy condition (23) should imply drastic simplification of the profile of function H, so we have reasons to suggest the conjecture (23) $\Rightarrow$ (24).

6. Symplectic schemes for linear Hamiltonian systems

By a linear Hamiltonian system here we mean the system with a quadratic form $H(z) = \phi_A(z) = (z'Az)/2$, $A' = A$. as the Hamiltonian. Then the system becomes

$$\frac{dz}{dt} = J\nabla\phi_A(z) = JAz \tag{25}$$

with phase flow

$$g_H^t = exp(tJA) = I + \sum_{k=1}^{\infty} \frac{t^k}{k!}(JA)^k \tag{26}$$

as linear symplectic operators. All the constructions of symplectic schemes in 4 and in 7,8 below can be applied, but here we present the approach of symplectic approximations to the exponential [1,2,9]. The Taylor series in (27) can be arbitrarily truncated, but the do not provide, apart from some rare exceptions, symplectic approximations. The natural way is looking for rational and especially the Padé approximations of the exponential function $exp(\lambda)$.

$$exp(\lambda) - \frac{P_{m,l}(\lambda)}{Q_{m,l}(\lambda)} = O(|\lambda|^{m+l+1}), \quad |\lambda| \sim 0 \tag{27}$$

$P_{m,l}$ and $Q_{m,l}$ is of degree m, l respectively. It can be proved that, the matrix Padé transform $P_{m,l}(\tau JA)/Q_{m,l}(\tau JA)$ is symplectic iff the Padé approximant is diagonal, i.e., $m = l$; then $P_{m,m}(\lambda) = P_m(\lambda)$, $Q_{m,m}(\lambda) = P_m(-\lambda)$, and

$$P_0(\lambda) = 1, \quad P_1(\lambda) = 2+\lambda, \quad P_2(\lambda) = 12+6\lambda+\lambda^2, \quad P_m(\lambda) = 2(2m-1)P_{m-1}(\lambda)+\lambda^2 P_{m-2}\lambda) \tag{28}$$

We get the *Padé symplectic* scheme

$$P_m(-\tau JA)\hat{z} = P_m(\tau JA)z, \quad \hat{z} = G^\tau z = \frac{P_m(\tau JA)}{P_m(-\tau JA)}z. \tag{29}$$

which is of order $2m$. For $m = 1$ we get the time-centered Euler scheme (8).

It is known that every linear system (26) in R^{2n} with $H = \phi_A$ is completely integrable in the sense that there exists n invariant quadratic forms, $\phi_{A_1} = \phi_A, \cdots, \phi_{A_n}$ which are functionally independent and mutually commuting $\{\phi_{A_i}, \phi_{A_j}\} = 0, \quad i, j = 1, \cdots, n$. The *Padé symplectic schemes* have the remarkable property that they posess *the same set of invariant quadratic forms* as that of the original linear Hamiltonian system. If the original linear phase flow is not only symplectic but also unitary, then the Padé schemes are also unitary, so they are useful for quantum mechanical applications.

In general, the linear symplectic schemes for linear Hamiltonian system, even when they do not preserve the quadratic invariants (including the energy) of the original system, we still can construct for any linear symplectic transition operator G_H^τ (in fact even for any $G \in Sp(2n)$) a set of n independent, mutually commuting quadratic forms ϕ_{B_i} which are invariant under G. this set is near to the corresponding set of the original when τ is small, so they behave well in the aspects of orbital structure and stability.

7. Symplectic leap-frog schemes

The concept of symplecticity for multi-level schemes for Hamiltonian system is more complicated then the 2-level case . We discuss only the 3-level schemes. For system (3) with Hamiltonian H and phase flow operator g_H^t, we have the following exact relations between the solution at 3 successive moments, we fix $\tau > 0$ and t:

$$z(t) \to z(t+\tau) = g_H^\tau z(t)$$
$$z(t) \to z(t-\tau) = g_H^{-\tau} z(t)$$

Then we have

$$z(t+\tau) - z(t-\tau) = \delta_H^\tau z(t), \quad \delta_H^\tau = g_H^\tau - g_H^{-\tau} \tag{30}$$

where δ_H^τ is the exact difference operator for the Hamiltonian system, and can be shown to be infinitesimally symplectic, i.e., its Jacobian matrix belongs to $sp(2n)$ everywhere.

We take the general form of 3-level leap-frog schemes as

$$z^{k+1} - z^{k-1} = \Delta_H^\tau z^k \tag{31}$$

here the coefficients chosen before z^{k+1} and z^{k-1} are justified on grounds of symmetry. The *leap-frog* scheme (31) is called *symplectic* if the operator Δ_H^τ is *infinitesimally symplectic*. So our problem is to seek for approximations Δ_H^τ to δ_H^τ preserving infinitesimal symplecticity.

For linear Hamiltonian system (25, 26) we have

$$\delta_H^\tau = exp(\tau JA) - exp(-\tau JA) = 2sinh(\tau JA) = 2\sum_{j=1}^{\infty} \frac{(-1)^j}{(2j-1)!}(\tau JA)^{2j-1} \tag{32}$$

here all the truncations are infinitesimally symplectic, so we get symplectic leap-frog schemes

$$\text{order } 2: \quad z^{k+1} - z^{k-1} = 2\tau JAz^k$$

$$\text{order } 2m: \quad z^{k+1} - z^{k-1} = 2(\sum_{j=1}^{m} \frac{(-1)^j}{(2j-1)!}(\tau JA)^{2j-1})z^k.$$

Similarly, for non-linear Hamiltonian system, from the Lie series (6) of phase flow we get

$$(\delta_H^\tau z)_i = 2\sum_{j=1}^{\infty} \frac{\tau^{2j-1}}{(2j-1)!}\{\{\cdots\{\{z_i^k, H\}, H\}\cdots, H\}, H\} \tag{33}$$

whose truncations provide *symplectic leap-frog* schemes

$$\text{order } 2: \quad z^{k+1} - z^{k-1} = 2\tau J\nabla H(z^k) \tag{34}$$

$$\text{order } 2m: \quad z_i^{k+1} - z_i^{k-1} = \delta_{H,m}^\tau z^k \tag{35}$$

where $\delta_{H,m}^\tau z$ is the m-term truncation of (33).

For these schemes we can establish the conservation property as: If $F(z) = z'Bz$ is a quadratic invariant under g_H^t, then follows the " cross invariance" under the scheme (34)

$$(z^{k+1})'Bz^k = (z^k)'Bz^{k-1}.$$

All the symplectic leap-frog schemes are explicit. The simplest 2nd order scheme (34) is one of the rare cases of existing methods which are symplectic and seems to be the oldest one. It is interesting to note that, according to linearized analysis, the stability domain for (34) is only a segment on the imaginary axis in the complex plane, it is absolutely unstable when applied to asymptotically stable systems. However, it works well with Hamiltonian systems, which correspond to the critical case with pure imaginary characteristic exponents and are never asymptotically stable owing to the symplecticity of dynamics.

8. Fractional Transforms and Generating Functions

A matrix α of order $4n$ is called a *Darboux matrix* if

$$\alpha' J_{4n}\alpha = \tilde{J}_{4n} \quad J_{4n} = \begin{pmatrix} 0 & -I_{2n} \\ I_{2n} & 0 \end{pmatrix}, \quad \tilde{J}_{4n} = \begin{pmatrix} J_{2n} & 0 \\ 0 & -J_{2n} \end{pmatrix},$$
$$\alpha = \begin{pmatrix} a & b \\ c & d \end{pmatrix}, \quad \alpha_{-1} = \begin{pmatrix} a_1 & b_1 \\ c_1 & d_1 \end{pmatrix}, \tag{36}$$

Each Darboux matrix induces a (linear) *fractional transform* between symplectic and symmetric matrices

$$\sigma_\alpha: \quad Sp(2n) \to sm(2n)$$
$$\sigma_\alpha(S) = (aS+b)(cS+d)^{-1} = A, \quad \text{for} \quad |cS+d| \neq 0 \tag{37}$$

with inverse transform $\sigma_\alpha^{-1} = \sigma_{\alpha^{-1}}$

$$\sigma_\alpha^{-1}: \quad sm(2n) \to Sp(2n),$$
$$\sigma_\alpha^{-1}(A) = (a_1A+b_1)(c_1A+d_1)^{-1} = S, \quad \text{for} \quad |c_1A+d_1| \neq 0. \tag{38}$$

The above machinery can be extended to generally non-linear operators in R^{2n}. Let $Symp(2n)$ denote the totality of symplectic operators, and $symm(2n)$ denote the totality of symmetric operators. Each $f \in symm(2n)$ corresponds, at least locally a real function ϕ (up to a constant) such that $f(w) = \nabla\phi(w)$. Then we have

$$\sigma_\alpha: \quad Symp(2n) \to symm(2n),$$
$$\sigma_\alpha(g) = (ag+b)(cg+d)^{-1} = \nabla\phi, \quad \text{for} \quad |cg_z+d| \neq 0, \tag{39}$$

or alternatively

$$ag(z) + bz = (\nabla\phi)(cg(z) + dz), \tag{40}$$

where ϕ is called the *generating function* of Darboux type α for the symplectic operator g. Then

$$\begin{aligned}&\sigma_\alpha^{-1}: \qquad symm(2n) \to Symp(2n),\\ &\sigma_\alpha^{-1}(\nabla\phi) = (a_1\nabla\phi + b_1)(c_1\nabla\phi + d_1)^{-1} = g, \quad \text{for} \quad |c_1\phi_{ww} + d_1| \neq 0\end{aligned} \tag{41}$$

or alternatively

$$a_1\nabla\phi(w) + b_1(w) = g(c_1\nabla\phi(w) + d_1 w). \tag{42}$$

where g is called the symplectic operator of Darboux type α for the generating function ϕ.

For the study of symplectic difference algorithms we may narrow down the class of Darboux matrices to the subclass of *normal Darboux matrices*, i.e., those satisfying $a + b = 0$, $c + d = I_{2n}$. The normal Darboux matrices α can be characterized as

$$\begin{aligned}\alpha &= \begin{pmatrix} a & b \\ c & d \end{pmatrix} = \begin{pmatrix} J & -J \\ \frac{1}{2}(I + JB) & \frac{1}{2}(I - JB) \end{pmatrix}, \quad B' = B\\ \alpha^{-1} &= \begin{pmatrix} a_1 & b_1 \\ c_1 & d_1 \end{pmatrix} = \begin{pmatrix} \frac{1}{2}(JBJ - J) & I \\ \frac{1}{2}(JBJ + J) & I \end{pmatrix}.\end{aligned} \tag{43}$$

The fractional transform induced by normal Darboux matrix establishes a 1-1 correspondence between *symplectic operators near identity* and *symmetric operators near nullity*. Then the determinantal conditions could be taken for granted. Those B listed in 4 correspond to the most important normal Darboux matrices.

For each Hamiltonian H with its phase flow g_H^t and for each normal Darboux matrix α, we get the *generating function* $\phi = \phi_H^t = \phi_{H,\alpha}^t$ of *normal Darboux type* α for the *phase flow* of H by

$$\nabla\phi_{H,\alpha}^t = (ag_H^t + b)(cg_H^t + d)^{-1}, \quad \text{for small} \quad |t|. \tag{44}$$

$\phi_{H,\alpha}^t$ satisfies the *Hamilton-Jacobi* equation

$$\frac{\partial}{\partial t}\phi = -H(a_1\nabla\phi(w) + b_1 w) = -H(c_1\nabla\phi(w) + d_1 w) \tag{45}$$

and can be expressed by Taylor series in t

$$\phi(w,t) = \sum_{k=1}^{\infty}\phi^{(k)}(w)t^k, \quad |t| \quad \text{small}, \tag{46}$$

The coefficients can be determined recursively

$$\phi^{(1)}(w) = -H(w), \quad \text{and for} \quad k \geq 0, \quad a_1 = \frac{1}{2}(JBJ - J):$$

$$\phi^{(k+1)}(w) = \frac{-1}{k+1}\sum_{m=1}^{k}\frac{1}{m!}\sum_{i_1,\cdots,i_n=1}^{2n} H_{z_{i_1}\cdots z_{i_m}}(w) \sum_{\substack{j_1+\cdots+j_m=k\\ j_i\geq 1}} (a_1\nabla\phi^{(j_1)}(w))_{i_1}\cdots(a_1\nabla\phi^{(j_m)}(w))_{i_m}. \tag{47}$$

Let ψ^τ be a truncation of $\phi_{H,\alpha}^\tau$ up to certain power t^m, say. Using inverse transform σ_α^{-1} we get the symplectic operator

$$G^\tau = \sigma_\alpha^{-1}(\nabla\psi^\tau), \quad |\tau| \quad \text{small}$$

which depends on τ, H, α (or equivalently B) and the mode of truncation. It is a symplectic approximation to the phase flow g_H^τ and can serve as the transition operator of a symplectic difference scheme (for the Hamiltonian system (3))

$$\hat{z} = G^\tau z \tag{48}$$

which, by (39,40,41) can be written as

$$J\hat{z} - Jz = \nabla\psi^\tau(c\hat{z} + (I-c)z), \quad c = \frac{1}{2}(I + JB). \tag{49}$$

Thus, using the machinery of phase flow generating functions we have constructed, for each H and each normal Darboux matrix, an hierarchy of symplectic schemes by truncation. The simple symplectic schemes in 4 corresponds to the lowest truncation. The conservation properties of all these higher order schemes are the same as stated in 4.

9. Numerical experiments and discussions

Numerical experimentation shows strong evidences of superior performance of symplectic schemes over the non-symplectic ones. Most of the conventional schemes are non-symplectic when applied to Hamiltonian systems. Unavoidably they introduce, above all, artificial dissipation and other parasitic effects which are alien to Hamiltonian dynamics, leading eventually to grave, qualitative distortions.

On the other hand, symplectic schemes tend to purify (sometimes even simplify) the algorithms, inhibiting artificial dissipations and all other non-Hamiltonian distortions. We present a comparative study for the harmonic oscillator, Fig 1,2 and for a nonlinear oscillator, Fig 3,4. For some other result, see [4]. The non-symplectic Runge-Kutta scheme yields persistent spiraling of phase trajectories with artificial creation of attractors (impossible for Hamiltonian system), irrespective of the higher order of the scheme and the smallness of step sizes. The symplectic scheme always give clear-cut invariant elliptical trajectories even for low order of schemes and large step sizes. The striking contrast is, at least, in the the aspects of global and structural preservation and long-term tracking capabilities.

The most important characteristic property of Hamiltonian dynamics is the intricate *coexistence* of *regular* and *chaotic* motions, as implied by the celebrated KAM theorem [14]. The advantage of symplectic schemes consists primarily in the essential fidelity to the original in this aspecct. The behavior of symplectic schemes should be and could be understood and analysed in the context of KAM theory on preservation and break-down of invariant tori. There are evidences showing that an analog of KAM theorem should hold for symplectic symplectic difference schemes, we have a related result for a simple special case.

Symplectic schemes for Hamiltonian systems do not imply non-Hamiltonian perturbations, but they do inevitably imply Hamiltonian perturbations. An example is the break-down of invariant tori when the scheme is applied to integrable or near integrable systems. The effect is indiscernible for small step size, but becomes pronounced with large step sizes, see Fig 5. Another example is the *artificial chaos* in the form of Arnold diffusion. We computed by explicit symplectic schemes (21) of order 2, the Hamiltonian system H_k (22) with k-fold rotational symmetry in 1 degree of freedom, suggested by Sagdeev, Zaslavsky *et al* [18] in the context of chaotic stream lines of Lagrangian turbulence, see also [19-21]. Fig 6 gives some results for $k = 4$, when initial point is chosen on the original separatrix, the diffusion shows a remarkable combination of *regularity* and *chaoticity*, almost rectilinear motion along the "highways" and random turning at the "crossroads". Fig 7,8 give some results for $k = 3, 5$. For $k = 5$ the orbit is reminiscent of Alhambra fresco decoration or Penrose tiling or quasi-crystal with 10-fold symmetry. It is worth to note the connections of symplectic schemes with these widely different topics.

Fig.1. Harmonic oscillator $H = (p^2 + a^2q^2)/2, \quad a = 2.$
Runge-Kutta method, order 4, non-symplectic.
stepsize=τ, total number of steps=N , number of steps per plot =M.
Single orbit is computed in all cases with the same initial point.
Upper-left: τ=0.5, N=2000, M=1, Upper-right: τ=0.1, N=2000, M=1.
Lower-left: τ=0.3, N=2000, M=1, Lower-right: τ=0.1, N=200,000, M=100.
For large stepsize 0.5 (upper-left), the orbit spirials quickly around and towards the origin with a wrong formation of an artificial attractor there. The dissipation effect diminishes with smaller stepsize 0.3 (lower-left), but still pronounced. At very small stepsize 0.1 (upper-right), the orbit appears to be well-behaved as a closed ellipse when total number of steps is limited to 2000 as before. However, the artificial dissipation eventually becomes pronounced again when computed to twenty thousands steps (lower-right).

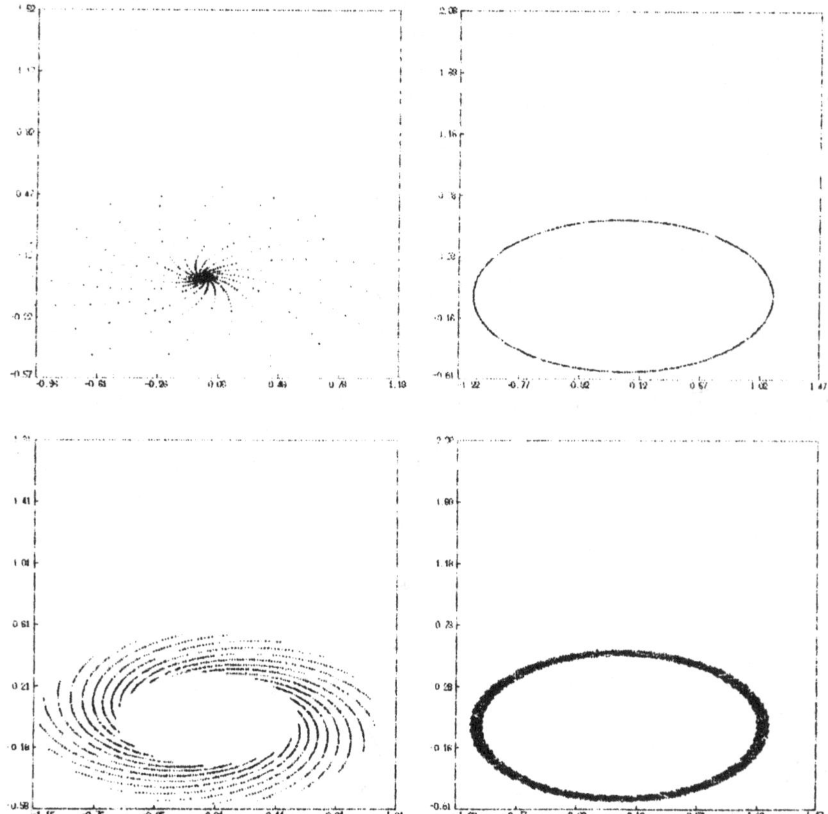

Fig.2. Harmonic oscillator $H = (p^2 + a^2q^2)/2, \quad a = 2.$
Explicit symplectic composite scheme, order 2
stepsize=τ, total number of steps=N , number of steps per plot =M.
Single orbit is computed in all cases with the same initial point.
Upper-left: τ=0.5, N=2000, M=1, Upper-right: τ=0.1, N=2000, M=1.
Lower-left: τ=0.3, N=2000, M=1, Lower-right: τ=0.1, N=1,000,000, M=500.

The arrangement of calculation is just the same as in Fig.1 for comparison. In all cases the symplectic computing gives always clear-cut invariant ellipses even for the very long run of a million steps (upper-right). The artificial dissipation is fully inhibited. The computational stability and the ability for long time tracking and the preservation of the structure of the phase portrait is truely remarkable. We note that the specific symplectic scheme used here does not preserve the quadratic energy exactly but it does possess an invariant quadratic form whose level lines are ellipses very close to and practically indistinguishable from the original ellipses.

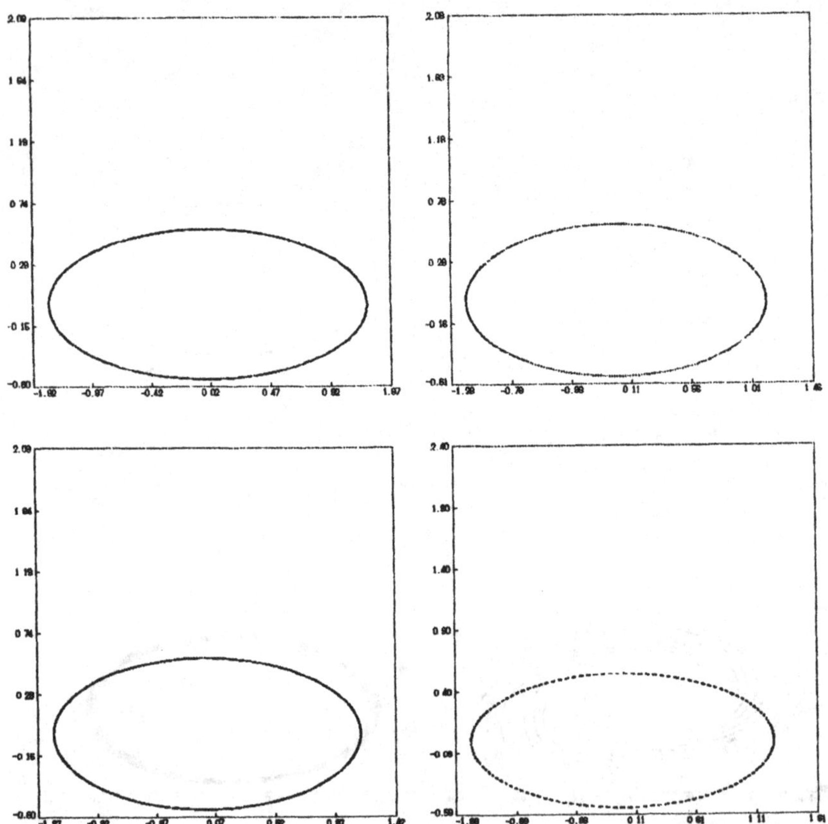

Fig.3. Nonlinear oscillator $H = (p^2 + a^2q^2 + a^4q^4/12)/2, \quad a = 2.$
Runge-Kutta method, order 4, non-symplectic.
stepsize=τ, total number of steps=N , number of steps per plot =M.
Single orbit is computed in all cases with the same initial point.
Upper-left: τ=0.5, N=2000, M=1, Upper-right: τ=0.1, N=2000, M=1.
Lower-left: τ=0.3, N=2000, M=1, Lower-right: τ=0.1, N=200,000, M=100.

This nonlinear oscillator is the harmonic one plus a nonlinear perturbation. At low energy values the invariant orbits are only nearly elliptical in form. For larger stepsizes 0.5 (upper-left), and 0.3 (lower-left), the spiraling structures of the orbits are much more complicated then the corresponding linear cases in the left hand side of Fig.1. At very small stepsize 0.1 (upper-right and lower-right), the situation is again similar to the linear case with the eventual cummulative artificial dissipation noticeable in the long run. Articifcial dissipation is inevitable for both linear and nonlinear cases in non-symplectic schemes.

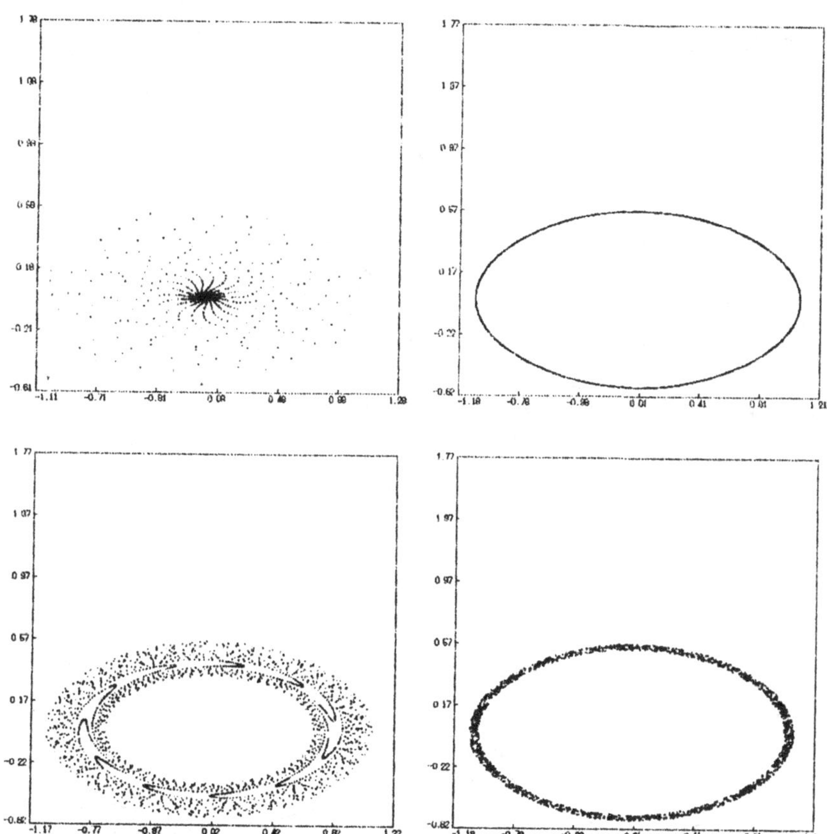

Fig.4. Nonlinear oscillator $H = (p^2 + a^2q^2 + a^4q^4/12)/2, \quad a = 2.$
Explicit symplectic composite scheme, order 2.
stepsize=τ, total number of steps=N , number of steps per plot =M.
Single orbit is computed in all cases with the same initial point.
Upper-left: τ=0.5, N=2000, M=1, Upper-right: τ=0.1, N=2000, M=1.
Lower-left: τ=0.3, N=2000, M=1, Lower-right: τ=0.1, N=1,000,000, M=500.

Here again the symplectic computing gives always clear-cut invariant closed orbits including the run of a million steps without any dissipation. The computed orbits are only nearly elliptical but hardly discernible from the ellipses in Fig.2. The apparent identity of all the orbits in Fig.1 and 2 shows the stability and robustness of the method. From Fig.1-4 one sees the sharp contrast between the performances of the two methods. The symplectic method is far superior at least in the aspects of computational stability, global and structural preservation and long-term simulation. Incidentally the specific 2nd order symplectic method adopted is 4 times faster than the 4th order Runge-Kutta method.

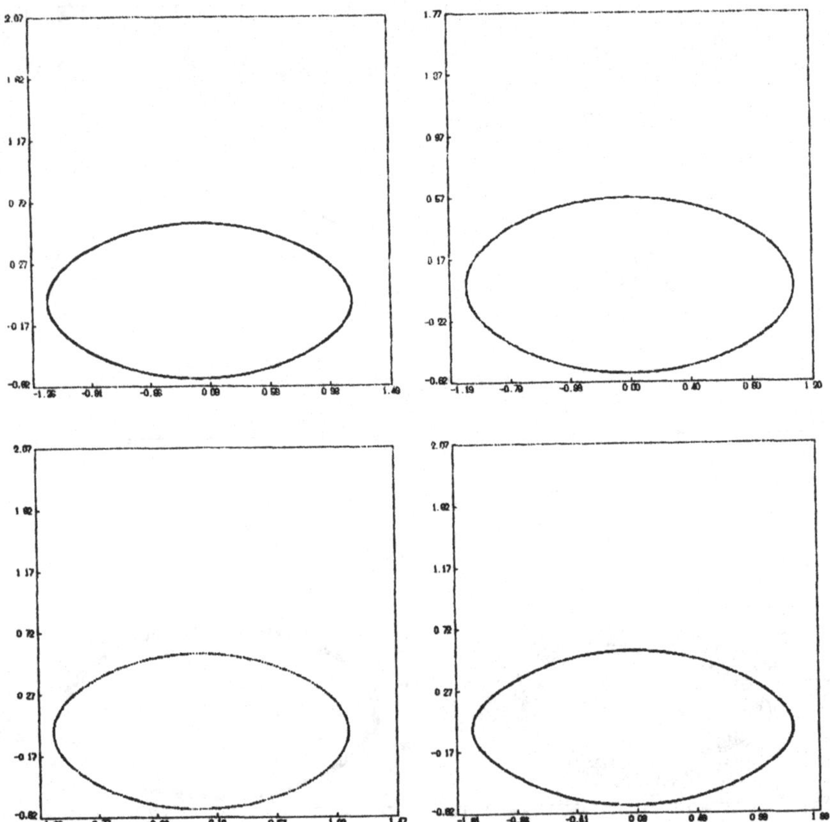

Fig.5. Preservation and breakdown of invariant tori for nonlinear oscillator $H = (p^2 + a^2q^2 + a^4q^4/12)/2, \quad a = 2$.
Explicit symplectic composite scheme, order 2.
stepsize=τ, total number of steps=N , number of steps per plot =M.
Five orbits are computed in all cases with the same set of five initial points.
Upper-left: τ=0.5, N=2000, M=1, Upper-right: τ=0.1, N=2000, M=1.
Lower-left: τ=0.3, N=2000, M=1, Lower-right: τ=0.1, N=1,000,000, M=500.
The five initial points are (0.5, 0.25), (0.65, 0.535), (0.375, 0.55), (0.685, 0.65), (0.495, 0.375). For smaller stepsize 0.3671 (upper-left), all the invariant tori are preserved practically, the effect of segmentation is almost inperceptible. However, for larger stepsizes (upper-left, lower-left, lower-right), while some invariant tori are preserved, the breakdown becomes pronounced. Some invariant tori are severely segmented and some are broken down into islands. This phenomenon is typical in the perturbation theory of Hamiltonian dynamics.

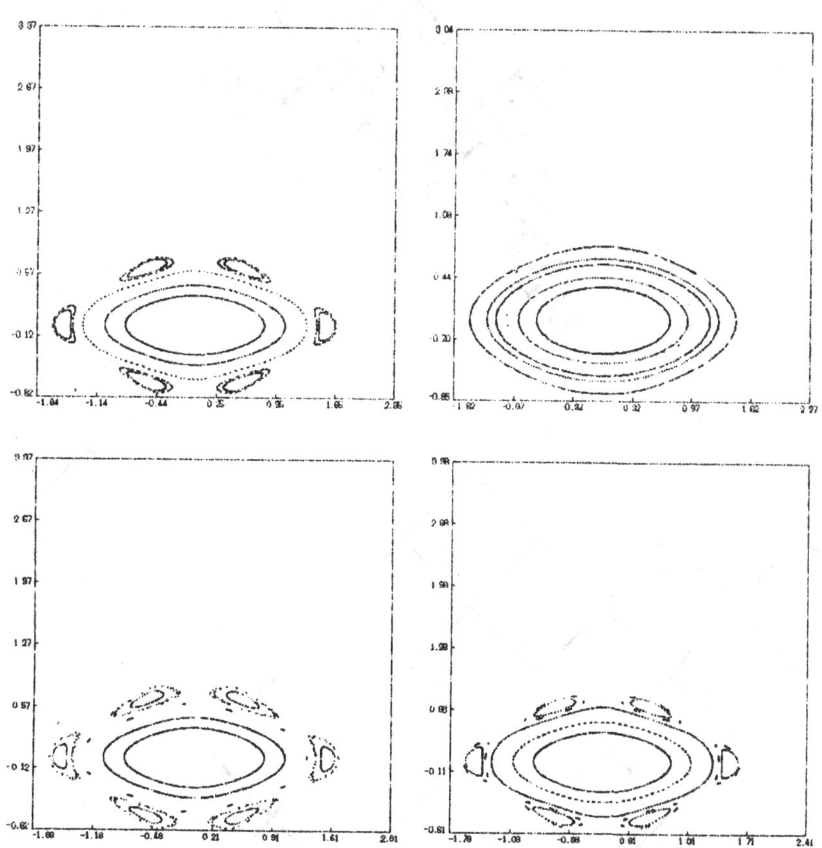

Fig.6. Hamiltonian with 4-fold rotational symmetry, using H_k in (22), $k = 4$.
Explicit symplectic composite scheme, order 2 (21).
stepsize=τ, total number of steps=N , number of steps per plot =M.

Upper-part: τ=1.3, N=10,000, M=1,

Lower-left: τ=0.20000002, N=40,000, M=1, Lower-right: τ=0.2, N=40,000, M=1.

In the symplectic scheme for system with Hamiltonian H_k, the initial point, chosen on or very near to a separatrix, moves within an Arnold web which is a network of canals with the separatrices as its skeleton. The canals become thinner with smaller τ. The separatrix network of H_4 is a square lattice. For larger τ=1.3 (upper part) the diffusion pattern is more wide-spread and coarse. For small τ the motion is almost rectilinear along the lattice plus random turnings at the crossings, see the cases of τ=0.2 (lower-right) and τ=0.20000002 (lower-left), which, in addition, show the extreme sensitivity of the diffusion pattern to the stepsize τ.

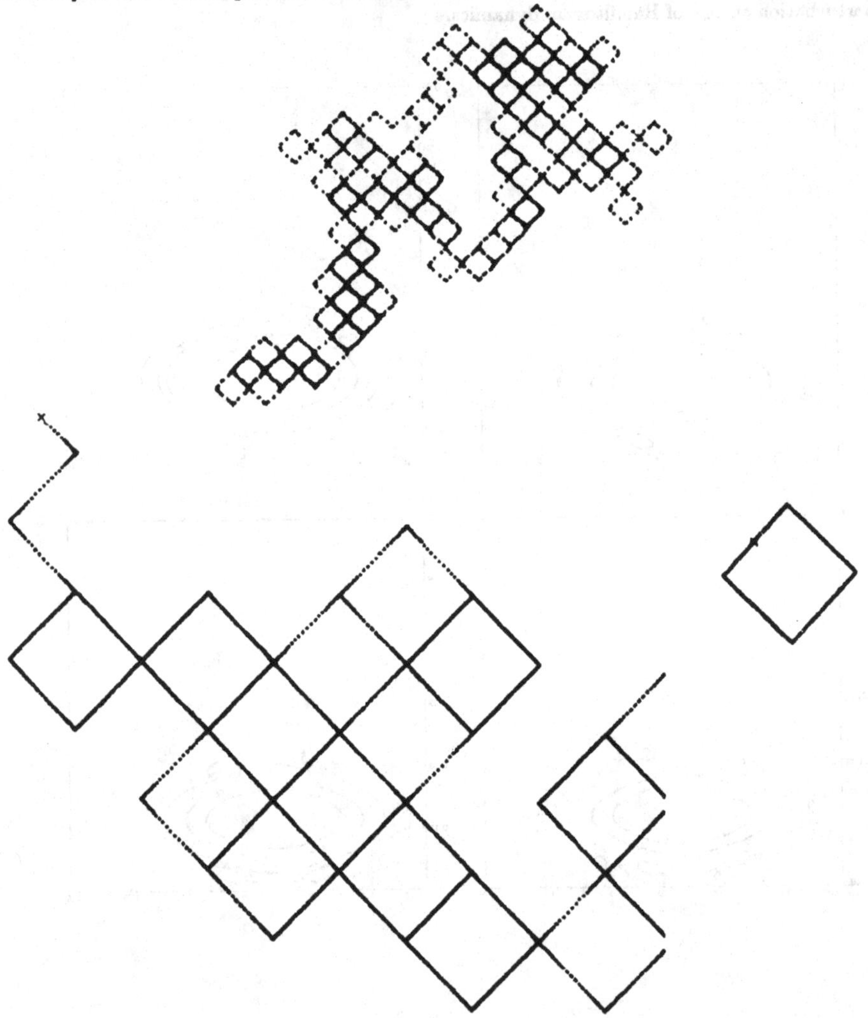

Fig.7. Hamiltonian with 3-fold rotational symmetry, using H_k in (22), $k = 3$.
Explicit symplectic composite scheme, order 2 (21).
stepsize=τ, total number of steps=N , number of steps per plot =M.
Single orbit is computed in all cases.

Upper-left: τ=1.4, N=4000, M=2, Upper-right: τ=1.04, N=10000, M=1.

Lower-left: τ=0.7002, N=10000, M=2, Lower-right: τ=0.7, N=25,000, M=5.

The separatrix network of H_3 (which is proportional to H_6) is a periodic lattice composed of triangles and hexagons. For large τ=1.4 (upper-left) we get a very coarse web; then at τ=1.04 (upper-right) the diffusion path is still not fully straightened. For smaller τ=0.7002 (lower-left) the diffusion is quite uniform and wide-spread. However, with slightly different τ=0.7 (lower-right), the diffusion behavior becomes grossly different. In an early stage up to 14000 steps the point is confined on an initial triangle, the it moves out to a far away hexagon and stays there again for a long time and then in certain way repeats such oscillating visits between these two entities.

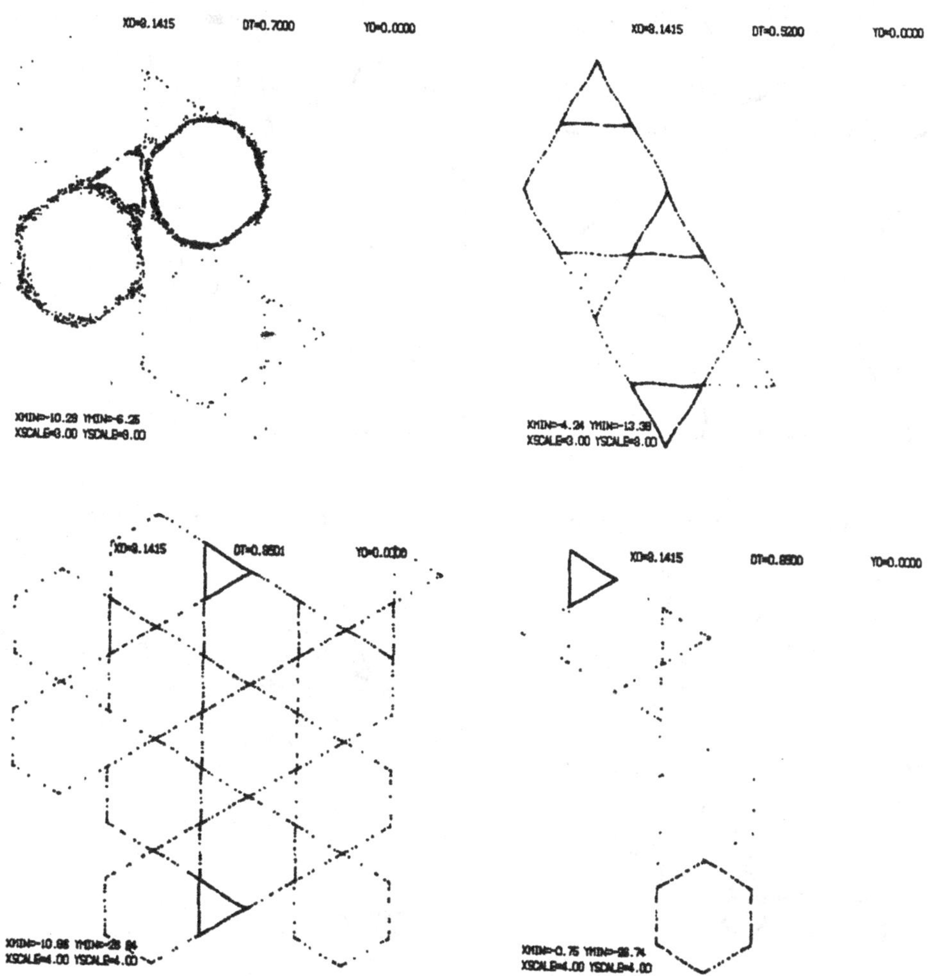

Fig.8. Hamiltonian with 5-fold rotational symmetry, using H_k in (22), $k = 5$.
Explicit symplectic composite scheme, order 2 (21).
stepsize=τ, total number of steps=N , number of steps per plot =M.
Single orbit is computed in all cases.

Upper-left: τ=0.5, N=100,000, M=10, Upper-right: τ=0.56, N=50,000, M=5.

Lower-left: τ=0.701, N=20,000, M=1, Lower-right: τ=0.701, N=100,000, M=10.

The separatrix network of H_5 (proportional to H_{10}) is not periodic but quasiperiodic and not connected, looks like a Penrose tiling with 10-fold symmetry. For small values τ=0.5 (upper-left) and τ=0.56 (upper-right) we get trapped diffusion, resulting in rosette patterns with 10-fold symmetry. For larger τ=0.701 (lower left and right, for evolution) we get the spread-out pattern (see also [21]) which looks like certain Moslem decoration and the electron diffraction pattern of a 2D decagonal quasicrystal, cf. [22].

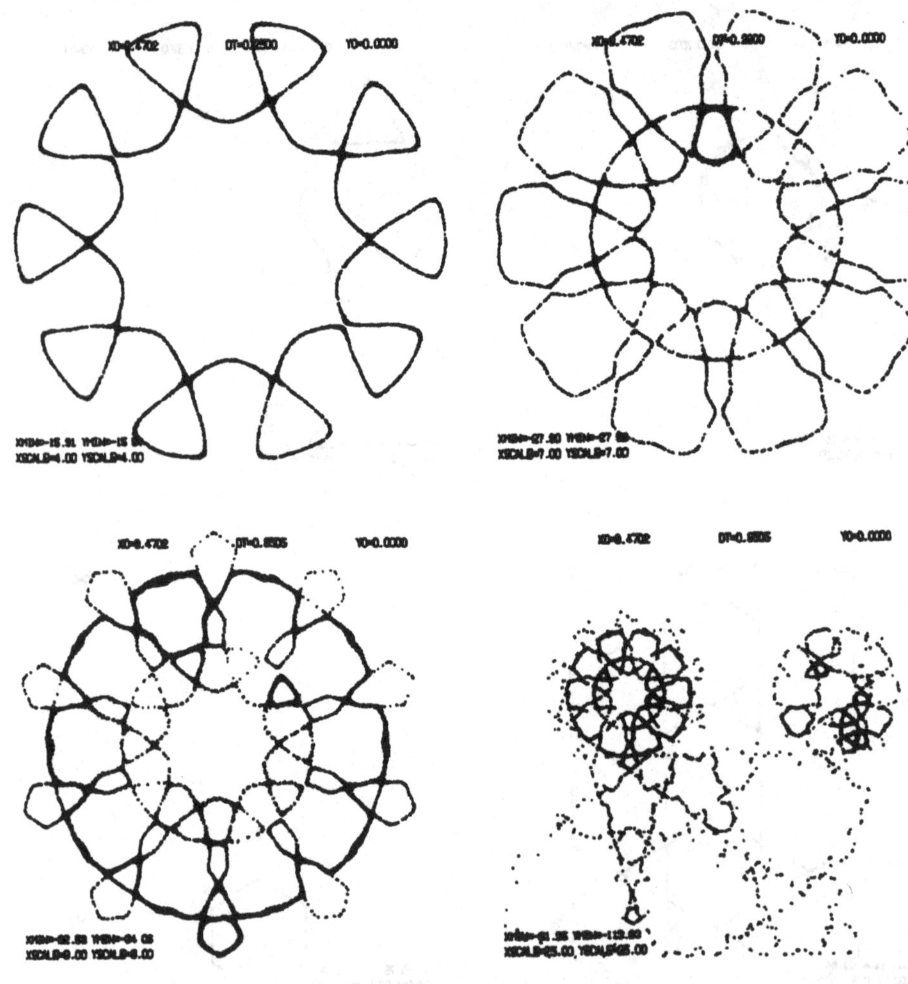

Acknowledgements This research program is supported by the National Natural Science Foundation of China. The author expresses his gratitude to his collaborators M.Z. Qin, H.M. Wu, Z. Ge, D.L. Wang, Y.H. Wu, M.Q. Zhang for close and fruitful cooperation, and in particular to M.Z. Qin for his involvement in the work and supplying his unpublished numerical results and to M.Q. Zhang for discussion and assistance in preparing the manuscript. Thanks are also due to Prof. R. Spigler for the invitation and support for the author's participation of Venice-1 Symposium 1989 to present this paper.

References

[1] Feng Kang, Proc. 1984 Beijing Symposium on Differential Geometry and Differential Equations, Ed. Feng Kang, Science Press, 42-58.

[2] Feng Kang, Jour. Comp. Math. 4(1986), 279-289.

[3] Feng Kang, Proc. 10th Inter. Conf. Numer. Methods in Fluid Dynamics, Ed. F.G. Zhuang, Lect. Notes in Phys. 264, Springer, 1986, 1-7.

[4] Feng Kang, Qin Meng-zhao, Proc. Conf. Numer. Methods for PDE's, Ed. Zhou You-lan, Lect. Notes in Math. 1297, Springer, 1987, 1-37.

[5] Ge Zhong, Feng Kang, Jour. Comp. Math. 6(1988), 88-97.

[6] Feng Kang, Wu Hua-mo, Qin Meng-zhao, Wang Dao-liu, Jour. Comp. Math. 7(1989), 72-96.

[7] Li Chun-wang, Qin Meng-zhao, Jour. Comp. Math. 6:2 (1989).

[8] Wu Yu-hua, Computer Math. & Appl. 15:12(1989), 1o41-1050.

[9] Qin Mengzhao, Math. Meth. in Appl. Sci. 11(1989), 543-557.

[10] W.R. Hamilton, Mathematical Papers, v. 2, Cambridge, 1940.

[11] J.L. Synge, Scripta Math. 10(1944), 13-24.

[12] .F.Klein, Entwickelung der Mathematik im 19 Jahrhundert, Teubner, 1928.

[13] E. Schroedinger, Scripta Math. 10(1944), 92-94.

[14] V.I. Arnold, Mathematical Methods of Classical Mechanics, Nauka, 1974.

[15] R. Feynman, Lectures on Physics, Addison-Wesley, 1963.

[16] A.V. Pogorelov, Dokl. Akad. Nauk. SSSR, 111(1956), 757-759, 945-947.

[17] P. Hartman, L. Nirenberg, Amer. Jour. Math. 81(1959), 901-920.

[18] A.A. Chernikov, R.Z. Sagdeev, G.M.Zaslavsky, Physica D33(1988), 65-76.

[19] V.I. Arnold, Physica D33(1988), 21-25.

[20] V.I. Arnold, Dokl. Akad. Nauk. SSSR, 156(1964), 9-12.

[21] V.V. Beloshapkin, R.Z. Sagdeev, G.M. Zaslavsky, *et al,* Nature, 337(1989), 543-557.

[22] K.K. Fung, *et al,* Phys. Rev. Lett. 56(1986), 2060.

THE SPEED OF WAVEFORM METHODS FOR ODES[1]

C. W. GEAR and FEN-LIEN JUANG
Department of Computer Science
University of Illinois at Urbana-Champaign

ABSTRACT. This paper considers iterative solution techniques for ordinary differential equations and investigates the speed of convergence of various processes. The iterates are defined as solutions of a sequence of differential equations. The solutions are the "waveforms." Waveform iterations are superlinearly convergent so a measure of the speed of convergence is defined and this is used to compare the value of various waveform methods. This measure is the rate of increase of the order of accuracy. The speed of the waveform Gauss Seidel method depends on the numbering of the equations. The numbering of the equations corresponds to a numbering of the directed graph specifying equation dependencies. We show how to compute the rate of order increase from the structure of the numbered graph and hence the optimum numbering, that is, the one which maximizes the speed of convergence.

1 Introduction

Waveform methods were first proposed [4, 5] in the context of VLSI circuit simulation where they were used to solve differential-algebraic equations (DAEs). In this paper we examine their effectiveness for the solution of ordinary differential equations (ODEs), which are a special case of DAEs.

The standard approach of solving ODE systems is based on three techniques applied in order:

1. Using implicit integration methods to discretize the system of differential equations. (If the equations are stiff, stiffly stable methods must be used.)
2. Using a functional iteration or modified Newton method to solve the system of non-linear algebraic equations obtained at each time point of the discretization. (If the equations are non stiff, only one functional iteration is needed, if stiff, an average of slightly more than one Newton iteration is needed.)
3. Using a direct method to solve the system of linear algebraic equations generated by Newton's method. (If the equations are non stiff, this last step is not necessary.)

[1]Work supported in part by US DOE under grant DOE DEFG02-87ER25026 and by NSF under grant DMS 87-03226

R. Spigler (ed.), *Applied and Industrial Mathematics*, 37–48.

As the size of the ODE system grows, the standard approach can become inefficient. One reason is that large systems usually contain variables that change at very different rates. The direct application of integration methods forces one to discretize all the variables identically and the discretization must be fine enough to accurately reflect the behavior of the most rapidly changing variable. If each variable in the system could use the largest possible timestep that would accurately reflect its behavior, i.e. if we could use different stepsizes for different variables, then the efficiency of the simulation could be improved greatly. Approaches that allow different stepsizes for different components in solving systems of ordinary differential equations are called *Multirate Integration Methods* [1, 2, 7].

In contrast, waveform methods apply the iteration first to define, by a sequence of differential equations, a sequence of functions of time ("waveforms") which converge to the solution of the differential equations. The discretization of the resulting differential equations is done as a second step. Waveform methods can result in systems of ODEs which are mutually decoupled. This not only reduces communication requirements (in parallel processing) but also permits simple implementation of multirate integration.

In this paper we consider the first step of the process, the iterative solution of the ODEs, and investigate the speed of convergence of various iterative processes. The numerical method used can affect the convergence of the iteration; furthermore, the numerical method used should take advantage of the error expected in each iterate to control its step size and order. However, we do not consider the discretization of the differential equations which define the iterates in this paper: rather we focus on the iteration itself. Since all waveform iterations are superlinearly convergent (in the absence of discretization), the main difficulty is measuring the speed of convergence. This paper introduces a measure of the speed of convergence and compares the value of various waveform methods using this measure. The speed of the waveform Gauss-Seidel (WGS) method depends on the ordering of the equations. The measure permits us to define the optimum ordering, that is, the one which maximizes the speed of convergence. Many of the details appear in the doctoral thesis [3]. In particular, proofs of all explicitly stated theorems appear in that thesis and are not repeated here.

1.1. THE BASIC IDEA OF WAVEFORM RELAXATION

Waveform relaxation is a family of iterative methods that are applied to solve systems of ordinary differential equations. One of its basic ideas is to partition a big system into loosely coupled subsystems and to solve each subsystem independently. The coupling between subsystems is neglected in the sense that at each iteration sweep each subsystem is solved by using values of previous iterates of other subsystems. The iterative process is continued until satisfactory convergence is obtained for each subsystem. At each iteration sweep of waveform relaxation, each subsystem can be discretized differently according to its behavior. At the first iteration, a zero-th order (constant) extrapolation from the initial value is usually used to approximate the values of variables in other subsystems; at later iterations, interpolation is used to approximate the needed values. A *waveform* is a continuous representation of a solution component.

We will consider the autonomous system of ordinary differential equations

$$\dot{\mathbf{u}} = F(\mathbf{u}), \qquad \mathbf{u}(0) = \mathbf{u}_0 \tag{1}$$

where $\mathbf{u} \in R^n$, and $F : R^n \to R^n$. The simplest iterative method is the Picard method.

This takes the form

$$\dot{u}^{[k+1]} = F(u^{[k]}), \qquad u^{[k+1]}(0) = u_0$$

starting from, usually, $u^{[0]}(t) \equiv u_0$.

In many cases, the system will also be considered as the m coupled subsystems

$$\begin{aligned} \dot{u}_1 &= f_1(u_1, u_2, \ldots, u_m), & u_1(0) = u_{1,0} \\ &\vdots \\ \dot{u}_m &= f_m(u_1, u_2, \ldots, u_m), & u_m(0) = u_{m,0} \end{aligned}$$

where $u_i \in R^{n_i}$, $\mathbf{u} = (u_1^T, u_2^T, \ldots, u_m^T)^T$, $f_i : R^n \to R^{n_i}$, $F = (f_1^T, f_2^T, \ldots, f_m^T)^T$, $1 \le i \le m$, and $\sum_{i=1}^m n_i = n$. Two common forms of waveform relaxation treat these subsystems independently. In the waveform Jacobi (WJ) method, each subsystem is integrated using the previous value for all other subsystems, that is,

$$\dot{u}_i^{[k+1]} = f_i(u_1^{[k]}, u_2^{[k]}, \ldots, u_{i-1}^{[k]}, u_i^{[k+1]}, u_{i+1}^{[k]}, \ldots, u_m^{[k]}), \qquad u_i^{[k+1]}(0) = u_{i,0}.$$

The obvious attraction of this method is that each subsystem can be integrated independently of the others in parallel. In the WGS method, the subsystems are integrated in sequence using the most recent values of the other subsystems, that is,

$$\dot{u}_i^{[k+1]} = f_i(u_1^{[k+1]}, u_2^{[k+1]}, \ldots, u_{i-1}^{[k+1]}, u_i^{[k+1]}, u_{i+1}^{[k]}, \ldots, u_m^{[k]}), \qquad u_i^{[k+1]}(0) = u_{i,0}.$$

Both of these are particular examples of the general idea of *splitting*. In splitting we select an arbitrary function $G(\mathbf{v}, \mathbf{u})$ and trivially rewrite (1) as

$$\dot{\mathbf{u}} - G(\mathbf{u}, \mathbf{u}) = F(\mathbf{u}) - G(\mathbf{u}, \mathbf{u}) \tag{2}$$

With this we can use the iteration

$$\begin{aligned} \dot{\mathbf{u}}^{[k+1]} - G(\mathbf{u}^{[k+1]}, \mathbf{u}^{[k]}) &= F(\mathbf{u}^{[k]}) - G(\mathbf{u}^{[k]}, \mathbf{u}^{[k]}) & (3) \\ \mathbf{u}^{[k+1]}(0) &= \mathbf{u}_0 & (4) \end{aligned}$$

In this notation the i-th component of G in the WJ method is given by

$$[G(\mathbf{v}, \mathbf{u})]_i = f_i(u_1, \ldots, u_{i-1}, v_i, u_{i+1} \ldots, u_m)$$

while for WGS it is

$$[G(\mathbf{v}, \mathbf{u})]_i = f_i(v_1, \ldots, v_i, u_{i+1}, \ldots, u_m).$$

Note that the splitting $G = 0$ corresponds to the Picard method, while the splitting $G(\mathbf{v}, \mathbf{u}) = F(\mathbf{v})$ corresponds to integrating the original ODE in the first iteration.

We want to choose a splitting to accomplish several objectives: we want fast convergence, and for this, $G(\mathbf{v}, \mathbf{u})$ should, in some senses, be like $F(\mathbf{v})$; and we would also like the ODE (3) to be easy to integrate (by chosing a very simple function G). The Picard method yields the simplest integration: it is only a quadrature. However, its convergence is slow unless F is almost independent of $\mathbf{u}$. WJ and WGS require slightly more complex integrations, but they are simpler than the original problem because it has been reduced to a number of simpler subsystems. The important characteristic of these two splittings is that no comunication from other subsystems is needed during the integration of a single subsystem; it can happen prior to the integration. The goal of fast convergence is achieved by methods like waveform Newton in which we choose

$$G(\mathbf{v},\mathbf{u}) = \frac{\partial F}{\partial \mathbf{u}}\mathbf{v}.$$

For this, $G(\mathbf{v},\mathbf{u})$ "looks like" $F(\mathbf{v})$ in that their first derivatives are identical at $\mathbf{u}$. Note that the error in successive iterates of a waveform method, $\epsilon^{[k]} = \mathbf{u}^{[k]} - \mathbf{u}$ satisfies

$$\dot{\epsilon}^{[k+1]} + G_{\mathbf{v}}\epsilon^{[k+1]} = (F_{\mathbf{u}} - G_{\mathbf{v}})\epsilon^{[k]} - O(\epsilon^{[k]} + \epsilon^{[k+1]})^2. \quad (5)$$

In waveform Newton, the first term on the right-hand side vanishes.

The fast convergence properties of waveform Newton are offset by the greater cost of each iteration. First there is the expensive computation of $\partial F/\partial \mathbf{u}$ at each step. Second, when the system of ODEs is very large and we try to integrate them on a parallel processor, there will be extensive communication between subsystems which destroys the potential advantages of parallel execution.

In the next section we will introduce a measure of the rate of convergence of a waveform method. Clearly waveform relaxation can not converge unless the original system (1) has a solution. If we require that F is Lipschitz continuous with respect to $\mathbf{u}$, then a unique solution for the system exists. We will also assume continuity of as many derivatives as necessary for our analysis. In [8] it is shown that the waveform relaxation algorithm is a contraction mapping in an exponentially scaled norm and in [6] waveform relaxation is proved to converge superlinearly on any finite intervals. In the following chapter we will look at waveform relaxation from a different point of view. Instead of discussing the convergence property of waveform relaxation, we will discuss how the order of accuracy of successive approximate solutions increases.

2 Accuracy Increase in Waveform Relaxation

Different splittings (2) yield different waveform methods with different convergence properties. All waveform relaxation methods converge superlinearly on any finite interval so it is not possible to use a measure like rate of convergence to compare different splittings. In this section we consider, instead, the rate of increase in the order of successive approximations. Consider Picard applied to the simple problem $y' = y$, $y(0) = 1$ starting from the approximation $y^{[0]}(t) = 1$. The k-th iterate is $y^{[k]}(t) = 1 + t + t^2/2! + \ldots + t^k/k!$. Each successive iterate has one additional correct term in its power series. This is not peculiar to simple problems. Consider the scalar Riccati equation;

$$\dot{u} = u - 2u^2, \quad u(0) = 3. \quad (6)$$

The exact solution to this equation is

$$u = \frac{1}{2 - \frac{5}{3}e^{-t}}.$$

It's Taylor series expansion is:

$$\begin{aligned} u(t) &= 3 - 15t + \frac{165t^2}{2} - \frac{905t^3}{2} + \frac{19855t^4}{8} - \frac{108901t^5}{8} + \\ & \frac{3583811t^6}{48} - \frac{137595781t^7}{336} + O(t)^8. \end{aligned}$$

The first three iterates by the Picard method starting with $u^{[0]} = 3$ are:

$$\begin{aligned} u^{[1]}(t) &= u(t) + \frac{-165t^2}{2} + O(t)^3 \\ u^{[2]}(t) &= u(t) + \frac{605t^3}{2} - O(t)^4 \\ u^{[3]}(t) &= u(t) + \frac{-6655t^4}{8} + O(t)^5. \end{aligned}$$

Neither is it peculiar to the Picard method. Consider the waveform method applied to the scalar Riccati equation above with the splitting $G(\mathbf{v}, \mathbf{u}) = \mathbf{v}$. The iteration is

$$\dot{u}^{[k+1]} - u^{[k+1]} = -2(u^{[k]})^2, \qquad u^{[k+1]}(0) = 3.$$

The first three approximate solutions are:

$$\begin{aligned} u^{[1]}(t) &= u(t) - 90t^2 + O(t)^3 \\ u^{[2]}(t) &= u(t) + 360t^3 + O(t)^4 \\ u^{[3]}(t) &= u(t) - 1080t^4 + O(t)^5. \end{aligned}$$

From the above we see that exactly one additional correct term is picked up in each iteration. The reason is evident from a consideration of the error term, $\epsilon^{[k]}(t) = \mathbf{u}^{[k]}(t) - \mathbf{u}(t)$, which satisfies (5). If the partial derivatives in that equation are evaluated at suitable points near the solution, the higher order terms can be ignored in that equation, so we find that

$$\epsilon^{[k+1]} = \int_0^t \hat{G}(\tau)\epsilon^{[k]}(\tau)d\tau$$

where $\hat{G}$ is the Greens function for the left hand side of (5). Clearly, if $\epsilon^{[k]}(t)$ is a power series starting with t^{k+1} then $\epsilon^{[k+1]}(t)$ will be a power series starting with t^{k+2}. This leads us to define the order of accuracy of an approximation as follows. Let $u_i(t)$ be the ith component of the exact solution and $z_i(t)$ be the ith component of an approximate solution.

Definition 2.1 *If $z_i(t) - u_i(t) = O(t)^{M_i+1}$ over a fixed, finite interval $[0, T]$, then the order of accuracy, $N(z_i)$, of $z_i(t)$ is M_i for $1 \le i \le n$. The order of accuracy of $\mathbf{z}(t)$, denoted by $N(\mathbf{z})$, is defined as $\min_{1 \le i \le n} N(z_i)$.*

The basic theorem of accuracy increase in waveform relaxation is:

Theorem 2.2 *Suppose that the exact solution of (1) can be written as*

$$\mathbf{u}(t) = \sum_{i=0}^{M} \mathbf{a}_i t^i + O(t)^{M+1} \tag{7}$$

that is, $\mathbf{u}(t) \in C^M$. Given $\mathbf{z}(t)$, an approximate solution, which satisfies $\mathbf{z}(0) = \mathbf{u}_0$, define

$$\mathbf{R}(t) \doteq \mathbf{z}(t) - \mathbf{u}(t) = \sum_{i=N(\mathbf{z})+1}^{M} \mathbf{b}_i t^i + O(t)^{M+1}. \tag{8}$$

Let $\mathbf{y}(t)$ *be the solution to the following system*

$$\dot{\mathbf{y}} - G(\mathbf{y}, \mathbf{z}) = F(\mathbf{z}) - G(\mathbf{z}, \mathbf{z}), \tag{9}$$

where $G : R^{2n} \to R^n$, F *and* G *are sufficiently smooth and define*

$$\mathbf{E}(t) \doteq \mathbf{y}(t) - \mathbf{u}(t) = \sum_{i=N(\mathbf{y})+1}^{M} \mathbf{c}_i t^i + O(t)^{M+1}. \tag{10}$$

Then, if $F(\mathbf{z})$ *and* $G(\mathbf{y}, \mathbf{z})$ *are sufficiently differentiable,*

$$N(\mathbf{y}) \geq N(\mathbf{z}) + 1.$$

Theorem 2.2 gives an inequality. Normally this is an equality unless there is cancellation, sparsity, or a special nature of the problem or splitting. Cancellation is exploited in the waveform Netwon method, while sparsity will be exploited below in the WGS method. A problem with a special character is $\dot{y} = t^{n-1}y$. It increases in accuracy by n at each iteration starting from $t_0 = 0$, although that accuracy increase does not occur from other starting points.

Let us consider the Riccati example with the waveform Newton method. We use the splitting $G(u^{[k+1]}, u^{[k]}) = (1 - 4u^{[k]})u^{[k+1]}$. The iteration is

$$\dot{u}^{[k+1]} - (1 - 4u^{[k]})u^{[k+1]} = 2(u^{[k]})^2, \quad u^{[k+1]}(0) = 3.$$

The first three approximate solutions are

$$\begin{aligned} u^{[1]}(t) &= u(t) + 150t^3 + O(t)^4 \\ u^{[2]}(t) &= u(t) + \frac{45000t^7}{7} + O(t)^8 \\ u^{[3]}(t) &= u(t) + \frac{2700000000t^{15}}{49} + O(t)^{16}. \end{aligned}$$

In this scheme we see that more than one correct term is picked up after each iteration. Actually, the number of correct terms in the Taylor expansion of an approximation almost doubles after each iteration because of the quadratic convergence of the Newton iteration.

This behavior is a particular case of Theorem 2.3 below which relates the order increase to the "closeness" of $G(\mathbf{v}, \mathbf{u})$ to $F(\mathbf{v})$. We define

$$Q_{\mathbf{u}}(\mathbf{v}) = G(\mathbf{v}, \mathbf{u}) - F(\mathbf{v}).$$

If Q is zero, then the "splitting" leads to the solution in one iteration. If $Q_{\mathbf{u}}(\mathbf{v})$ is relatively insensitive to changes in $\mathbf{v}$ near $\mathbf{v} = \mathbf{u}$, we get rapid convergence, as shown by

Theorem 2.3 *If in addition to the assumptions in Theorem 2.2 we assume that*

$$Q_{\mathbf{u}}(\mathbf{v}) - Q_{\mathbf{u}}(\mathbf{u}) = O(\mathbf{v} - \mathbf{u})^q.$$

Then in Theorem 2.2

$$N(\mathbf{y}) \geq q(N(\mathbf{z}) + 1).$$

In particular, if $q = 2$*, that is,* $\partial F/\partial \mathbf{u} = \partial G(\mathbf{v}, \mathbf{u})/\partial \mathbf{v}$ *at* $\mathbf{v} = \mathbf{u}$*, this iteration scheme is the Waveform Newton method and it converges quadratically, i.e., we get orders of accuracy 0, 2, 6, 14, 30, ..., when starting with a constant zeroth iterate.*

Note that $q \geq 1$ for any smooth F and G, but the case $q = 2$ is the only other "practical" one. As noted, for large systems even it is not practical because of the cost of computation of the Jacobian and the communication involved. For the WJ method, $q = 1$ so there is little that can be done to get faster order increase. In the next section we will examine WGS.

3 Accuracy Increase in Waveform Gauss Seidel

The accuracy increase in WGS is dependent on the numbering of the equations. In this section we will show how the average rate of increase is determined from the structure of the dependency digraph and its numbering.

We first reconsider Theorem 2.2 when subsystems are integrated sequentially.

Theorem 3.1 *Consider the equation for the ith component after partitioning,*

$$\dot{u}_i = f_i(u_1, \ldots, u_i, \ldots, u_m), \quad u_i(0) = u_{i,0}. \tag{11}$$

The equation to be solved after applying WGS scheme is

$$\dot{u}_i^{[k+1]} - f_i(u_1^{[k+1]}, \ldots, u_{i-1}^{[k+1]}, u_i^{[k+1]}, u_{i+1}^{[k]}, \ldots, u_m^{[k]}) = 0, \quad u_i^{[k+1]}(0) = u_{i,0}. \tag{12}$$

Assume that

$$\left.\begin{array}{lllll} E_j^{[k+1]} & \doteq u_j^{[k+1]} - u_j & = & O(t)^{N_j^{[k+1]}} & \textit{for } j \leq i, \\ E_j^{[k]} & \doteq u_j^{[k]} - u_j & = & O(t)^{N_j^{[k]}} & \textit{for } j > i. \end{array}\right\} \tag{13}$$

and all the $E_j^{[k]}$'s and $E_j^{[k+1]}$'s are sufficiently smooth. Then

$$N_i^{[k+1]} \geq \min(N_1^{[k+1]}, \ldots, N_{i-1}^{[k+1]}, N_{i+1}^{[k]}, \ldots, N_m^{[k]}) + 1, \tag{14}$$

with equality unless there is cancellation.

This theorem assumes that all variables appear in all equations. If variable j appears in the equation for variable i only if $j \in I_i$ where I_i is a subset of $[1, \ldots, m]$, then (14) can be replaced by

$$N_i^{[k+1]} \geq \min_{j \in I_i}(N_j^{[k+H(i-j)]}) + 1, \tag{15}$$

where $H(i-j) = 1$ if $i > j$ and 0 otherwise. (A similar result holds for WJ with H identically zero.)

It is instructive to consider some simple examples. The simple harmonic oscillator,

$$\begin{array}{rlr} y' = & z, & y(0) = 0 \\ z' = & -y, & z(0) = 1 \end{array}$$

can be solved by the WJ method to get the iterates

$$y^{[1]} = t, \qquad z^{[1]} = 1,$$
$$y^{[2]} = t, \qquad z^{[2]} = 1 - \frac{t^2}{2!},$$
$$y^{[3]} = t - \frac{t^3}{3!}, \qquad z^{[3]} = 1 - \frac{t^2}{2!}.$$

We see the usual order increase of one per step (partially masked by the oddness and eveness of the solutions in this case). If WGS is used, we get the iterates

$$y^{[1]} = t, \qquad z^{[1]} = 1 - \frac{t^2}{2!}$$
$$y^{[2]} = t - \frac{t^3}{3!}, \qquad z^{[2]} = 1 - \frac{t^2}{2!} + \frac{t^4}{4!}$$
$$y^{[3]} = t - \frac{t^3}{3!} + \frac{t^5}{5!}, \qquad z^{[3]} = 1 - \frac{t^2}{2!} + \frac{t^4}{4!} - \frac{t^6}{6!}$$

with an order increase of two per step after the first step.

If we consider the system of three equations

$$\begin{aligned} \dot{y}_1 &= f_1(y_1, y_3) \\ \dot{y}_2 &= f_2(y_1, y_2) \\ \dot{y}_3 &= f_3(y_2, y_3) \end{aligned}$$

and compute the order increase on three successive iterations we get

	Iteration No					
Equation	0	1	2	3	4	...
y_1	0	1	4	7	10	...
y_2	0	2	5	8	11	...
y_3	0	3	6	9	12	...

Order of Accuracy

which shows an accuracy increase of 3 per iteration after the first. Three is the number of equations, but more importantly, it is the length of the only cycle in the dependency graph for these equations. The *dependency graph* for a set of ordinary differential equations is defined as a graph containing a node for each variable and a directed edge from node i to node j if variable numbered i appears in the equation defining variable j, that is, if $\partial F_j/\partial u_i$ is non zero. Hence, the graph displays the sparsity structure of the Jacobian. For the problem above, the graph is

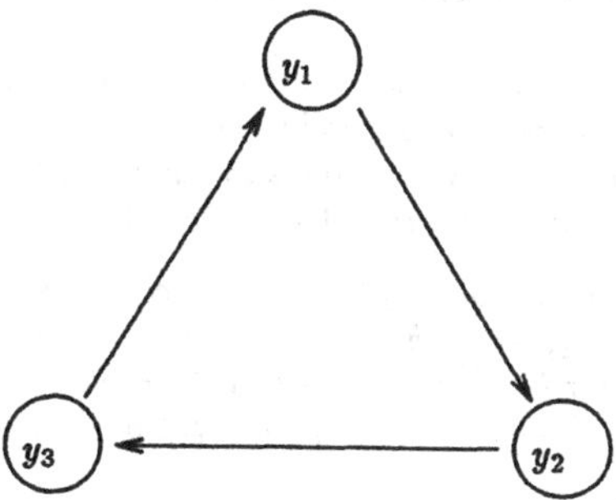

Its structure is independent of the numbering, but the numbering affects the average accuracy increase. Consider the same example with a different numbering:

$$\begin{aligned}\dot{z}_1 &= f_1(z_1, z_2)\\ \dot{z}_2 &= f_3(z_3, z_2)\\ \dot{z}_3 &= f_2(z_1, z_3)\end{aligned}$$

This leads to the order increase pattern

Equation	Iteration No					
	0	1	2	3	4	...
z_1	0	1	2	4	5	...
z_2	0	1	3	4	6	...
z_3	0	2	3	5	6	...

Order of Accuracy

which shows an accuracy increase of 3 every 2 iterations. The dependency digraph for this exmple is

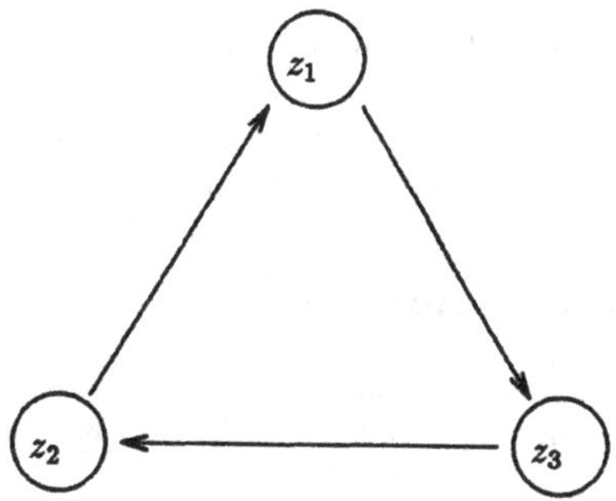

which shows a cycle of length 3, but this time the numbering is "out of order" around the cycle.

In the last example, the two iterations that were required to achieve an order increase of three (which is the cycle length) were determined by the number of times that the numbering of nodes decreased as the cycle was traversed once in the direction specified by the edges. This is also the number of *ascending chains* in the cycle, where

Definition 3.2 *An ascending chain of length l in a cycle $\tilde{A}$ is a sequence of nodes with numerical ordering $j_0, j_1, \ldots, j_{l-1}$, such that (1) $j_0 < j_1 < \cdots < j_{l-1}$, (2) there exists an edge from node j_i to node j_{i+1} for $i = 0, 1, \ldots, l-2$ in the cycle, and (3) no ascending chain in cycle $\tilde{A}$ contains $\{j_0, j_1, \ldots, j_{l-1}\}$ as a subsequence (in other words, the ascending chain is as long as possible).*

It follows from the definition that any cycle can be decomposed into a mutually exclusive set of ascending chains. With the equality assumption in Theorem 3.1, we know that after one WGS iteration, each node, j_i, in an ascending chain can not have order of accuracy more than one greater than the order of accuracy of its predecessor node, j_{i-1}, in the chain at this iteration, while the first node in an ascending chain can not have order of accuracy greater than one plus the order of accuracy, prior to the iteration, of its predecessor, the last node in the chain that precedes it. (If no other node except its predecessor in the cycle is connected to a node k, it will achieve exactly this order increase.) This implies that

Theorem 3.3 *If a cycle $\tilde{A}$ of length C consists of d ascending chains, then, after the first iteration, the accuracy increase at the internal nodes of $\tilde{A}$ due to d WGS iterations is bounded by C.*

In particular, when $d = 1$, that is all the internal nodes of a cycle are solved in cyclic order, the accuracy increase in one WGS iteration is then bounded by the cycle length.

We are interested in the average rate of accuracy increase, so we define the *average accuracy increase* of a node to be $\lim_{k\to\infty} p_k/k$ where p_k is the order of the node after the k-th iteration. We define the average accuracy increase of a numbered graph to be the minimum average accuracy increase of any of its nodes. The previous result says that the average accuracy increase is bounded by C/d for a cycle, hence we know that the average accuracy increase of a graph is bounded by $\min_i(C_i/d_i)$ where the minimum is taken over all cycles, $i \in I$, of length C_i with number of ascending chains d_i. We call a cycle whose C/d ratio achieves this minimum a *minimum cycle*. The key result, shown in [3] is

Theorem 3.4 *Suppose a minimum cycle in the dependency graph of a given system is of length C and has d ascending chains. Then the waveform Gauss-Seidel method applied to this system has average accuracy increase C/d.*

This tells us what the average is, but not how to number the graph to maximize that average. Finding such a numbering appears to be NP-hard, so a heuristic approach will almost certainly have to be used. The result above suggests using heuristics that attempt to maximize the length of ascending chains are appropriate, particularly those in short cycles.

4 Parallelism and Extensions

Although the waveform Gauss-Seidel is usually thought of as a serial method, there are several ways in which it can be used for parallel computation. In one approach, several nodes can be integrated in parallel. In this approach for a p processor system, up to p nodes can be numbered with each number in the sequence so long as there are no branches between nodes with the same number. At each step of a single waveform Gauss-Seidel sweep, all nodes with the same number can be integrated in parallel. Since they are mutually independent, the order result of Theorem 3.1 applies, and hence all results in this paper apply. This approach amounts to "collapsing" several nodes into a single node and then observing that there are independent equations within that node.

In another approach, the integration of later nodes can be staggered in time. The first node is integrated over a small interval of time, $[t_0, t_1]$. Its output is then ready to be used for the integration of the second node over that interval while the first node is integrated over a second interval, $[t_1, t_2]$, and so on. In the m-th step, the i-th node is being integrated over the $(m+1-i)$-th time interval for $i = 1, \ldots, m$. Its inputs for that time interval from earlier variables with lower i values have already been computed. In the $(m+1)$-st step, the first node begins the second iteration of the first interval while each of the others advance.

The results above assumed that there was no undeclared structure or cancellation so that the lower bound on order increase predicted by the theorems was the actual value. If each node represents a collection of equations, they may have a sparse structure. It is convenient to think of the collection of equations itself as a directed graph, where the number of nodes is the number of equations. We will call this the *full graph*. The grouping of equations into subsystems is equivalent to the grouping of nodes of the full graph into supernodes. Each supernode is a subgraph of the full graph, and the edges of the supernode graph are the edges of the full graph joining nodes grouped in different supernodes. When a single supernode is integrated, each equation in the supernode is integrated simultaneously.

Suppose that the incoming edges to a particular supernode, P, come from nodes $i \in I_P$ in the full graph with orders θ_i prior to the integration. Suppose that the outgoing edges from the supernode P leave node j in the full graph and have orders ϕ_j after the integration. Let the minimum path length from node i to node j *using only incoming edges to P plus edges in the subgraph for supernode P* be L_{ij}. Then we have

$$\phi_j \geq \min_{i \in I_p}(L_{ij} + \theta_i),$$

where equality holds unless there is numerical cancellation or a Jacobian element is zero. The proof of this is similar to proofs of the earlier theorems.

References

[1] Gear, C. W., "Automatic multirate methods for ordinary differential equations", *Proc. IFIP* 1980, 717-722, North-Holland Publishing Company.

[2] Gear, C. W. and Wells, D. R., "Multirate linear multistep methods", *BIT* 24 (1984), 484-502.

[3] Juang, F. L., "Waveform methods for ordinary differential equations," PhD Thesis, Department of Computer Science, University of Illinois at Urbana-Champaign, November, 1989.

[4] Lelarasmee, E., "The waveform relaxation methods for the time domain analysis of large scale nonlinear dynamical systems", Ph.D. dissertation, University of California, Berkeley.

[5] Lelarasmee, E., Ruehli, A. E., and Sangiovanni-Vincentelli, A. L., "The waveform relaxation method for time-domain analysis of large scale integrated circuits", *IEEE Trans. on CAD of IC and Sys.* Vol. 1, No. 3, pp. 131-145, July 1982.

[6] Nevanlinna, O., "Remarks on Picard-Lindelöf iteration", REPORT-MAT-A254, Helsinki University of Technology, Institute of Mathematics, Finland, December 1987.

[7] Wells, D. R., "Multirate linear multistep methods for the solution of systems of ordinary differential equations", Report No. UIUCDCS-R-82-1093, Department of Computer Science, University of Illinois at Urbana-Champaign, 1982.

[8] White, J., Odeh, F., Sangiovanni-Vincentelli, A.L., and Ruehli, A., "Waveform relaxation: theory and practice", Memorandum No. UCB/ERL M85/65, 1985, Electronics Research Laboratory, College of Engineering, University of California, Berkeley.

DIFFUSIVELY COUPLED DYNAMICAL SYSTEMS

J.B. KELLER
Departments of Mathematics and Mechanical Engineering
Stanford University
Stanford, CA 94305

ABSTRACT. The reaction-diffusion equation $u_\tau = -f(u) + \varepsilon^2 \Delta u$ is considered for small values of ε. When $\varepsilon = 0$ the equation reduces to an ordinary differential equation or a dynamical system at each x. This system may have one, two or a finite number of attracting rest points, a connected manifold of rest points, two manifolds of rest points, an attracting limit cycle, or a family of periodic orbits. In each of these cases the behavior of $u(x, \tau, \varepsilon)$ is described for small values of ε. It involves flow by curvature or curve shortening in the case of two rest points, diffusion to a harmonic map in the case of a manifold of rest points, etc.

1. Introduction

A dynamical system can be described by a state vector $u(\tau)\varepsilon R^m$ which is governed by an ordinary differential equation

1.1. $$\frac{du}{d\tau} = -f(u), \qquad u(0) = g\,.$$

Suppose that one such system is located at each point x in a domain Ω in R^n, with state vector $u(x, \tau)$ and initial state $g(x)$. If these systems are coupled together by a weak diffusive interaction, their evolution will obey a parabolic partial differential equation such as

1.2. $$u_\tau = -f(u) + \varepsilon^2 \Delta u, \qquad u(x, 0) = g(x)\,.$$

In addition, at the boundary of Ω, u will satisfy some condition, for example

1.3. $$\partial_n u = 0, \qquad x \in \partial\Omega\,.$$

The equation (1.2) is called a reaction-diffusion equation because of its occurrence in the description of chemical reactions. Then u represents the vector of concentrations of various chemical species at x, τ; $-f(u)$ specifies the rate of change of these concentrations due to chemical reactions, and $\varepsilon^2 \Delta u$ determines the rate of change of u due to diffusion with the small diffusion coefficient ε^2. Equation (1.2) also occurs in population dynamics. Then u represents the vector of number densities of various populations at x, τ; $-f(u)$ describes their rates of increase due to birth, death, predation, etc., and $\varepsilon^2 \Delta u$ represents the effect

R. Spigler (ed.), Applied and Industrial Mathematics, 49–56.

of migration. In both the chemical and population interpretations, (1.3) represents the condition at an impermeable boundary.

The question we consider is "How is the evolution of these dynamical systems modified by the small diffusive coupling?" The answer depends upon the nature of the function $f(u)$. We shall describe the results obtained for various classes of functions $f(u)$ by a number of authors.

In section 2 we show that if $f(u)$ is such that (1.1) has a single globally attracting fixed point U_1, then the solution $u(x,t)$ of (1.2) will tend to U_1 at every x, and the rate of approach will be only slightly modified by diffusion. However, if (1.1) has two or more attracting fixed points, say U_1 and U_2, that in some part of Ω, u may tend to U_1 and in another part u may tend to U_2. Consequently a front will develop across which u changes rapidly from U_1 to U_2. In section 3 we show that this front will move in a manner which can be determined by boundary layer analysis. The result is that the front moves normal to itself with velocity εc_0 where c_0 depends upon $f(u)$. If $f(u) = V_u(u)$, c_0 is proportional to $|V(u_2) - V(u_1)|$ and directed toward the side with the greater value of $V(u)$. When $c_0 = 0$ the front moves at a much slower speed proportional to ε^2 times the mean curvature of the front. This is called flow by curvature. In section 4 we show that this flow will usually lead to the disappearance of the front, so that u will tend to U_1 or to U_2 throughout Ω. However for some special domains Ω, the front may tend to a fixed position, and then u will tend to a nearly piecewise constant function (Rubinstein, Sternberg, and Keller [1]).

In section 5 we consider $f(u)$ such that (1.1) has a smooth manifold M of attracting fixed points. Then, at each point x of Ω the solution u will tend quickly to some point on M. But then on a longer time scale diffusion will act, while the values of u remain on M. Ultimately, equilibrium will be attained and the limit $u(x,\infty)$ will be a harmonic map from Ω to M (Rubinstein, Sternberg and Keller [2]).

If (1.1) has two disconnected smooth manifolds M_1 and M_2 of attracting fixed points, the preceding description applies to M_1 and M_2. In addition there will be a front in Ω separating points where u tends to a value in M_1 from those where u tends to M_2. This front will move as in the case of two fixed points [2].

When (1.1) has an attracting limit cycle, at each x the solution u will tend quickly to that limit cycle. However, the phase with which $u(x,t)$ describes the limit cycle will depend upon x. Then on a longer time scale, diffusion will act to modify the phase and ultimately bring all points into the same phase (Neu [3]). When (1) has a $k+1$-dimensional manifold, e.g. a torus, composed of attracting orbits, at each x u will tend quickly to one of these orbits. Then the phase and the k parameters labelling the orbit will evolve slowly in time due to diffusion.

Some of the other phenomena which can occur are described in the books of Fife [4], [5].

2. One attracting fixed point

A natural classification of the problem (1.2), (1.3) is based upon the motion of the uncoupled systems. For them $\varepsilon = 0$ so (1.2) becomes the ordinary differential equation (1.1) for each x with $g = g(x)$ and the boundary equation (1.3) must be omitted.

Suppose $f(u)$ is such that (1.1) has a single globally attracting fixed point U_1. Then for every initial value g, the solution of (1.1) will tend to U_1 as $\tau \to \infty$. As a consequence it is to be expected that, for ϵ sufficiently small, the solution $u(x,t)$ will tend to the constant U_1 as $\tau \to \infty$. This expectation can be verified formally by writing $u(x,t,\varepsilon)$ as an expansion in powers of ε^2

2.1. $$u(x,\tau,\varepsilon) = u_0(x,\tau) + \varepsilon^2 u_,(x,\tau) + 0(\varepsilon^4) .$$

By substituting (2.1) into (1.2) and equating terms in ε^0 we find that u_0 satisfies (1.1), and therefore $u_0(x,\tau) \to U_1$ as $\tau \to \infty$. However this solution $u_0(x,\tau)$ will not satisfy (1.3) in general. It will do so if $\partial_n g(x) = 0$ for $x \varepsilon \partial w$ but then u_1 will not satisfy (1.3).

We conclude that (2.1) holds only in the interior of w and a different expansion of $u(x,\tau,\varepsilon)$ holds for x near ∂w. We call it a boundary layer expansion. It can be found by first introducing the stretched variable $\nu = n/\varepsilon$, where $n(x)$ is the distance from x to ∂w, and $n-1$ coordinates y orthogonal to $n(x)$. Then for x near $\partial\Omega$ we write

2.2 $$u(x,\tau,\varepsilon) = w_0(\nu,y,\tau) + \varepsilon^2 w_1(\nu,y,\tau) + 0(\varepsilon^4) .$$

Then from (1.2) and (1.3) we find that w_0 satisfies the equations

2.3. $$\begin{aligned} &\partial_\tau w_0 = \partial_\nu^2 w_0 - f(w_0)\,, \quad w_0(\nu,y,0) = g(0,y)\,,\\ &\partial_\nu w_0(0,y,\tau) = 0\,, \quad \lim_{\nu\to\infty} w_0(\nu,y,\tau) = u_0(0,y,\tau)\,. \end{aligned}$$

The solution $w_0(\nu,y,\tau)$ of (2.3) tends to U_1 as $\tau \to \infty$.

This result and similar considerations for further terms show that when (1.1) has a single globally attracting fixed point U_1, the solution $u(x,t,\varepsilon)$ of (1.2) and (1.3) tends to U_1 as $t \to \infty$. The solution can be represented by an interior expansion (2.1) and a boundary layer expansion (2.2), and the terms in these expansions can be found successively by solving simpler problems.

When the boundary condition (1.3) is replaced by the condition $u(x,\tau) = 0$ for $x\varepsilon\partial w$, then w_0 tends to a non-constant limit as $\tau \to \infty$. Thus in this case the boundary layer persists. Then the ultimate steady solution is asymptotic to the constant U_1 in the interior of Ω with a boundary layer near $\partial\Omega$. Within this layer u changes rapidly from U_1 to the value zero at $\partial\Omega$. (Owen, Sternberg and Rubinstein [6].)

3. Two or more attracting fixed points

Next we suppose that (1.1) has two or more attracting fixed points $U_1, U_2, \cdots U_K$ with $K \geq 2$. Corresponding to each U_j there is a basin of attraction $B_j \subset R^m$ such that when $u(0)\varepsilon B_j$ then $u(\tau) \to U_j$ as $\tau \to \infty$. Thus Ω is divided into subdomains within each of which the values of $g(x)$ are all in the same B_j. On a subdomain boundary, which we shall call a front, $g(x)$ lies on the boundary between two of the B_j. Within a subdomain where $g(x)\varepsilon B_j$ it follows that $u_0(x,\tau) \to U_j$ as $\tau \to \infty$.

Since $u_0(x,\tau)$ tends to two different values on the two sides of a front, i.e. a boundary between subdomains, the expansion (2.1) cannot be valid on and near a front. Instead a boundary layer expansion will hold there. Furthermore, the front will not remain stationary, but it will move in a manner which must be determined.

To describe the boundary layer and motion of the front we follow Rubinstein, Sternberg and Keller [1]. We introduce introduce the two time variables $t = \varepsilon\tau$ and $\eta = \varepsilon t = \varepsilon^2\tau$, and a function $\varphi(x,t,\eta)$. Then we write

3.1. $$u(x,\tau,\varepsilon) = u^0(z,x,\tau,t,\eta) + \varepsilon u^1(z,x,\tau,t,\eta) + 0(\varepsilon^2)$$

where $z = \varepsilon^{-1}\varphi(x,t,\eta)$. Substituting (3.1) into (1.2), and equating coefficients of ε^0 and ε^1 yields equations for u^0 and u^1. We assume that both u^0 and u^1 tend to traveling waves as $\tau \to \infty$:

3.2.
$$u^0(z,x,\tau,t,\eta) \sim Q(z - c\tau, x, t, \eta)\,,$$
$$u^1(z,x,\tau,t,\eta) \sim P(z - c\tau, x, t, \eta) \quad \text{as } \tau \to \infty\,.$$

Upon substituting (3.2) into the equation for u^0 we obtain an ordinary differential equation for Q as a function of $z - ct$:

3.4
$$-(\nabla\varphi)^2 Q'' + (\varphi_t - c)Q' + f(Q) = 0$$

We require that $Q \to U_i$ as $z \to -\infty$ and $Q \to U_j$ as $z \to +\infty$ where U_i and U_j are the limiting values of u on the two sides of the front. To find it we introduce $s = (z - ct)/|\nabla\varphi|$ and set $R(s) = Q(z - ct)$ in (3.4) to get

3.5.
$$R_{ss} - \frac{(\varphi_t - c)}{|\nabla\varphi|} R_s - f(R) = 0$$

We assume that (3.5) has a solution tending to U_i and U_j at $\pm\infty$ for a unique value c_0 of the coefficient of R_s:

3.6.
$$(c - \varphi_t)/|\nabla\varphi| = c_0$$

The left side of (3.6) is the normal velocity of a level set of $\psi(x,t,\eta) = \varphi(x,t,\eta) - ct$, and R is a function of $\varepsilon^{-1}\psi/|\nabla\psi|$. Thus the level sets of R and therefore of u, move normal to themselves at the constant speed c_0.

In the special case in which $f(u) = V_u(u)$, where V is a scalar function, we multiply (3.5) by R_s and integrate from $-\infty$ to $+\infty$ to obtain

3.7.
$$c_0 = [V]\left(\int_{-\infty}^{\infty} R_s^2\,ds\right)^{-1}$$

Here $[V] = V(u_j) - V(u_i)$. This equation shows that the front moves toward the side with the larger value of V. It also shows that when $[V] = 0$ then $c_0 = 0$, so that the front does not move on the t time scale.

Returning to general $f(u)$, we consider the case in which $c_0 = 0$. Then we use (3.3) in the equation for u^1 to get an equation for P. From this equation we derive the relation

3.8.
$$\varphi_\eta/|\nabla\varphi| = \kappa_\varphi$$

where κ_φ is the mean curvature of the level set $\varphi =$ constant. Thus when $c_0 = 0$ the level sets of φ, and therefore of R and of u, move normal to themselves on the η timescale with velocity $-\kappa_\varphi$. This motion of a surface is called flow by curvature. It was derived before in a less formal manner by Allen and Cahn [7]. Bronsard and Kohn [8] have proved that (3.8) is valid for the spherically symmetic case, and de Mottoni and Schatzman [9] have proved it in general.

Finally we consider the intersection of a front, defined by $\psi = 0$, with $\partial\Omega$. By applying the boundary condition (1.3) to the solution $u \sim R(\psi(x,t,\eta)/\varepsilon|\nabla\psi|)$, which is valid as $\tau \to \infty$, we conclude that $\partial_n\psi(x,t,\eta) = 0$ for $x\varepsilon\partial\Omega$ and $\psi = 0$. This shows that the front must intersect the boundary normally when $\partial_n u = 0$ on $\partial\Omega$. However when the bounday condition is $u = 0$ on $\partial\Omega$, and $f(u) = V_u(u)$, then the front must intersect the boundary at a certain constant angle which depends upon the function $V(u)$ and the values U_i and U_j on the two sides of the front. (Owen, Sternberg and Rubinstein [6].)

When $c_{ij} \neq 0$ for every pair (i,j), the system cannot have a limiting steady state with a front. Then the only steady state is $u = U_i$ throughout Ω for some i. However the system may not tend to a steady state, but may tend to a nearly piecewise constant state with continually moving fronts. For example, suppose there are three attracting fixed points U_1, U_2, U_3 with $c_{12} = c_{23} = c_{31}$ and that Ω is the exterior of a circle in R^2. Then one can imagine a spiral wave with three involutes of the circle as fronts, separating regions where $u = U_1, U_2$, or U_3, with the fronts continuously evolving.

If $c_{ij} = 0$ for some pair i,j then there can be steady states with fronts for some special domains Ω, as we shall see.

4. Flow by curvature and curve shortening

The flow by curvature of a curve embedded in the plane R^2 has been studied by Gage and Hamilton [10] and by Grayson [11]. In that case the equation (3.8) can be reformulated as an equation for a single level curve $x = \gamma(x,t)$, $0 \leq s \leq L(t)$, where s is arclength and t is used instead of η. Then (3.8) implies that γ satisfies

4.1. $$\gamma_t = \gamma_{ss}\ , \ 0 \leq s \leq L(t)\ .$$

In [10] it is shown that the length $L(t)$ of the curve decreases according to

4.2. $$\frac{dL(t)}{dt} = -\int_0^{L(t)} \kappa^2(s,t)ds\ .$$

Here κ is the curvature of the curve. Because of (4.2) flow by curvature of a plane curve is also referred to as curve shortening. Gage and Hamilton [10] proved that a closed convex curve evolving according to (4.1) will shrink to a point in a finite time. Grayson [11] has extended this result to any simple closed curve. Furthermore they have shown that $\gamma(s,t)$ lies in the interior of $\gamma(s,0)$ for all $t > 0$.

The preceding results also apply to any curved front in a simply connected domain Ω in R^2, provided that the initial curve does not intersect $\partial\Omega$. If it does intersect $\partial\Omega$ then it must do so orthogonally, as was pointed out after (3.8). In that case Rubinstein, Sternberg and Keller [1] showed that if two smooth curves $\gamma_1(s,t)$ and $\gamma_2(s,t)$ both evolve according to (4.1) and do not intersect at $t = 0$, then they never intersect. This comparison result is based upon a Hopf-type corner lemma for parabolic equations, which they prove.

This result can be used to study possible equilibrium or time independent fronts $\gamma_e(s)$. Since the front velocity is proportional to its curvature, an equilibrium front $\gamma_e(s)$ must have curvature zero. Therefore, γ_e must be a straight line segment, and it must also be orthogonal to $\partial\Omega$ at its endpoints. The following conclusion about γ_e is proved in [1] using the comparison lemma. *If γ_e is of locally maximal length among neighboring straight line segments with endpoints on $\partial\Omega$, then it is unstable under the flow (4.1). On the other hand,*

if γ_e is strictly shorter than any nearby curve with endpoints on $\partial\Omega$, then it is stable under this flow.

This result is related to the earlier work of Sternberg [12], and Kohn and Sternberg [13], on the local minimizers of the functional $F_\varepsilon(u)$ defined by

4.3. $$F_\varepsilon(u) = \int_\Omega \left\{\varepsilon^{-1}V[u(x)] + \frac{\varepsilon}{2}|\nabla u(x)|^2\right\} dx$$

These minimizers are the possible equilibrium solutions of the equation $u_t = -\delta F_\varepsilon(u)/\delta u$, which is just (1.2) with $f(u) = V_u(u)$. They assumed that V has two minimizers u_1 and u_2 with $V(u_1) = V(u_2)$, with $u\varepsilon R^1$ or $u\varepsilon R^2$. They showed that then as $\varepsilon \to 0$, any local minimizer of $F_\varepsilon(u)$ converges to a piecewise constant function $u_0(x)$ with values u_1 and u_2. In addition, the interfaces across which $u_0(x)$ jumps are local minimizers of surface area. Our analysis shows how such a $u_0(x)$ can result from the solution of (1.2) as $t \to \infty$ and $\varepsilon \to 0$.

5. A connected manifold of equilibria

Now we consider the case in which (1.1) has a smooth compact connected d- dimensional manifold M of rest points in R^m, i.e. points at which $f(u) = 0$. We also assume that as $\tau \to \infty$, every solution $u(\tau)$ of (1.1) tends to some point $u(\infty)\varepsilon M$. Then following Rubinstein, Sternberg and Keller [2], we write the solution of (1.2) and (1.3) in the form

5.1. $$u(x,\tau,\eta,\varepsilon) = u^0(x,\tau,\eta) + \varepsilon^2 u^1(x,\tau,\eta) + 0(\varepsilon^4)\ .$$

By using (5.1) in (1.2) we find that u^0 satisfies (1.1). Therefore as $\tau \to \infty$ our assumption implies that u^0 tends to a point $u^0(x,\infty,\eta)\varepsilon M$. When $\eta = 0$ we have from (1.2) that $u^0(x,0,0) = g(x)$, and we denote by $w(x)$ the limit of this solution:

5.2. $$u^0(x,\infty,0) = w(x)\varepsilon M\ .$$

Thus during the initial period, at every point $x\varepsilon\Omega$ the solution tends to some point $w(x)\varepsilon M$.

Next we derive the equation for u^1 by using (5.1) in (1.2) and retaining terms of order ε^2 to get

5.3. $$u^1_\tau + f_u(u^0)u^1 = -u^0_\eta + \Delta u^0\ .$$

Then we make the assumption that as $\tau \to \infty$, u^1 tends to a limit $u^1(x,\infty,\eta)$ and we obtain from (5.3)

5.4. $$f_u(u^0)u^1(x,\infty,\eta) = -u^0_\eta + \Delta u^0\ .$$

Let $l_1(u^0),\ldots,l_d(u^0)$ be an orthonormal basis of the tangent space to M at u^0. These are the left null vectors of $f_u(u^0)$, so (5.4) yields

5.5. $$l_i(u^0)\cdot[-u^0_\eta + \Delta u^0] = 0, \quad i = 1,\ldots,d\ .$$

Since $u^0(x,\infty,\eta)\varepsilon M$, (5.5) describes a diffusion mapping Ω into the manifold M. The initial value of u^0 is given by (5.2). When the initial function $g(x)$ satisfies $\partial_n g(x) = 0$ for $x\varepsilon\partial\Omega$ it follows that

5.6. $$\partial_n u^0(x,\infty,\eta) = 0, \quad x\varepsilon\partial\Omega\ .$$

This is the boundary condition for the diffusion (5.5).

When $u^0(x,\infty,\eta)$ has a limit $u^0(x,\infty,\infty)$ as $\eta \to \infty$ then (5.5) and (5.6) become equations for this limit:

5.7. $$l_i(u^0) \cdot \Delta u^0 = 0\,, \quad i = 1, \cdots, d\,,$$

5.8. $$\partial_n u^0 = 0\,, \quad x \varepsilon \partial\Omega\,.$$

A solution $u^0(x)$ of (5.7) and (5.8) is called a harmonic map of Ω into M, so (5.5) and (5.6) may be called a diffusion to a harmonic map. Such diffusions have been studied in order to characterize harmonic maps. Coron and Ghidaglia [14] have used a reaction-diffusion equation with a small parameter to examine these diffusions.

The preceding analysis is formal. Therefore in [2] it is supplemented by a proof that $u(x,\tau,\varepsilon)$ does converge to $u^0(x,\eta)$ as $\varepsilon \to 0$ under rather strong hypotheses on $f(u)$. Further results of this type were obtained by Struwe and Chen [15].

The special case in which M is the circle S^1 is considered in [2]. It is shown how to construct the diffusion $u^0(x,\eta)$ and the limiting harmonic map when Ω is multiply connected. Some similar solutions were found by Neu [16].

6. Two manifolds of equilibria

We now suppose that (1.1) has two manifolds of equilibria, M_1 and M_2, separated by an $(m-1)$-dimensional oriented hypersurface S. Suppose that solutions from all initial points on one side of S tend to M_1 and those from points on the other side of S tend to M_2. Then solutions from initial points on S remain on S, so the limit $u^0(x,\infty,\eta)$ will lie on M_1 or M_2 or S. Thus Ω consists of two subdomains Ω_1 and Ω_2 with $u^0(x,\infty,\eta)\varepsilon M_j$ for $x\varepsilon\Omega j$, $j = 1,2$.

In each subdomain the preceding equations (5.1)–(5.8) hold with d replaced by d_j, the dimension of M_j. The boundary between the two subdomains is a moving front at which the boundary layer analysis of section 3 applies. This leads to front motion with a velocity c determined by the points on M_1 and M_2 which are connected by the boundary layer. The case when $f(u) = V_u(u)$ and $V(u)$ is constant on M_1 and on M_2 is considered in [2]. Then c is proportional to the jump in V across the front. When this jump is zero, the front moves according to flow by curvature. The special case in which $V(u) = V(|u|)$ is spherically symmetric is treated in detail.

7. Limit cycles

Finally we consider the case in which (1.1) has a periodic solution. When the periodic solution is an attracting limit cycle, then for ε small, the solution of (1.2) will tend to that limit cycle with a phase $\varphi(x,t)$. Howard and Koppel [17] derived a first order partial differential equation governing the phase. Then Neu [3] derived a modification of the phase equation which included a diffusion term, but only for $x\varepsilon R^1$ and $u\varepsilon R^1$. Hagan [18] extended this analysis to $x\varepsilon R^n$ and $u\varepsilon R^m$. Then Koppel [19] proved that the equation for φ, which had been derived formally, was asymptotically correct for so-called λ, w systems, which are special forms of (1.2) with $u\varepsilon R^2$.

When (1.1) has a family of periodic solutions, we can expect the solution of (1.2) to tend to one of these solutions with some phase $\varphi(x,t)$. Both the phase and the particular periodic solution will vary gradually with x and t. Some results on this problem have been obtained by Rosales, Rubinstein, Sternberg, and Keller (unpublished).

References

[1] Rubinstein, Jacob, Sternberg, Peter, and Keller, Joseph B. (1989) 'Fast Reaction, Slow Diffusion, and Curve Shortening,' SIAM J. Appl. Math. 49, 116–133.

[2] Rubinstein, Jacob, Sternberg, Peter and Keller, Joseph B. (1989) 'Reaction-Diffusion Processes and Evolution to Harmonic Maps,' SIAM J. Appl. Math. 49, 1722–1733.

[3] Neu, John C., (1979) 'Chemical waves and the diffusive coupling of limit cycle oscillators,' SIAM J. Appl. Math. 36, 509–515.

[4] Fife, Paul C. (1979) Mathematical Aspects of Reacting and Diffusing Systems, Springer, New York.

[5] Fife, Paul C. (1988) 'Dynamics of internal layers and diffusive interfaces,' CBMS-NSF Regional Conference Series in Applied Mathematics 53, SIAM, Philadelphia.

[6] Owen, Nicholas, Sternberg, Peter, and Rubinstein, Jacob (1988) preprint.

[7] Allen, S.W., and Cahn, J.W. (1979) 'A microscopic theory for antiphase boundary motion and its application to antiphase domain coarsening,' Acta Metallurgica 27, 1085–1095.

[8] Bronsard, L., and Kohn, R. (1988) preprint.

[9] de Mottoni, P. and Schatzman, M. (1989) 'Évolution géométrique d'interfaces,' Compt. Rend. 309, 453–458.

[10] Gage, M., and Hamilton, R.S. (1986) 'The heat equation shrinking convex plane curves,' J. Differential Geometry 23, 69–96.

[11] Grayson, M. (1987) 'The heat equation shrinks embedded plane curves to round points,' J. Differential Geometry 26, 285–314.

[12] Sternberg, Peter (1988) 'The effect of a singular perturbation on nonconvex variational problems,' Arch. Rational Mech. Anal. 101, 209–240.

[13] Kohn, R.V., and Sternberg, Peter (1989) 'Local minimizers and singular perturbations,' Proc. Roy. Soc. Edinburgh Sect. A., 69–84.

[14] Coron, J.M., and Ghidaglia, J.M. (1989) 'Explosion en temps fini pour le flot des applications harmonique,' Compt. Rend. 308, 339–344.

[15] Struwe, M., and Chen, Y., (1990) 'Existence and partial regularity results for the heat flow for haromonic maps,' preprint.

[16] Neu, J. (1986) preprint.

[17] Howard, L.N. and Koppel, N. (1977) 'Slowly varying waves and shock structures in reaction-diffusion equations,' Stud. Appl. Math. 56, 95–145.

[18] Hagan, P. (1981) 'Target patterns in reaction-diffusion systems,' Adv. Appl. Math. 2, 400–416.

[19] Koppel, N. (1981) 'Target pattern solutions to reaction-diffusion equations in the presence of impurities,' Adv. Appl. Math. 2, 389–399.

DETERMINISTIC TURBULENCE

Peter D. Lax
Courant Institute, New York University

There are many theories of turbulence but they all fall into two categories: Probabilistic and deterministic. In one kind of probabilistic theory one deals with an ensemble of flows, dependent on a random field of some kind. The task for the theorist is:

a) To prove that the variables of physical interest – velocity, pressure, density – have expected values.

b) To derive laws governing the evolution of these expected values. These laws will involve expected values of functions of velocity, pressure, density and of their translates.

c) To find approximations to the exact equations of evolution, accurate enough for engineering purposes, under conditions that can be ascertained, which involve only a finite number of variables.

Another kind of probabilistic theory exploits the breakdown at some critical time of classical solutions, and the nonuniqueness of subsequent generalized solutions. In such a case the task of the theorist is to form some kind of Boltzmann-Gibbs type average over the ensemble of generalized solutions. The two further tasks b) and c) remain the same.

In a deterministic theory one looks at physically significant parameters that appear in the equations of flow, such as Reynolds number coefficient of heat conduction, etc. As one of these parameters approaches a critical value, solutions become more and more oscillatory, with shorter and shorter wave lengths and with amplitudes that do not tend to zero. Such as a sequence does

R. Spigler (ed.), Applied and Industrial Mathematics, 57–58.

not converge in the ordinary i.e. strong sense. In this case the task of the theorist is to show that these solutions converge in a weak sense, such as the sense of distributions. These weak limits play the role of expected values of variables of physical interest. The two further tasks b) and c) remain as before.

In this talk I will discuss some examples of turbulence that go the deterministic route. The critical parameter in these examples, whose approach to zero triggers off turbulence is dispersion, not viscosity. In this case turbulence occurs already in one-dimensional flows, whereas viscosity-controlled turbulence is quintessentially a three dimensional phenomenon.

The main example is a centered leap-frog like difference scheme proposed and tried out by von Neumann in 1944 for the one-dimensional equations of Lagrangean compressible flow. A series of carefully controlled numerical calculations carried out by Tom Hou and the author indicate that as Δx and Δt tend to zero, the approximate solutions converge uniformly in some regions of x, t space, and weakly in the complementary region. The weak limits do not satisfy the equations of compressible flow, contrary to what von Neumann conjectured.

The other two examples deal with Burgers' one-component caricatures of one-dimensional compressible flows. One is a study by the author and David Levermore of the zero-dispersion limit of the KdV equation, and the other a study by Jonathan Goodman and the author of a semidiscrete space-centered approximation of the one component equation. Both these examples are completely integrable, therefore can be handled rigorously. In particular, task c) can be carried out exactly by using modulated multiphase waves.

EXACT CONTROLLABILITY FOR DISTRIBUTED SYSTEMS. SOME TRENDS AND SOME PROBLEMS

J.L. LIONS
Collège de France et C.N.E.S.
3, rue d'Ulm
75005 Paris
France

1. Introduction

We consider a *distributed system*, i.e. a system whose *state* y is given, as a function of x (the space variable), t (the time) and v (the *control function*), by the solution of the Partial Differential Equation (PDE) :

$$\frac{\partial y}{\partial t} + \mathcal{A}y = \mathcal{B}v. \tag{1.1}$$

In (1.1) $\mathcal{A}$ is a Partial Differential Operator, which can be linear or non linear ; it is defined in a *bounded* open set Ω of $\mathbb{R}^n (n = 1, 2, 3$ in the applications).

The *Boundary conditons* depend on the structure of $\mathcal{A}$. We *do not* write the Boundary conditions in (1.1), but of course this will be made precise in the examples studied below.

The control function v spans a space $\mathcal{U}$ and in (1.1) the operator $\mathcal{B}$ maps $\mathcal{U}$ into a space such that (1.1) makes sense (again, this has to be made precise !).

The control function expresses that we can *act* on the system which is modelled through the operator $\frac{\partial}{\partial t} + \mathcal{A}$.

In the applications one can act only on *"small"* geometrical parts of Ω, i.e. either on *part of the boundary* (then v is a *Boundary control*) or on *part of the domain* (then v is said to be a *distributed control*).

In (1.1), y can be a *scalar* function or a *vector* function.

R. Spigler (ed.), Applied and Industrial Mathematics, 59–84.

The *examples* that we have in mind are :

$$(\frac{\partial^2}{\partial t^2} - \Delta)y = \mathcal{B}v \tag{1.2}$$

(the *wave operator* ; one writes (1.2) as a first order system in t to obtain representation (1.1)) ;

$$(\frac{\partial}{\partial t} - \Delta)y = \mathcal{B}v \tag{1.3}$$

(the *diffusion operator*)
and

$$\begin{aligned} &\frac{\partial y}{\partial t} + y\nabla y - \Delta y = \mathcal{B}v - \nabla\pi \\ &div\ y = 0, \end{aligned} \tag{1.4}$$

the *Navier-Stokes* equations.

Let us now introduce the notion of *Exact Controllability.*
Let us add to (1.1) the *initial condition*

$$y(0) = y^o \tag{1.5}$$

where $y(0)$ denotes the function $x \to y(x,0)$.

It is *assumed* that given v and given y^o (in suitable functions spaces) and with suitable Boundary conditions "built in" equation (1.1), (1.1) and (1.5) *uniquely define a state,* $y(x,t;v) = y(v)$.

Remark 1.1

This hypothesis does not make difficulties for Examples 1.2 and 1.3. It is still satisfied for system (1.4) if the space dimension is $n = 2$. Existence *and* Uniqueness of the solution of (1.4) if n=3 in an appropriate functional class is an open question. One has then to modify accordingly the formulation of the problem of Exact Controllability. Cf. Remark 1.3 below. □

Let T be *given* and let y^1 be *any* element in *the function space where we take* y^o.

We want to find v (*if it exists*) such that

$$y(T;v) = z^o. \tag{1.6}$$

In other words, we want to drive the system from state y^o to state z^o in the time interval T.

If this is possible for any couple y^o, z^o, one says that the system is *exactly controllable.*
A few remarks are in order.

Remark 1.2

Suppose that, given the state equation (1.2) (wave equation) (where now y actually denotes the couple $y, \frac{\partial y}{\partial t}$), one can act on the system either at the boundary Γ of Ω or on a subset σ of Ω. Then one wants to find v such that the system is driven from state $\{y^o, \frac{\partial y}{\partial t}(0) = y^1\}$ to

$$y(T) = z^o, \quad \frac{\partial y}{\partial t}(T) = z^1.$$

Due to the finite speed of wave propagation, it is obvious that Exact Controllability *can* be possible *only for T large enough.*
The situation is entirely different for (1.3). Again this will be made precise below. □

Remark 1.3

If one considers (1.4) for $n = 3$ (i.e. where existence *and* uniqueness is an open question, but where global (in time) existence is known), the question can be formulated as follows : given y^o and z^o, and given T, it is possible to find v such that there exists a solution of (1.4) such that one has (1.5) and (1.6) ? □

Remark 1.4

In the above formulation there is a *large flexibility* : we have, to some extent, the *choice of the function spaces* where we choose y^o, z^o on one hand and where we choose v on the other hand (provided they are compatible). At least this is true in the *linear situations*, where by using "weak" solutions one can choose very general spaces (and even spaces which are *not* distributions's spaces). The analogous is *not* true in the nonlinear case. □

Remark 1.5

If the system is exactly controllable for $T > T_*$, then there will be *infinitely many ways* to drive the system from u^o to z^o. Indeed let ε be chosen such that $T-\varepsilon > T_*$. We take $v = v_\varepsilon$ *arbitrarily* in the time interval $(0, \varepsilon)$. This drives the system from

y^o to $y^o_\varepsilon = y(\varepsilon)$. Then according to the exact controllability, one can choose $v = w_\varepsilon$ in the time interval (ε, T), driving the system from y^o_ε to z^o. Then

$$v = \begin{cases} v_\varepsilon & \text{for} \quad 0 < t < \varepsilon \\ w_\varepsilon & \text{for} \quad \varepsilon < t < T \end{cases}$$

drives the system from y^o to z^o.

It is then natural to ask for *the "best"* way to drive the system from y^o to z^o. We shall consider therefore the following problem :

if system (1.1) *is exactly controllable, find v such that*

$$\|v\|_{\mathcal{U}} = \min \tag{1.7}$$

among all possible v's (in a space $\mathcal{U}$) *driving the system from* y^o *to* z^o.

We have used in (1.7) the *norm* of the space $\mathcal{U}$. In the examples we have in mind, $\mathcal{U}$ will be chosen to be a *Hilbert space.* □

In the case of the *diffusion equations*, the notion of Exact Controllability is somewhat too restrictive, as we shall explain in Section 3 below.
It is then useful to introduce *another* problem : instead of driving the system from state y^o to state z^o, we want now to find controls v such that if $y(0) \in B^o$ then $y(T) \in C^o$ where B^o (resp. C^o) denotes the ball of center y^o (resp. z^o) and radius (in the norm of the space where the state is considered) β_o (resp.γ_o). *In short, we want to drive the system from* B^o *to* C^o (and again we add (1.7)).
For *linear* cases, it is easy to see that this is indeed possible. We address then the question of finding v satisfying (1.7).

All questions considered here are motivated by problems in solid mechanics or in fluid mechanics. For instance, the stabilization of flexible structures is related to the above questions, by considering the elasticity system, or the plate equations. We refer to J. LAGNESE [1], J. LAGNESE and J.L. LIONS [1].
For the Navier-Stokes system, these questions *may* be related to turbulence theory. Simular questions -although in a slightly different form- arise in the context of climatology : are they irreversible changes, or not ?

The problems of Exact Controllability are classical in the framework of systems described by *Ordinary* Differential Equations (O.D.E.'s).
For *linear* O.D.E.'s (i.e. for the case where in (1.1) $\mathcal{A}$ and $\mathcal{B}$ are *matrices*) an algebraic condition giving necessary and sufficient conditions for Exact Controllability is known (cf. e.g. R.W. BROCKETT [1]). The situation is much more

complicated in *nonlinear* situations, where no general answer seems to be known (possible ?). But very interesting results are known, for instance in the case of *bilinear* systems, of the type

$$\frac{\partial y}{\partial t} + \mathcal{A}y = \mathcal{B}(y, v) \tag{1.8}$$

where $\mathcal{B}$ is a bilinear operator in y and v (a situation which makes sense also in the case of distributed systems !). We refer to H.J. SUSSMAN and V. JURDJEVIC [1], R. HERMANN and A.J. KRENER [1]. We also refer to the interesting work of A.M. BLOCH and J.E. MARSDEN [1].

For systems described by P.D.E.'s several methods are available for *linear systems.* Using Fourier type expansions, the problem is transformed into a moment equation. Cf. D.L. RUSSELL [1] where methods using extension of solutions to the whole space are also introduced. A systematic method (H.U.M. = Hilbert Uniqueness Method) has been introduced by the Author (J.L. LIONS [1]), based on Uniqueness theorems and on the construction of "abstract" function spaces corresponding to these uniqueness theorems (classical or new), cf. J.L. LIONS [2], [3] for a systematic presentation of H.U.M. and cf. the Bibliography therein.
Very few results are known for *nonlinear* distributed systems. Some of them are indicated in Section 2 below.

In what follows we begin in Section 2 by a brief, but *self contained*, presentation of HUM for the *wave equation* with boundary control, and we proceed by a survey (without proofs) some recent results of Y. JINHAI and E. ZUAZUA.

In section 3 we study similar problems for the *heat equation.* The results presented here seem to be new. We use some variants of HUM and we also use the duality theory of W. FENCHEL [1] and T.R. ROCKAFELLAR [1] for convex functions (cf. also I.EKELAND and R. TEMAM [1]).

All methods presented in Sections 2 and 3 are *constructive.* They lead in a natural way to numerical algorithms, after introducing some further adequate approximations (and regularizations).
We refer to R. GLOWINSKI, C.H. LI, J.L. LIONS [1], R. GLOWINSKI [1], where *hyperbolic* situations are studied. Numerical work is in progress for *parabolic* situations, with the goal to arrive at *numerical algorithms* for studying Exact Controllability for the Navier-Stokes equations. This leads to open problems, some of them being indicated in Section 4.

Similar problems can be considered for other classes of linear or non linear equations ; Schroedinger Equations for instance. They are not studied here (H.U.M applies, in the linear cases).* After the oral presentation of this paper, Prof. OCKENDON raised the question of exact controllability type problem for the *backward* heat equation (where one wants do "drive" the system from the state y^o to a ball C^o). This can indeed to be treated, but it requires further developments that will be presented elsewhere.

2. Hyperbolic systems

2.1. Let Ω be a bounded open set in $\mathbb{R}^n$, with (smooth) boundary Γ.

In the domain $\Omega \times (O,T)$, we consider the *wave equation*

$$y" - \Delta y = 0 \tag{2.1}$$

(where $y" = \dfrac{\partial^2 y}{\partial t^2}$). We suppose that

$$y(0) = 0, \quad y'(0) = 0 \qquad (y' = \frac{\partial y}{\partial t}) \tag{2.2}$$

(i.e. we start from the initial state 0, something that does not restrict the generality in the *linear* situation). We assume that we can act on the system *through the boundary* :

$$y = v \quad \text{on} \quad \Gamma \times (0,T). \tag{2.3}$$

Remark 2.1

One can, more generally, consider the case where

$$\begin{aligned} y = v \quad &\text{on} \quad \Gamma_o \times (0,T), \quad \Gamma_o \subset \Gamma \\ y = 0 \quad &\text{on} \quad \Gamma \backslash \Gamma_o \times (0,T) \end{aligned} \tag{2.4}$$

We return to that later on. □

We want to drive the system, at a *given* time T (large enough), to a *given state*

$$y(T) = z^o, \quad y'(T) = z^1. \tag{2.5}$$

One has to make precise the *functions spaces* where v, z^o, z^1 are considered. In the *linear* case this can be done in infinitely many ways. We present here *one* situation, probably the most natural one. We assume that

$$v \in L^2(\Sigma), \quad \Sigma = \Gamma \times (O,T). \tag{2.6}$$

---

* There are forthcoming papers on this subject, by A. BENSOUSSAN, by G. LEBEAU and by E. ZUAZUA.

Then the *weak* solution of (2.1), (2.2), (2.3) is such that

$$y(T) \in L^2(\Omega), \quad y'(T) \in H^{-1}(\Omega) \tag{2.7}$$

(where $H^{-1}(\Omega)$ = dual of $H_0^1(\Omega), H_0^1(\Omega)$ = SOBOLEV Space of functions φ such that $\varphi, \dfrac{\partial \varphi}{\partial x_1}, ..., \dfrac{\partial \varphi}{\partial x_n} \in L^2(\Omega)$ and such that $\varphi = 0$ on Γ).

The *proof* of (2.7) is based on the transposition method as in J.L. LIONS and E. MAGENES [1] and on the following regularity theorem : let φ be the solution of

$$\varphi'' - \Delta\varphi = 0 \quad \text{in} \quad \Omega \times (0,T), \tag{2.8}$$

$$\varphi(0) = \varphi^o \quad , \varphi'(0) = \varphi^1 \quad , \varphi^o \in H_0^1(\Omega), \varphi^1 \in L^2(\Omega) \tag{2.9}$$

$$\varphi = 0 \quad \text{on} \quad \Sigma. \tag{2.10}$$

Then

$$\frac{\partial \varphi}{\partial \nu} \in L^2(\Sigma) \tag{2.11}$$

and the mapping $\{\varphi^o, \varphi^1\} \to \dfrac{\partial \varphi}{\partial \nu}$ is continuous from $H_0^1(\Omega) \times L^2(\Omega) \to L^2(\Sigma)$ (cf. J.L. LIONS [4] for a proof).

The Exact Controllability problem is now, in a precise form :
among all possible v's in $L^2(\Sigma)$ such that one has (2.1), (2.2), (2.3), (2.5), where $z^o \in L^2(\Omega)$ and $z^1 \in H^{-1}(\Omega)$ *(if* such v's do exist !), *find the control such that*

$$\|v\|_{L^2(\Sigma)} = \min. \tag{2.12}$$

□

2.2. The HUM Solution

We introduce a *penalized* problem. We consider the functional

$$\frac{1}{2}\int_\Sigma v^2 d\Sigma + \frac{1}{2\varepsilon}\int\int_{\Omega\times(O,T)} (y'' - \Delta y)^2 dx dt \tag{2.13}$$

where y satisfies (2.2), (2.3), (2.5). The $inf\cdot$ of (2.13) is achieved at $v_\varepsilon, y_\varepsilon$. Let us set

$$\rho_\varepsilon = \frac{1}{\varepsilon}(y''_\varepsilon - \Delta y_\varepsilon). \tag{2.14}$$

One has

$$\int_\Sigma v_\varepsilon \hat{v} d\Sigma + \int\int_{\Omega\times(O,T)} \rho_\varepsilon(\hat{y}" - \Delta\hat{y})dxdt = 0, \tag{2.15}$$

where

$$\hat{y} = \hat{v} \quad \text{on} \quad \Sigma, \quad \hat{y}(0) = \hat{y}'(0) = 0, \quad \hat{y}(T) = \hat{y}'(T) = 0.$$

It follows from (2.15) that

$$\begin{aligned} &\rho"_\varepsilon - \Delta\rho_\varepsilon = 0 \quad \text{in} \quad \Gamma \times (0,T), \\ &\rho_\varepsilon = 0 \quad \text{on} \quad \Sigma \end{aligned} \tag{2.16}$$

and

$$\frac{\partial\rho_\varepsilon}{\partial\nu} + v_\varepsilon = 0 \quad \text{on} \quad \Sigma. \tag{2.17}$$

Let us assume, for the time being, that *we do have exact controllability.* Then, as $\varepsilon \to 0$, ρ_ε satisfies (2.16) *and*

$$\frac{\partial\rho_\varepsilon}{\partial\nu} = -v_\varepsilon \quad \text{remains in a bounded set of} \quad L^2(\Sigma). \tag{2.18}$$

We use these informations by introducing a *new norm* "built" on these informations. Let us consider the solution ρ of

$$\begin{aligned} &\rho" - \Delta\rho = 0 \quad \text{in} \quad \Omega \times (0,T), \\ &\rho(T) = \rho^o, \quad \rho'(T) = \rho^1 \quad \text{where} \\ &\rho^o \quad \text{and} \quad \rho^1 \quad \text{are given smooth functions in} \quad \Omega, \quad \rho^o = 0 \quad \text{on} \quad \Gamma, \\ &\rho = 0 \quad \text{on} \quad \Sigma \end{aligned} \tag{2.19}$$

and let us define

$$\|\{\rho^o, \rho^1\}\|_F = \left(\int_\Sigma \left(\frac{\partial\rho}{\partial\nu}\right)^2 d\Sigma\right)^{\frac{1}{2}}. \tag{2.20}$$

If T is large enough, we have defined in this way a Hilbertian norm. Indeed if $\|\{\rho^o, \rho^1\}\|_F = 0$, then

$$\frac{\partial\rho}{\partial\nu} = 0 \quad \text{on} \quad \Sigma. \tag{2.21}$$

It then follows from the Holmgren Uniqueness theorem that if $T >$ diameter (Ω), then $\rho = 0$, and therefore

$$\|\{\rho^o, \rho^1\}\|_F = \{0,0\}.$$

Returning to ρ_ε, we see that, as $\varepsilon \to 0$,

$$\{\rho_\varepsilon(T), \rho'_\varepsilon(T)\} \quad \text{remains in a bounded set of} \quad F. \tag{2.22}$$

We can then let $\varepsilon \to 0$ (using weak solutions, defined by transposition) and we obtain for $\rho = lim \cdot \rho_\varepsilon$, $\quad y = lim \cdot y_\varepsilon$, *the optimality system :*

$$\begin{aligned}
&\rho" - \Delta\rho = 0 \quad \text{in} \quad \Omega \times (0,T), \\
&\rho(T) = \rho^o \quad \rho'(T) = \rho^1, \\
&\rho^o = 0 \quad \text{on} \quad \Sigma, \\
&y" - \Delta y = 0 \quad \text{in} \quad \Omega \times (0,T), \\
&y(0) = 0, \quad y'(0) = 0 \\
&y = -\frac{\partial\rho}{\partial\nu} \quad \text{on} \quad \Sigma,
\end{aligned} \tag{2.23}$$

and

$$y(T) = z^o, \quad y'(T) = z^1. \tag{2.24}$$

Of course, at this stage, everything is still *formal*, since we do *not* know if the problem admits a solution, so that (2.18) is unknown.
But if we can solve *directly* (2.23), (2.24) -*where the unknowns are* $\{\rho^o, \rho^1\}$-, then the above procedure will be justified.

Let us consider *the operator* Λ *defined* by

$$\Lambda\{\rho^o, \rho^1\} = \{y'(T), -y(T)\}. \tag{2.25}$$

If $\{\hat\rho^o, \hat\rho^1\}$ is another couple and if $\hat\rho$ is the corresponding solution, we verify that

$$< \Lambda\{\rho^o, \rho^1\}, \{\hat\rho^o, \hat\rho^1\} >= \int_\Sigma \frac{\partial\rho}{\partial\nu}\frac{\partial\hat\rho}{\partial\nu} d\Sigma. \tag{2.26}$$

It follows easily from (2.20) and (2.26) that if $T >$ diameter Ω, Λ *is an isomorphism from* F *onto its dual* F'.
Therefore if z^o, z^1 are such that

$$\{z^1, -z^o\} \in F' \tag{2.27}$$

then the equation

$$\Lambda\{\rho^o, \rho^1\} = \{z^1, -z^o\} \tag{2.28}$$

admits a unique solution. □

The next step is to study the space F. One can prove (cf. L.F. HO [1]) that, if $T >$ diameter (Ω)

$$\left(\int_\Sigma \left(\frac{\partial \rho}{\partial \nu}\right)^2 d\Sigma\right)^{\frac{1}{2}} \geq c\|\{\rho^o, \rho^1\}\|_{H_0^1(\Omega)\times L^2(\Omega)}. \tag{2.29}$$

Since, by virtue of (2.11), one has also the inverse inequality, it follows that

$$F = H_0^1(\Omega) \times L^2(\Omega). \tag{2.30}$$

So that of course $F' = H^{-1}(\Omega) \times L^2(\Omega)$ and we have proven *the exact controllability for* $T >$ *diameter* (Ω) *and* $z^o \in L^2(\Omega)$, $z^1 \in H^{-1}(\Omega)$. □

Remark 2.2

It is quite clear that the above method -denoted by H.U.M.- is very general at least for *linear* situations. It relies on two main steps :

(i) use a *uniqueness* theorem for constructing new *norms* ;

(ii) obtain informations on the "abstract" space F.

This is studied systematically in J.L. LIONS [3]. Let us confine ourselves to the following remarks. □

Remark 2.3

If we consider now the situation (2.4) where the control is applied on $\Gamma_o \subset \Gamma$, the same considerations than above lead to the introduction of

$$\|\{\rho^o, \rho^1\}\|_F = \left(\int_{\Gamma_o \times (0,T)} \left(\frac{\partial \rho}{\partial \nu}\right)^2 d\Sigma\right)^{\frac{1}{2}}. \tag{2.31}$$

(In fact $F = F(\Gamma_o)$).

Then, according to Holmgren's uniqueness Theorem, we have a *norm* if

$$T > 2d(\Omega, \Gamma_0), d(\Omega, \Gamma_0) = \text{ greatest distance from } x \in \Omega \text{ to } \Gamma_0 \tag{2.32}$$

If T_0 is "very small", then F will be "very large", and therefore F' "very small".

A necessary and sufficient condition on Γ_0 in order to have

$$F = H_0^1(\Omega) \times L^2(\Omega) \tag{2.33}$$

is given by C. BARDOS, G. LEBEAU and J. RAUCH [1].

(The set Γ_0 should be "large enough" and $T >$ some $T(\Gamma_0)$).
If Γ_0 is such that there exist rays in Ω which can be reflected for arbitrarily long times *without* crossing Γ_0, then (2.33) is not true. One has

$$F \supset H_0^1(\Omega) \times L^2(\Omega) \quad \text{strictly.} \tag{2.34}$$

A complete *characterization* of $F(= F(\Gamma_0))$ is not known. □

2.3. Numerical approximation

Solving (2.28) is equivalent to finding $\{\rho^o, \rho^1\}$ which *minimizes*

$$\frac{1}{2} < \Lambda\{\rho^o, \rho^1\}, \{\rho^o, \rho^1\} > - < \{z^1, -z^o\}, \{\rho^o, \rho^1\} > \tag{2.35}$$

i.e.

$$\frac{1}{2} \int_\Sigma (\frac{\partial \rho}{\partial \nu})^2 d\Sigma - \int_\Omega (z^1 \rho^o - z^o \rho^1) dx, \tag{2.36}$$

where ρ is the solution of (2.19).
This is constructive.

Numerical algorithms based on (2.36) are introduced and studied in R. GLOWINSKI, C.M. LI and J.L. LIONS [1], where further regularization procedures are introduced. They can be partly avoided by using the technique introduced in R. GLOWINSKI [1]. □

Remark 2.4

Let us define

$$F(v) = \frac{1}{2} \int_\Sigma v^2 d\Sigma,$$

$$Lv = \{y(T), y'(T)\} \quad , \quad y = \quad \text{solution of (2.1),(2.2), (2.3),}$$

$$G(q) = \begin{cases} 0 \text{ if } q^o = z^o \quad , \quad q^1 = z^1 \quad , \quad q = \{q^o, q^1\}, \\ +\infty \quad \text{otherwise.} \end{cases}$$

Then problem (2.12) is equivalent to finding

$$inf \cdot_{v \in L^2(\Sigma)} [F(v) + G(Lv)]. \tag{2.37}$$

The *dual* problem, in the sense of FENCHEL and ROCKAFELLAR is given by

$$inf \cdot_{v \in L^2(\Sigma)} [F(v) + G(Lv)] = -inf_{\rho=\{\rho^o,\rho^1\}}[F^*(L^*\rho) + G^*(-\rho)] \tag{2.38}$$

where in general $F^*(v) = sup_{\hat{v}}[(v, \hat{v}) - F(\hat{v})]$.

The problem in the right hand side of (2.38) *coincides* with (2.36). [cf. J.L. LIONS [3], Vol. 1, Chapter VIII]. □

2.4. Other linear problems for the wave equation

Remark 2.5

One can consider the problem (2.1), (2.2), (2.3), (2.5) *constrained by further conditions* on v. For instance

$$\int_0^T t^j v(x,t)dt = 0 \quad \forall x \in \Gamma, \qquad 0 \leq j \leq N. \tag{2.39}$$

This is indeed possible (Y. JINHAI [1]). The key to the answer lies on precise estimates to obtain a characterization of F and on the following simple observation: let ρ be a solution of

$$\begin{aligned} &\rho'' - \Delta\rho = 0 \quad \text{in} \quad \Omega \times (0,T) \\ &\rho = 0 \quad \text{on} \quad \Gamma \times (O,T) \\ &\frac{\partial \rho}{\partial \nu} = a_o(x) + ta_1(x) + ... + t^N a_N(x) \quad \text{on} \quad \Gamma \times (O,T). \end{aligned} \tag{2.40}$$

Then, if $T >$ diameter $(\Omega), \rho = 0$. Indeed, if we introduce

$$\sigma = \frac{\partial^{N+1}}{\partial t^{N+1}}\rho \tag{2.41}$$

we have

$$\begin{aligned} &\sigma'' - \Delta\sigma = 0, \\ &\sigma = 0 \quad , \quad \frac{\partial \sigma}{\partial \nu} = 0 \quad \text{on} \quad \Gamma \times (O,T). \end{aligned}$$

Then $\sigma = 0$, so that

$$\rho = b_o(x) + tb_1(x) + ... + t^N b_N(x)$$

and $\rho" - \Delta\rho = 0$. Therefore $b_o = ... = b_N = 0$, so that we can apply H.U.M. We refer to JINHAI, loc. cit., for further results. □

Remark 2.6

Let us consider a space $G \subset L^2(\Omega) \times H^{-1}(\Omega)$. We shall say that we have *Extended* Exact Controllability (with respect to G) *if we can find* v *such that*

$$\{y(T), y'(T)\} \in G + \{z^o, z^1\}. \tag{2.42}$$

This condition is of course weaker than (2.5) and it can therefore be fulfilled *either* for T *smaller* than the one needed for (2.5) (for *all* couples $\{z^o, z^1\}$), *or* with functions v of *smaller* norm. Cf. J.L. LIONS [5].

Another variant, studied in Y. JINHAI [2], is to find v_h such that

$$\frac{1}{h}\int_{T-h}^{T} y(t)dt = z_h^o, \quad \frac{1}{h}\int_{T-h}^{T} y'(t)dt = z_h^1 \tag{2.43}$$

such that, as $h \to 0, \{z_h^o, z_h^1\} \to \{z^o, z^1\}$ in the $L^2(\Omega) \times H^{-1}(\Omega)$ topology and to study the convergence of v_h. □

2.5. Nonlinear problems

Suppose we want to address the same problems than before but for the *nonlinear equation*

$$y" - \Delta y + f(y) = 0, \tag{2.44}$$

where f is "suitably" given, such as

$$f(y) = \lambda \, sin \, y \tag{2.45}$$

or

$$f(y) = \lambda y^3. \tag{2.46}$$

Of course one has a serious difficulty due to the fact that we *cannot* use *general* function spaces, since if y is too "weak", $f(y)$ may be impossible to define.

We only want to indicate here, without going into the (non trivial) technical details, the following idea of E. ZUAZUA [1], [2]. One begins with the *linear* equation

$$\begin{aligned} &\rho'' - \Delta\rho = 0 \\ &\rho(T) = \rho^o, \quad \rho^1(T) = \rho^1, \quad \rho = 0 \quad \text{on} \quad \Sigma \end{aligned} \tag{2.47}$$

and *then* (compare to (2.23)) we introduce the *nonlinear* equation *

$$\begin{aligned} &y'' - \Delta y + f(y) = 0, \\ &y(0) = 0, y^1(0) = 0 \\ &y = -\frac{\partial\rho}{\partial\nu}. \end{aligned} \tag{2.48}$$

One has then to solve the *nonlinear equation*

$$M\{\rho^o, \rho^1\} = \{z^1, -z^o\}, \tag{2.49}$$

where

$$M\{\rho^o, \rho^1\} = \{y'(T), -y(T)\}. \tag{2.50}$$

One can then apply fixed points arguments. Cf. E. ZUAZUA, loc.cit., for examples of *type* (2.45) and Y. JINHAI [3] for example (2.46) in dimension 1 and for λ small enough. □

3. Diffusion equation

3.1. Exact controllability

Let Ω be an open set of $\mathbb{R}^n$, this time *not* necessarily bounded and let σ be an open subset of Ω. We assume that we can act on the system on σ, the case of a distributed system.

Remark 3.1

We could as well, in what follows, consider the case of a *boundary control.* But this introduces some further technical difficulties that we want to avoid here. □

If we denote by χ_σ the characteristic function of σ, we have

$$y' - \Delta y = v\chi_\sigma \quad \text{in} \quad \Omega \times (0, T), \tag{3.1}$$

* Of course in the *nonlinear* case, one should consider general initial data $y(0) = y^o, y'(0) = y^1$.

$$y(0) = 0 \quad \text{in} \quad \Omega \tag{3.2}$$

$$y = 0 \quad \text{on} \quad \Sigma = \Gamma \times (0, T). \tag{3.3}$$

Given $T > 0$ and given z^o in a suitable function space, we want to find v such that

$$y(T) = z^o \tag{3.4}$$

and such that

$$\|v\|_{L^2(\sigma\times(0,T))} = min. \tag{3.5}$$

□

Let us proceed in the same way as in the hyperbolic case. We begin with a penalized problem : given $\varepsilon > 0$, we define the functional

$$\frac{1}{2}\|v\|^2_{L^2(\sigma\times(O,T))} + \frac{1}{2\varepsilon}\|y' - \Delta y - v\chi_\sigma\|^2_{L^2(\Omega\times(O,T))}, \tag{3.6}$$

where y is subject to (3.2), (3.3), (3.4).

The inf.of (3.6) is achieved for $v_\varepsilon, y_\varepsilon$.

We introduce

$$\rho_\varepsilon = \frac{1}{\varepsilon}(y'_\varepsilon - \Delta y_\varepsilon - v_\varepsilon\chi_\sigma). \tag{3.7}$$

We have

$$\int\int_{\sigma\times(O,T))} v_\varepsilon \hat{v} dx dt + \int\int_{\sigma\times(O,T))} \rho_\varepsilon(\hat{y}' - \Delta\hat{y} - \hat{v}\chi_\sigma) dx dt = 0 \tag{3.8}$$

$$\forall \hat{v}, \hat{y} \quad \text{such that} \quad \hat{y} = 0 \quad \text{on} \quad \Gamma \times (0,T),\ \hat{y}(0) = \hat{y}(T) = 0.$$

Then

$$\begin{aligned} -\rho'_\varepsilon - \Delta\rho_\varepsilon = 0 \quad &\text{in} \quad \Omega \times (0,T), \\ \rho_\varepsilon = 0 \quad &\text{on} \quad \Gamma \times (0,T) = \Sigma, \end{aligned} \tag{3.9}$$

and

$$\rho_\varepsilon = v_\varepsilon \quad on \quad \sigma \times (0,T) \tag{3.10}$$

Let us assume now that problem (3.1),...,(3.5) admits a solution (i.e. that we have Exact Controllability). Then we know that, as $\varepsilon \to 0$,

$$\|v_\varepsilon\|_{L^2(\sigma\times(0,T))} \leq \quad \text{constant}. \tag{3.11}$$

This fact leads to the introduction of the following new Hilbert space. Let ρ^o be given, say a smooth function which is 0 in Γ, and let ρ be the solution of

$$-\rho' - \Delta\rho = 0 \quad ,\rho(T) = p^o \quad ,\rho = 0 \quad \text{on} \quad \Sigma. \tag{3.12}$$

We then define

$$\|\rho^o\|_F = \left(\int\int_{\sigma\times(O,T))} \rho^2 dxdt\right)^{\frac{1}{2}}. \tag{3.13}$$

We define in this way a *norm*. Indeed if $\|\rho^o\|_F = 0$, then $\rho = 0$ on $\sigma \times (O,T))$ and therefore $\rho \equiv 0$; hence $\rho^o = 0$.

We then define F as the completion of smooth functions with 0 boundary values for the norm (3.13). It is a *Hilbert space* which is "very large". In particular it contains elements which are *not* distributions (outside $\bar{\sigma}$).

It now follows from (3.11) and by the *definition* of F that

$$\rho_\varepsilon(0) \quad \text{remains in a bounded set of} \quad F. \tag{3.14}$$

We obtain in this way the *formal* optimality system (it is formal since we do *not* know at this stage if (3.1),...(3.4) admits a solution)

$$\begin{aligned} &-\rho' - \Delta\rho = 0 \\ &\rho(T) = \rho^o \quad , \quad \rho = 0 \quad \text{on} \quad \Sigma \end{aligned} \tag{3.15}$$

$$\begin{aligned} &y' - \Delta y = \rho\chi_\sigma \\ &y(0) = 0 \quad , \quad y = 0 \quad \text{on} \quad \Sigma \end{aligned} \tag{3.16}$$

The problem is "reduced" to finding ρ^o such that (3.4) holds true.

If we define the linear operator Λ by

$$\Lambda\rho^o = y(T), \tag{3.17}$$

(3.4) writes

$$\Lambda\rho^o = z^o. \tag{3.18}$$

We proceed now as in the hyperbolic case, at least formally. We consider $\hat{\rho}^o$ and we denote by $\hat{\rho}$ the corresponding solution of (3.15).
If we multiply (3.16) by $\hat{\rho}$ we obtain

$$< \Lambda\rho^o, \hat{\rho}^o > = \int\int_{\sigma\times(O,T))} \rho\hat{\rho}dxdt. \tag{3.19}$$

It follows from (3.19) that Λ *is an isomorphism* from F *onto* F' = dual space of F. We also notice that

$$\Lambda^* = \Lambda. \tag{3.20}$$

Therefore(3.18) *admits a unique solution if and only if*

$$z^o \in F'. \tag{3.21}$$

Since F is "big", F' is "small", so that (3.21) is a strong restriction on z^o. □

Remark 3.2

That some restriction should be imposed on z^o is quite clear, since y solution of (3.1), (3.2), (3.3) will always be (in particular) C^∞ outside $\bar{\sigma}$. □

Remark 3.3

Solving (3.18) is equivalent to minimizing

$$\frac{1}{2} < \Lambda\rho^o, \rho^o > - < z^o, \rho^o > \tag{3.22}$$

i.e.

$$\frac{1}{2}\int\int_{\sigma\times(O,T))} \rho^2 dxdt - \int_\Omega z^o \rho^o(x)dx. \tag{3.23}$$

□

Remark 3.4

Let us introduce

$$F(v) = \frac{1}{2}\int\int_{\sigma\times(O,T))} v^2 dxdt,$$

$$G(q) = \begin{cases} 0 & \text{if} \quad q = z^o \\ +\infty & \text{otherwise} \end{cases}$$

$$Lv = y(T). \tag{3.24}$$

Then problem (3.1),...,(3.5) is equivalent to finding

$$inf_v[F(v) + G(Lv)]. \tag{3.25}$$

Using FENCHEL-ROCKAFELLAR duality theorem (as in Section 2), we have

$$inf_v[F(v) + G(Lv)] = -inf_{\rho^o}[F^*(L^*\rho^o) + F^*(-\rho^o)]. \tag{3.26}$$

We verify that

$$L^*\rho^o = -\rho\chi_\sigma, \tag{3.27}$$

$$G^*\rho^o = (\rho^o, z^o)$$

so that

$$F^*(L^*\rho^o) + G^*(-\rho^o) = \frac{1}{2}\int\int_{\sigma\times(O,T))} \rho^2 dxdt - \int_\Omega \rho^o z^o dx,$$

which shows that (up to the sign) *the problem of minimizing* (3.23) *is the dual of* (3.1),...,(3.5). □

Remark 3.5

The formulation in Remark 3.3 is *not* entirely constructive, since we do *not* have a direct characterization of F'.
It is this fact that leads to another (related) problem, that we consider now.

3.2. Approximate controllability

We introduce

$$B = \quad \text{ball in} \quad L^2(\Omega) \quad \text{of center} \quad z^o \quad \text{and radius} \quad \beta. \tag{3.28}$$

It is easily seen that, when v spans $L^2(\sigma \times (O,T)), y(T)$ spans a *dense* subset of $L^2(\Omega)$, so that there always exist elements v such that

$$y(T) \in B. \tag{3.29}$$

The problem we want now to consider is *to find*

$$inf \cdot \|v\|_{L^2(\sigma\times(0,T))} \tag{3.30}$$

among all v's that (3.29) *holds true.*[We assume $\|z^o\|_{L^2(\Omega)} > \beta$ otherwise, $v = 0$ is the obvious solution].

This is the problem of approximate controllability.

We begin again with the penalized functional (3.6) where (3.4) is replaced by (3.29). Let us denote by $v_\varepsilon, y_\varepsilon$ the couple where the $inf\cdot$ is achieved and let us introduce ρ_ε by (3.7). We have this time

$$\begin{aligned}&\int\int_{\sigma\times(O,T))} v_\varepsilon(\hat{v} - v_\varepsilon)dxdt + \\ &\int\int_{\Omega\times(O,T))} \rho_\varepsilon[(\frac{\partial}{\partial t}) - \Delta)(\hat{y} - y_\varepsilon) - (\hat{v} - v_\varepsilon)\chi_\sigma]dxdt \geq 0\end{aligned} \tag{3.31}$$

for every couple $\hat{v}, \hat{y}$ satisfying $\hat{y} = 0$ on Σ, $\hat{y}(0) = 0$ and $\hat{y}(T) = b \in B$. It follows that we have (3.9), (3.10) *and* a condition on $\rho_\varepsilon(T) = \rho_\varepsilon^o$:

$$(\rho_\varepsilon(T), b - y_\varepsilon(T) \geq 0, \quad \forall b \in B. \tag{3.32}$$

Inequalities (3.32) are possible if and only if $y_\varepsilon(T)$ belongs to the boundary of B

$$\|y_\varepsilon(T) - z^o\|_{L^2(\Omega)} = \beta \tag{3.33}$$

and

$$\rho_\varepsilon(T) = \rho_\varepsilon^o = \lambda_\varepsilon(z^o - y_\varepsilon(T)), \lambda_\varepsilon \geq 0. \tag{3.34}$$

But if we denote by σ_ε the solution of

$$\begin{aligned} &-\sigma_\varepsilon' - \Delta\sigma_\varepsilon = 0 \quad \text{in} \quad \Omega \times (0,T)) \\ &\sigma_\varepsilon(T) = z^o - y_\varepsilon(T) \quad \text{in} \quad \Omega, \quad \sigma_\varepsilon = 0 \quad \text{on} \quad \Sigma \end{aligned} \tag{3.35}$$

then

$$\begin{aligned} &\rho_\varepsilon = \lambda_\varepsilon\sigma_\varepsilon \quad \text{and} \\ &\|\rho_\varepsilon\|_{L^2(\sigma\times(0,T))} = \|v_\varepsilon\|_{L^2(\sigma\times(0,T))} = \\ &\lambda_\varepsilon\|\sigma_\varepsilon\|_{L^2(\sigma\times(0,T))} \leq \quad \text{constant}. \end{aligned} \tag{3.36}$$

Since one has (3.33) it follows that $\|\sigma_\varepsilon\|_{L^2(\sigma\times(0,T))} \geq \delta > 0$ so that (3.36) implies

$$0 \leq \lambda_\varepsilon \leq \delta^{-1} \quad \text{constant}.$$

Therefore ρ_ε^o *is bounded* in $L^2(\Omega)$ and one can pass to the limit. The optimality system is now

$$\begin{aligned} &-\rho' - \Delta\rho = 0, \\ &\rho(T) = \rho^o, \quad \rho|_\Sigma = 0, \\ &y' - \Delta y = \rho\chi_\sigma, \\ &y(0) = 0, \quad y|_\Sigma = 0. \end{aligned} \tag{3.37}$$

where

$$\begin{aligned} &\rho^o = \lambda(z^o - y(T)), \lambda > 0, \\ &\|z^o - y(T)\|_{L^2(\Omega)} = \beta. \end{aligned} \tag{3.38}$$

□

Remark 3.6

Let us check *directly* that (3.37), (3.38) *admits a unique solution.*

1) One considers first the system

$$\begin{aligned} & -\sigma' - \Delta\sigma = 0, \\ & \eta' - \Delta\eta = \sigma\chi_\sigma, \\ & \sigma(T) + \eta(T) = z^o, \quad \eta(0) = 0, \\ & \sigma = 0,\ \eta = 0 \quad \text{on} \quad \Sigma. \end{aligned} \tag{3.39}$$

This system admits a unique solution. Indeed with the notations of Section 3.1, $\eta(T) = \Lambda(\sigma(T))$ to that

$$(\Lambda + I)(\sigma(T)) = z^o. \tag{3.40}$$

Equation (3.40) defines in a unique fashion $\sigma(T) \in L^2(\Omega)$ and (3.39) admits a unique solution.

2) We necessarily have

$$\rho = \lambda\sigma,\ y = \lambda\eta$$

so that the last equation in (3.38) becomes

$$\|z^o - \lambda\eta(T)\|_{L^2(\Omega)} = \beta, \quad \lambda > 0 \tag{3.41}$$

which uniquely defines λ. □

Remark 3.7

One can also apply duality theory. We introduce F and L as in Remark 3.4 and we define now G by

$$G(q) = \begin{cases} 0 & \text{if} \quad \|q - z^o\| \le \beta, \quad \text{i.e.} \quad q \in B, \\ +\infty & \text{otherwise.} \end{cases}$$

Then

$$\begin{aligned} G^*(\rho^o) &= \sup(\rho^o, q),\ q \in B \\ &= (\rho^o, z^o) + \beta\|\rho^o\|_{L^2(\Omega)} \end{aligned}$$

so that the *dual problem* of (3.30) is now

$$inf\ \cdot_{\rho^o \in L^2(\Omega)} [\frac{1}{2} \int\int_{\sigma\times(0,T)} \rho^2 dxdt - \int_\Omega z^o \rho^o dx + \beta\|\rho^o\|_{L^2(\Omega)}]. \tag{3.42}$$

This is now *constructive.* □

4. Controllability of Navier-Stokes equations. Open problems

Let us consider the Navier-Stokes equation in $\Omega \times (0,T), \Omega \subset \mathbb{R}^n, n = 2$ or $n = 3$:

$$y' + (y\nabla)y - \mu\Delta y = -\nabla\pi + v\chi_\sigma \tag{4.1}$$

subject to

$$div\ y = 0 \tag{4.2}$$

$$y(0) = y^o. \tag{4.3}$$

In (4.1), $\mu > 0, \pi =$ pressure, $y =$ velocity, σ is an open subset of Ω.

Let us denote by $\mathcal{V}$ the space of vector functions φ which are smooth, with compact support in Ω and which satisfy div $\varphi = 0$ in Ω.
We denote by V (resp. H) the closure of $\mathcal{V}$ in $(H^1(\Omega))^n$ (resp. $(L^2(\Omega))^n$).
We assume that

$$y^o \quad \text{is given in} \quad H \tag{4.4}$$

and that

$$v \in (L^2(\sigma \times (0,T)))^n. \tag{4.5}$$

In these conditions it is known that there exists a solution y such that

$$y \in L^2(0,T;V) \cap L^\infty(O,T;H), \tag{4.6}$$

$$y' \in L^2(0,T;V'_{\frac{n}{2}}) \tag{4.7}$$

where

$$V'_{\frac{n}{2}} = \quad \text{dual space of} \quad V_{\frac{n}{2}}$$

$$V_{\frac{n}{2}} = \quad \text{closure of } \mathcal{V} \quad \text{in} (H^{\frac{n}{2}}(\Omega))^n$$

(If $n = 2, V_{\frac{n}{2}} = V$).

If $n = 2$, the solution is *unique*. If $n = 3$, it is not known if uniqueness in the class (4.6), (4.7) is true or not. Cf. J.L. LIONS [6], Chapter 1, Section 6. The uniqueness in the above class for dimension 2 has been proven, following the basic work of J. LERAY [1], [2], by G. PRODI and the A. [1]. Uniqueness in a smaller class was proven by O.A. LADYZENSKAYA [1].

It follows from (4.6), (4.7) and J.L. LIONS-E. MAGENES [1] (Lemme 8.3, Chap. 3), that y is (after possible modification on a set of measure 0 on $(0,T)$) continuous from $[0,T] \to H$, provided with its weak topology.

Let B be the ball

$$\|z - z^o\|_H \leq \beta \quad ; \quad z^o \quad \text{given in} \quad H. \tag{4.8}$$

The problem of *approximate controllability* analogous to the one in Section 3.2 is now :

Problem 4.1. *:Is it possible to find a couple* $\{v, y\}$ *such that* $v \in (L^2(\sigma \times (0,T)))^n$ *and* y *satisfying* (4.1), (4.2), (4.3), (4.6), (4.7) *and such that*

$$y(T) \in B. \tag{4.9}$$

Remark 4.1

The statement makes sense since (4.6), (4.7) implies the continuity of $t \to y(t)$ from $[0,T] \to H$ weakly. If $n = 2$ a simpler statement of problem 4.1 is : to find (if possible) v such that *the* solution $y(t) = y(t; v)$ of (4.1), (4.2), (4.3), (4.6), (4.7) satisfies (4.9). □

Problem 4.1 is an *open question.* More generally the structure of the *reachability set*, i.e. of the set of all $y(T)$'s in H corresponding to (4.1),...,(4.7), is not known. It is, in particular, not known whether this set is dense of not in H.

Remark 4.2

Another formulation of similar questions is : find v and y which satisfy (4.1), (4.2), (4.3), (4.6), (4.7) and which minimize the functional

$$\frac{1}{2} \int\int_{\sigma \times (0,T)} |v|^2 dx dt + \frac{k}{2} \|y(T) - z^o\|_H^2 \tag{4.10}$$

where k is given > 0.

The inf. of the functional (4.10) is a function of μ, say $\phi(\mu)$. It would be very interesting to prove (?) that the function $\mu \to \phi(\mu)$ *decreases* as $\mu \to 0$. □

Remark 4.3

In a *very* formal way, one could expect that we get closer and closer from some kind of exact controllability (or approximate exact controllability) as μ decreases to 0. □

Remark 4.4

In the *linear* situation, where one drops the term $(y \cdot \nabla)y$ in (4.1), the methods of Section 3 apply to all the above problems, with some technical difficulties which will be presented elsewhere. □

Remark 4.5

As we already indicated in the introduction, the question was raised by Prof. OCKENDON, to study similar problems for the *backward* heat equation. More precisely, given a ball B in $L^2(\Omega)$, it is possible to find a couple $\{v, y\}$ such that

$$\begin{aligned} &y' + \Delta y = v\chi_\sigma \quad \text{in} \quad \Omega \times (0,T), \\ &y(0) = 0 \quad \text{in} \quad \Omega, \quad y = 0 \quad \text{on} \quad \Gamma \times (0,T) \end{aligned} \tag{4.11}$$

and such that

$$y \in L^2(0,T; H_0^1(\Omega)) \tag{4.12}$$

such that

$$y(T) \in B. \tag{4.13}$$

We shall return to this question on another occasion. Some of the methods introduced in Section 3 extend to this situation. □

Remark 4.6

One can consider questions similar to those introduced above with boundary controls. One considers

$$\begin{aligned} &y' + (y\nabla)y - \mu\Delta y = -\nabla\pi \\ &div\ y = 0 \quad \text{in} \quad \Omega \times (0,T) \end{aligned} \tag{4.14}$$

$$y = v \quad \text{on} \quad \Gamma \times (0,T) = \Sigma \tag{4.15}$$

where v is given in $(L^2(\Sigma))^n$ such that

$$\begin{aligned} &\int_\Gamma (v\nu)d\Gamma = 0 \quad \text{a.e. in} \quad t, \\ &\nu = \quad \text{unitary normal to} \quad \Gamma. \end{aligned} \tag{4.16}$$

Except for the linear case (where $(y\nabla)y$ is dropped), questions approximate controllability are open. □

Remark 4.7

Ideas of control theory *can* be applied to numerical algorithms for Navier-Stokes equations. This is being developped by R. GLOWINSKI, J. PERIAUX and the A. □

References

J. BALL and M. SLEMROD.
[1] Non harmonic Fourier Series and Stabilization of Distributed Semi linear Control Systems. C.P.A.M. XXXII (1979), p. 555-587.

A.M. BLOCH and J.E. MARSDEN.
[1] Controlling Homoclinic Orbits (Reprint May 1989, To appear).

R.W. BROCKETT
[1] *Finite Dimensional Linear Systems.* Wiley and Sons, 1970.

I. EKELAND et R. TEMAM
[1] *Analyse convexe et Problèmes variationnels.* Dunod-Gauthier Villars, 1974.

W. FENCHEL
[1] On conjugate convex functions. Can. J. Math. 1 (1949), p. 73-77.

R. GLOWINSKI
[1] C.R.A.S. Paris 1990. Séminaire Collège de France, Décembre 1989.

R. GLOWINSKI, C.H. LI, J.L. LIONS
[1] A numerical approach to the exact boundary controllability of the wave equation (I). Jap. J. of Appl. Math. (II) to appear.

R. HERMANN and A.J. KRENER
[1] Nonlinear controllability and observability. IEEE Trans. Aut. Contr. 22, 5 (1977), p. 725-740.

L.F. HO
[1] Observabilité frontière de l'équation des ondes. C.R.A.S. Paris, 302, 1986, p. 444-446.

Y. JINHAI.
[1] Contrôlabilité exacte pour l'équation des ondes avec contraintes sur le contrôle. To appear.

2] Contrôlabilité exacte "moyennée" pour l'équation des ondes. To appear.

3] Contrôlabilité exacte de systèmes hyperboliques non linéaires. To appear.

O.A. LADYZENSKAYA.

1] Solution "in the large" of the boundary value problem for Navier-Stokes equations for the case of two space variables. Doklady Akad. Nauk. SSSR, 123, (1958), p. 427-429.

J. LAGNESE

1] *Boundary Stabilization of Thin Plates.* SIAM Studies in Applied Math., 1989.

J. LAGNESE and J.L. LIONS

] *Modelling, Analysis and Control of Thin Plates.* R.M.A. Vol. 6, Masson, 1988.

J. LERAY.

] Etude de diverses équations intégrales non linéaires et de quelques problèmes que pose l'hydrodynamique.
J.M.P.A. XII (1933), p.1-82.

] Essai sur le mouvement plan d'un liquide visqueux que limitent des parois.
J.M.P.A. XIII (1934), p. 331-418.

J.L. LIONS

] Contrôlabilité exacte des systèmes distribués. C.R.A.S. Paris 302 (1986), p. 471-475.

Exact Controllability, Stabilization and Perturbation for distributed systems, SIAM Rev. 30 (1988), p. 1-68.

Contrôlabilité exacte, Perturbation et stabilisation de systèmes distribués. Tome 1, Contrôlabilité exacte. Tome 2, Perturbations. Masson, Paris, Collection R.M.A. Vol. 8 et 9 (1988).

*Contrôle des systèmes distribués singuliers.*Collection MMI, Gauthier-Villars, 1983.

Sur la contrôlabilité exacte élargie, in Partial Differential Equations and the Calculus of Variations. Vol. II, Essays in Honor of Ennio De Giorgi. Birkhauser, 1989, p. 705-727.

Quelques Méthodes de résolution des problèmes aux limites non linéaires. Dunod, Gauthier-Villars, 1969.

J.L. LIONS et E. MAGENES

Problèmes aux limites non homogènes et applications. Vol. 1 et 2, Paris, Dunod, 1968.

J.L. LIONS et G. PRODI.

[1] Un théorème d'existence et unicité dans les équations de Navier-Stokes en dimension 2. C.R.A.S. t.248 (1959), p. 3519-3521.

T.R. ROCKAFELLAR

[1] Duality and Stability in extremum problems involving convex functionals. Pac.J. Math. 21 (1967), p. 167-187.

D.L. RUSSELL

[1] Controllability and Stabilization theory for linear partial differential equatins. Recent progress and open questions. SIAM Rev. 20, (1978), p. 639-739.

H.J. SUSSMAN and V. JURDJEVIC

[1] Controllability of nonlinear systems. J. Diff. Eqns. 12 (1972), p. 95-116.

E. ZUAZUA.

[1] Contrôlabilité exacte de systèmes d'évolution non linéaires. C.R.A.S. Paris, 306 (1988), p. 129-132.

[2] Exact controllability for the semi-linear wave equation. J.M.P.A., 1990.

BEGINNING OF WEAKLY ANISOTROPIC TURBULENCE

V.P.MASLOV
MIEM, B.Vuzovsky 3/12
109028,Moscow,USSR

ABSTRACT. Multi-phase rapidly oscillating asymptotic solutions of the Navier-Stokes equations are considered in the case of large Reynolds numbers. Regularised equations describing the evolution of the one-phase solution are obtained as well as equations for two-phase solutions generalizing the Rayleigh equation in the nonlinear case. Equations for attractors are given and a new type of instability for quasi-two-dimensional flows is considered.

1.ONE-PHASE SOLUTION

In papers [1,2,3] the notion of incorrectness for nonlinear equations was defined and regularization of Euler's equations was considered. If we consider the Navier-Stokes equations

$$\frac{\partial u_i}{\partial t} + u_k\frac{\partial u_i}{\partial x_k} + \frac{\partial P}{\partial x_i} = h^2\,\Delta u_i\,,\ i = 1,2,3,\quad x\in\Omega\subset R^3, \tag{1}$$

$$\frac{\partial u_k}{\partial x_k} = 0,\quad u|_{t=0} = V\Big(\frac{S^0(x)}{h},x\Big),\quad u|_{\partial\Omega} = 0, \tag{2}$$

then the problem will be, of course, well posed (correct) in the sense of definition given in these papers. Here $S^0\in C^\infty$, $V(\tau,x)$ is a function from C^∞ almost periodic with respect to τ, $S^0|_{\partial\Omega} = 0$, $\langle\nabla S^0, V(\tau,x) - (V)_{av}\rangle = 0$, $h = 1/\sqrt{Re}$,

R. Spigler (ed.), Applied and Industrial Mathematics, 85–99.

$$(f(\tau, x, t))_{av} = \lim_{T\to\infty} \frac{1}{T}\int_0^T f(\tau, x, t)\, d\tau$$

However, if the Reynolds number Re is sufficiently large, then the so called asymptotic incorrectness appears which means, roughly speaking, that any small perturbation (of order h) results in changing the solution by a value of order $O(1)$ during the time $t \geq \delta > 0$. Thus, actually, this notion is equivalent to the notion of instability. The system of equations for the regularized solutions $v = (u)_{av}$ of the Navier-Stokes equations according to the papers [2,3] has the form

$$\frac{dv_i}{dt} = -\frac{\partial P}{\partial x_i} - \frac{\partial}{\partial x_k}(a_i a_k)_{av}, \qquad \frac{\partial v_k}{\partial x_k} = 0 \tag{3}$$

$$v|_{t=0} = (V)_{av}, \qquad v^{\perp}|_{\partial\Omega} = 0,$$

where $i=1,2,3$, and the functions $a_i(\tau,x,t)$ almost periodic in τ satisfy the equations

$$\frac{da_i}{dt} + a_k\frac{\partial}{\partial x_k}(v_i + a_i) + \lambda\frac{\partial a_i}{\partial \tau} = |\nabla S|^2\frac{\partial^2 a_i}{\partial \tau^2} - \frac{\partial S}{\partial x_i}P + \frac{\partial}{\partial x_k}\left(a_i a_k\right)_{av}, \tag{4}$$

$$\frac{\partial \lambda}{\partial \tau} + \frac{\partial a_k}{\partial x_k} = 0, \qquad \lambda|_{\tau=0} = 0, \quad \frac{\partial S}{\partial t} + v_k\frac{\partial S}{\partial x_k} = 0, \tag{5}$$

$$a|_{t=0} = V(\tau,x) - (V)_{av}, \qquad S|_{t=0} = S^0(x), \tag{6}$$

$$a^{\parallel}|_{\tau=0,\, x\in\partial\Omega} = -v^{\parallel}|_{x\in\partial\Omega}. \tag{7}$$

Here

$$\frac{d}{dt} = \frac{\partial}{\partial t} + v_k \frac{\partial}{\partial x_k},$$

$$P = \frac{1}{|\nabla S|^2}\left\{\left(a_i a_k - (a_i a_k)_{av}\right)\frac{\partial^2 S}{\partial x_i \partial x_k} - 2a_k \frac{\partial S}{\partial x_i}\frac{\partial v_i}{\partial x_k}\right\},$$

$f^{\perp}$, $f^{\parallel}$ are the vector f components normal and tangent to $\partial\Omega$ respectively.

In order to derive the Reynolds equation (3) and auxiliary equations (4), (5), we substitute the asymptotic solution of problem (1),(2) in the form of the expansion

$$u = V_0\left(\frac{s(x,t,h)}{h},x,t\right) + h\,V_1\left(\frac{s(x,t,h)}{h},x,t\right) = \ldots.$$

$$P = P_0(x,t) + h\,\mathcal{P}_1\left(\frac{s(x,t,h)}{h},x,t\right) + \ldots \tag{8}$$

$$s = S_0(x,t) + hS_1(x,t) + \ldots$$

where P_0, $S_j \in C^{\infty}$, $V_j(\tau,x,t)$, $\mathcal{P}_k(\tau,x,t)$ are smooth functions almost periodic in τ, and the notations are used

$$v_j = \left(V_j\right)_{av}, \quad a_j = V_j - \left(V_j\right)_{av}.$$

We substitute (8) into equation (1). After easy calculations we have

$$\frac{1}{h}\left\{\frac{\partial s}{\partial t} + \langle V_0, \nabla s\rangle\right\}\frac{\partial V_0}{\partial \tau} + \frac{\partial V_0}{\partial t} + \langle V_0, \nabla\rangle V_0 + \nabla P_0 + \left(\frac{\partial s}{\partial t} + \langle V_0, \nabla s\rangle\right)\frac{\partial V_1}{\partial \tau} +$$

$$+ \langle V_1, \nabla s\rangle\frac{\partial V_0}{\partial \tau} + \nabla s\frac{\partial \mathcal{P}_1}{\partial \tau} - |\nabla s|^2\frac{\partial^2 V_0}{\partial \tau^2} = \mathcal{O}(h) \tag{9}$$

$$\frac{1}{h}\left\langle \nabla s, \frac{\partial V_0}{\partial \tau}\right\rangle + \left\langle \nabla s, \frac{\partial V_1}{\partial \tau}\right\rangle + \left\langle \nabla, V_0\right\rangle = \mathcal{O}(h)$$

By equating to zero the terms of order $\mathcal{O}(1/h)$ in (9), we get

$$\frac{\partial S_0}{\partial \tau} + \left\langle v_0, \nabla S_0\right\rangle = 0 \tag{10}$$

and the condition

$$\left\langle a_0, \nabla S_0\right\rangle = 0. \tag{11}$$

By equating to zero the terms of order $\mathcal{O}(1)$ in (9) we see

$$\frac{\partial V_0}{\partial t} + \left\langle V_0, \nabla\right\rangle V_0 + \nabla P_0 + \left[\frac{\partial S_1}{\partial t} + \left\langle V_0, \nabla S_1\right\rangle + \left\langle V_1, \nabla S_0\right\rangle\right]\frac{\partial V_0}{\partial \tau} +$$

$$+ \nabla S_0 \frac{\partial \mathcal{P}_1}{\partial \tau} = \left|\nabla S_0\right|^2 \frac{\partial^2 V_0}{\partial \tau^2} \tag{12}$$

$$\frac{\partial}{\partial \tau}\left\{\left\langle V_0, \nabla S_1\right\rangle + \left\langle V_1, \nabla S_0\right\rangle\right\} + \left\langle \nabla, V_0\right\rangle = 0 \tag{13}$$

By averaging equations (12), (13), we obtain (3), where the index 0 is omitted by the function P_0 and the vectors v_0, a_0. According to (3), we can rewrite equations (12),(13) as follows

$$\frac{\partial a_0}{\partial t} + \left\langle v_0 + a_0\right\rangle a_0 + \left\langle a_0, \nabla\right\rangle v_0 + \nabla S_0 \frac{\partial \mathcal{P}_1}{\partial \tau} + \left[\frac{\partial S_1}{\partial t} + \left\langle V_0, \nabla S_1\right\rangle + \left\langle V_1, \nabla S_0\right\rangle\right]\frac{\partial a_0}{\partial \tau} =$$

$$= \left|\nabla S_0\right|^2 \frac{\partial^2 a_0}{\partial \tau^2} + \left(\left\langle a_0, \nabla\right\rangle a_0 + \left[\frac{\partial S_1}{\partial t} + \left\langle V_0, \nabla S_1\right\rangle + \left\langle V_1, \nabla S_0\right\rangle\right]\frac{\partial a_0}{\partial \tau}\right)_{av}, \tag{14}$$

$$\frac{\partial}{\partial \tau}\left\{\left\langle a_0, \nabla S_1\right\rangle + \left\langle a_1, \nabla S_0\right\rangle\right\} + \left\langle \nabla, a_0\right\rangle = 0 \tag{15}$$

We integrate (15) over the variable τ

$$\langle a_0, \nabla S_1 \rangle + \langle a_1, \nabla S_0 \rangle = -\int_0^\tau \langle \nabla, a_0 \rangle d\tau' + C, \tag{16}$$

where the constant of integration C can be obtained by means of the condition

$(a_0)_{av} = (a_1)_{av} = 0,$, i.e.

$$C(x,t) = \left(\int_0^\tau \langle \nabla, a_0 \rangle \, d\tau' \right)_{av} . \tag{17}$$

It is also evident that

$$\langle a_0, \nabla S_1 \rangle + \langle a_1, \nabla S_0 \rangle \Big|_{\tau=0} = C . \tag{18}$$

We define the function S_1 as a solution of the equation

$$\frac{\partial S_1}{\partial t} + \langle v_0, \nabla S_1 \rangle = - \langle \nabla v_1, \nabla S_0 \rangle - C . \tag{19}$$

Then (14),(15) have the form

$$\frac{\partial a_0}{\partial t} + \langle v_0 + a_0, \nabla \rangle a_0 + \langle a_0, \nabla \rangle v_0 + \nabla S_0 \frac{\partial \mathbb{P}_1}{\partial \tau} + \lambda \frac{\partial a_0}{\partial \tau} =$$

$$= |\nabla S_0|^2 \frac{\partial^2 a_0}{\partial \tau^2} + \left(\langle a_0, \nabla \rangle a_0 + a_0 \langle \nabla, a_0 \rangle \right)_{av} , \tag{20}$$

$$\frac{\partial \lambda}{\partial \tau} + \langle \nabla, a_0 \rangle = 0$$

where

$$\lambda = \langle a_0, \nabla S_1 \rangle + \langle a_1, \nabla S_1 \rangle - C .$$

Now we take into account that the solution of (20) must satisfy the condition (11). The system (11),(20) is not overdeterminate, since (20) contains an "extra" function $\mathcal{P}_1$. By multiplying (20) scalarly by ∇S_0, we have due to (10), (11)

$$|\nabla S_0|^2 \frac{\partial \mathcal{P}_1}{\partial \tau} =$$

$$= \langle a_0, \langle a_0, \nabla \rangle \nabla S_0 \rangle - 2\langle \nabla S_0, \langle a_0, \nabla \rangle v_0 \rangle - \left[\langle a_0, \langle a_0, \nabla \rangle \nabla S_0 \rangle\right]_{av} . \tag{21}$$

Denoting $P = \dfrac{\partial \mathcal{P}_1}{\partial \tau}$ and omitting the index 0, we obtain equations (4),(5) from (10), (20), (21).

Now we must pose initial and boundary conditions for (3)-(5). The initial conditions

$$S_0|_{t=0} = S^0(x), \quad (s - S_0)|_{t=0} = 0, \quad v_0|_{t=0} = (V)_{av}, \quad a_0|_{t=0} = V(\tau, x) - (V)_{av}$$

are obtained as a direct consequence of (2) and the asymptotic expansion (8). In order to pose the boundary conditions we first of all note that the solution (10) for $S^0|_{\partial\Omega} = 0$ and the function v_0 satisfying the impermeability condition $v_0^{\perp}|_{\partial\Omega} = 0$ vanishes identically on the boundary $\partial\Omega$. Further, by rewriting (18),(19) in the form

$$\frac{\partial S_1}{\partial t} + \langle V_0|_{\tau=0}, \nabla S_1 \rangle + \langle V_1|_{\tau=0}, \nabla S_0 \rangle = 0, \qquad S_1|_{t=0} = 0,$$

and using the condition of glueing, we can easily see that $S_1|_{\partial\Omega} = 0$ uniformly in t. Thus $s|_{\partial\Omega} = 0$ with accuracy $O(h^2)$, and we can neglect this error when defining the principal term asymptotics. Thus, the vector $\nabla S_0|_{\partial\Omega}$ is directed along a normal to $\partial\Omega$ and the condition (11) means, in particular, that $a_0^{\perp}|_{\tau=0, x\in\partial\Omega} = 0$. This and the condition of glueing (2) yield the condition (7).

The problem (3)-(7) defines the function $v_0(x,t)$, $a_0(\tau,x,t)$, $P_0(x,t)$, $S_0(x,t)$ completely, however the principal term of the velocity asymptotic, i.e. $V_0(S_0/h + S_1, x, t)$ remains to be undeterminate in this approximation, since the function S_1 is unknown. The equation (19) shows that S_1 depends on the correction v_1, i.e. on the lower term of the asymptotical expansion. This dependence means that the problem (1),(2) solution is asymptotically incorrect. Actually, if we replace the initial condition (2) by a value $O(h)$, i.e. we replace (2) by the condition

$$u|_{t=0} = V_0(\frac{S^0}{h}, x) + hv_1^0(x),$$

we obtain $v_1|_{t=0} = v_1^0(x)$ instead of $v_1|_{t=0} = 0$, i.e. the function v_1 varies by a value $O(1)$. For $t \geq \delta > 0$ this change yields, due to (19), a change of S_1, and thus $a_0(S_0/h + S_1, x, t)$ changes by a value of order $O(1)$. At the same time the regularized equations (3) are stable with respect to a small perturbation, since (3) are invariant to the change of $a_i(\tau,x,t)$ by $a_i(\tau + F, x,t)$ with an arbitrary smooth function $F(x,t)$ [2].

Further we shall need equations (3)-(5) written in the Lagrangian system of coordinates. We denote by $S_j(x)$ smooth functions such that the vectors $\nabla S^0(x)$ and $\nabla f_j(x)$ are mutually orthogonal, $j=2,3$. We define the functions $x^j = x^j(x,t)$ as the solution of the following problems

$$\frac{\partial x^j}{\partial t} + \langle v, \nabla x^j \rangle = 0,$$

$$x^1|_{t=0} = S^0(x), \quad x^i|_{t=0} = f_i(x), \quad i=2,3. \tag{22}$$

When $v \in C^\infty$, the Jacobian $\det\left|\frac{\partial x^j}{\partial x}\right|$ does not vanish for all t, so we can choose x^j as new coordinates. We denote by e_j, e^j the direction vectors of the new coordinate system

$$e_j = i_1 \frac{\partial x_1}{\partial x^j}, \quad e^j = i_1 \frac{\partial x^j}{\partial x_1},$$

where i_1 are the unit vectors of the Cartesian coordinate system, and we assume

$$g_{ik} = \langle e_i, e_k \rangle, \quad g^{ik} = \langle e^i, e^k \rangle, \quad g = \det(g_{ik}).$$

Further, we denote by F^j the coefficients of the vector F expansion with respect to the vectors e_1, e_2, e_3, and by $F^j_{,k}$ the covariant derivative of F^j with respect to the variable x^k

$$F = F^j e_j, \quad F^j = \langle F, e_j \rangle, \quad F^j_{,k} = \frac{\partial F^j}{\partial x^k} + \Gamma^j_{lk} F^l,$$

where $\Gamma^j_{lk} = \left\langle e^j, \frac{\partial e_l}{\partial x^k} \right\rangle$ is the Christoffel symbol of the second kind.

By passing to the Lagrangian coordinates in (3)-(7), we obtain the following problem

$$\frac{\partial v^j}{\partial t} + v^k v^j_{,k} + g^{jk} \frac{\partial P}{\partial x^k} = -\mathfrak{R}^j, \tag{23}$$

$$\frac{\partial}{\partial x^k}(\sqrt{g}\, v^k) = 0, \quad v^j\big|_{t=0} = \left(V^j\right)_{av}, \quad v^1\big|_{\partial\Omega} = 0,$$

where j=1,2,3, and the Reynolds tensions can be calculated by the formulas

$$\mathfrak{R}^1 = \Gamma^1_{lk}(a^k a^l)_{av}, \quad \mathfrak{R}^i = \frac{1}{\sqrt{g}} \frac{\partial}{\partial x^k} \sqrt{g}\,(a^k a^i)_{av} + \Gamma^i_{kl}(a^k a^l)_{av}, \quad i=2,3,$$

$a^1 = 0$, the functions $a^i(\tau, x^1, x^2, x^3, t)$, i=2,3, almost periodic in τ satisfy the following equations

$$\frac{\partial a^i}{\partial t} + a^k(2v^i_{,k} + a^i_{,k}) + \lambda \frac{\partial a^i}{\partial \tau} = g^{11} \frac{\partial^2 a^i}{\partial \tau^2} - g^{1i} p + \mathfrak{R}^i,$$

$$\frac{\partial \lambda}{\partial \tau} + \frac{1}{\sqrt{g}} \frac{\partial}{\partial x^k}(\sqrt{g}\, a^k) = 0, \qquad \lambda\big|_{\tau=0} = 0 \tag{24}$$

$$a^i\big|_{t=0} = V^i - \left(V^i\right)_{av}, \quad a^i\big|_{\tau=0,\partial\Omega} = -v^i\big|_{\partial\Omega},$$

$$P = -\frac{1}{g^{11}}\left\{2a^k v^i_{,k} + \Gamma^1_{lk}\left[a^k a^l - (a^k a^l)_{av}\right]\right\}.$$

We also note that if the initial value $V(\tau,x)$ decreases rapidly, $\lim_{\tau\to\infty} V(\tau,x) = v^0(x)$, the asymptotic solution (8) is a solution of "boundary layer" type, and equations (3),(4) are equivalent to the Prandtl equations. Actually, we set $u^{\parallel} = v^{\parallel} + a^{\parallel}$, $S = f(x^{\parallel})x^{\perp}$, where $x^{\perp}$ is the distance between the point x and $\partial\Omega$ along the normal, $f \neq 0$ is a smooth function of variables tangent to $\partial\Omega$, and take into account that similarly as in the theory of complex germ we have

$$(x^{\perp})^k \exp(-f(x^{\parallel})x^{\perp}/h) = O(h^k), \qquad k \geq 0$$

After the change $\xi = f(x^{\parallel})x^{\perp}$ we obtain the Prandtl equations for $u^{\parallel}$ in their usual form [4,5].

2. TWO-PHASE SOLUTIONS

Now we consider the two-phase situation which is a generalization of the preceding section. We asume that the initial conditions for (1) have the form

$$u|_{t=0} = V\left(\frac{S^0}{h}, x\right) + \varepsilon V^1\left(\frac{S^0}{h}, \frac{\Phi^0}{h}, x\right) \tag{25}$$

where V^1, $\Phi^0 \in C^\infty$, $\varepsilon \ll 1$, the function V^1 is almost periodic in τ and periodic in η with the period 2π, $\langle\nabla S^0, \nabla\Phi^0\rangle = 0$. By $\bar{f}(\tau,x,t)$ we shall denote the function $f(\tau,\eta,x,t)$ value averaged with respect to the variable η:

$$\bar{f}(\tau,x,t) = \frac{1}{2\pi}\int_0^{2\pi} f(\tau,x,t)\,d\eta.$$

We assume that the function $\langle V^1, \nabla S^0\rangle$ projection on the function $\chi\exp(i\eta)$, where χ is an eigenfunction of the Rayleigh equation

$$(\omega + \langle V,p\rangle)\left[(\nabla S^0)^2\frac{\partial^2}{\partial\tau^2} - p^2\right]\chi - (\nabla S^0)^2\langle V_{\tau\tau}, p\rangle\chi = 0, \tag{26}$$

corresponding to the maximal imaginary part of the equation (26) eigenvalue, is of order $O(1)$. Further, we assume that $\nabla\Phi^0$ is equal to such a value of p in (26), for which $\max \operatorname{Im} \omega(p)$ is achieved, since just such perturbations grow more rapid than the others.

We introduce the following notations: $u(\frac{S}{h},\frac{\Phi}{h},x,t)$, $p(\frac{S}{h},\frac{\Phi}{h},x,t)$ is the problem (1),(25) solution, $v=(\bar{u})_{av}$, $a=\bar{u}-v$, $w = u-a-v$, $\chi(\tau,\eta,x,t) = \langle w(\tau,\eta+G\tau,x,t),\nabla S\rangle$, $\theta(\tau,\eta,x,t)=\langle w(\tau,\eta+G\tau,x,t),\nabla\Phi\rangle$, $G=|\nabla S|^{-2}\langle\nabla S,\nabla\Phi\rangle$, $p_\perp=\nabla\Phi-G\nabla S$. Under the assumption $\langle a,\nabla S\rangle=0$, $dS/dt=0$ the functions χ and θ satisfy the Rayleigh equation with accuracy $O(h)$ for $t\geq\delta>0$ in general position and after regularization in the sense of [1] :

$$(\Phi_t+\langle v+a,\nabla\Phi\rangle)\left[(\nabla S)^2\frac{\partial^2}{\partial\tau^2}+p_\perp^2\frac{\partial^2}{\partial\eta^2}\right]\chi-(\nabla S)^2\langle a_{\tau\tau},\nabla\Phi\rangle\chi+$$

$$+p_\perp^2(\theta\chi_{\eta\eta}-\chi\theta_{\eta\eta})+(\nabla S)^2(\theta\chi_{\tau\tau}-\chi\theta_{\tau\tau})=0, \tag{27}$$

$$\chi_\tau+\theta_\eta=G\chi_\eta, \tag{28}$$

χ,θ are almost periodic in τ and periodic in η with the period 2π, the function χ satisfies the boundary condition

$\chi|_{\tau=0}=0$

We represent the solution of (27), (28) in the form of the Fourier series with respect to η :

$$\chi=\sum_n\chi e^{in\eta},\ \theta=\sum_n\theta_n e^{in\eta},\ \chi_0=\theta_0=0. \tag{29}$$

We multiply (27) by the function $\bar{\varphi}^*\exp(-i\eta)$, $\varphi^*=\bar{\varphi}(\bar{\omega}+\langle a,\nabla\Phi\rangle)^{-1}$, where φ is the Rayleigh equation solution,

$$\mathfrak{L}(\omega,\nabla\Phi,p_\perp)\varphi\equiv(\omega+\langle a,\nabla\Phi\rangle)\left[(\nabla S)^2\frac{\partial^2}{\partial\tau^2}-p_\perp^2\right]\varphi-(\nabla S)^2\langle a_{\tau\tau},\nabla\Phi\rangle\varphi=0 \tag{30}$$

integrate with respect to η from 0 to 2π, and average with respect to τ. We obtain the following equality

$$\lim_{T\to\infty}\frac{1}{2\pi T}\int_0^{2\pi}d\eta\int_0^T d\tau\left\{\chi\left[(\nabla S)^2\bar{\varphi}^*_{\tau\tau}-p_\perp^2\bar{\varphi}^*\right]e^{-i\eta}\right\}\times$$

$$\times(\Phi_t+\langle v_0,\nabla\Phi\rangle-\omega)+\lim_{T\to\infty}\frac{1}{2\pi T}\int_0^{2\pi}d\eta\int_0^T d\tau\,\bar{\varphi}^*\exp(-i\eta)\times$$

$$\times\left[(p_\perp^2)(\theta\chi_{\eta\eta}-\chi\theta_{\eta\eta})+(\nabla\Phi)^2(\theta\chi_{\tau\tau}-\chi\theta_{\tau\tau})\right]=0 \qquad (31)$$

By using (29), we obtain from (31)

$$\left(\frac{d\Phi}{dt}-\omega\right)\left(\chi_1\left[(\nabla S)^2\frac{\partial^2}{\partial\tau^2}-p_\perp^2\right]\bar{\varphi}^*\right)_{av}+$$

$$\left(\bar{\varphi}^*\left\{\sum_l p_\perp^2\left[\frac{i}{1-l}\frac{\partial\chi_{1-l}}{\partial\tau}+G\chi_{1-l}\right]\chi_l(1-2l)+\sum_l(\nabla S)^2\times\right.\right.$$

$$\left.\left.\left[\left(\frac{i}{1-l}\frac{\partial\chi_{1-l}}{\partial\tau}+G\chi_{1-l}\right)\frac{\partial^2\chi_l}{\partial\tau^2}-\chi_l\left(\frac{i}{1-l}\frac{\partial^2\chi_{1-l}}{\partial\tau^3}+G\frac{\partial^2\chi_{1-l}}{\partial\tau^2}\right)\right]\right\}\right)_{av}=0 \qquad (32)$$

It is easy to see that after summing in (32) all the summands containing the function G as a factor cancel out, and we finally have

$$\left(\frac{d\Phi}{dt}-\omega\right)\left(\chi_1\left[(\nabla S)^2\frac{\partial^2}{\partial\tau^2}-p_\perp^2\right]\bar{\varphi}^*\right)_{av}+$$

$$i\sum_l \left(\bar{\varphi}^* \left\{ p_\perp^2 \frac{1-2l}{1-l} \chi_l \frac{\partial \chi_{1-l}}{\partial \tau} + (\nabla S)^2 \frac{1}{1-l} \left(\frac{\partial \chi_{1-l}}{\partial \tau} \frac{\partial^2 \chi_l}{\partial \tau^2} - \chi_l \frac{\partial^3 \chi_{1-l}}{\partial \tau^3} \right) \right\} \right)_{av} = 0 \tag{33}$$

By equating to zero the real and imaginary parts in (33), we get the Hamilton-Jacobi equation for the phase Φ :

$$\frac{d\Phi}{dt} - \operatorname{Re}\omega + \operatorname{Re} A(x,t,\nabla\Phi) = 0 \tag{34}$$

and the equation

$$-\operatorname{Im}\omega + \operatorname{Im} A = 0 \tag{35}$$

which defines the attractor corresponding to the given perturbation. Here

$$A = i\sum_l \left(\bar{\varphi}^* \left\{ p_\perp^2 \frac{1-2l}{1-l} \chi_l \frac{\partial \chi_{1-l}}{\partial \tau} + \frac{(\nabla S)^2}{1-l} \left(\frac{\partial \chi_{1-l}}{\partial \tau} \frac{\partial^2 \chi_l}{\partial \tau^2} - \chi_l \frac{\partial^3 \chi_{1-l}}{\partial \tau^3} \right) \right\} \right)_{av} \times$$

$$\times \left\{ \left(\chi_1 \left[(\nabla S)^2 \, \partial^2/\partial\tau^2 - p_\perp^2 \right] \bar{\varphi}^* \right)_{av} \right\}^{-1}$$

If we assume that the imaginary part of ω is small, which holds for a boundary layer in a weakly Rayleigh unstable situation, we find χ by means of the perturbation theory, choosing as the zero approximation the solution of the Rayleigh equation $\chi^{(0)} = \alpha\varphi \exp(i\eta) + \overline{\alpha\varphi} \exp(-i\eta)$, where α defines the perturbation amplitude near the attractor. In this case (27),(28) yield

$$\chi = \alpha\varphi \exp(i\eta) + \alpha^2 \psi \exp(2i\eta) + \text{c.c.} + \dots$$

$$\theta = \alpha(i\varphi_\tau + G\varphi) + \text{c.c.} + \dots$$

where

$$\psi=-i\,(\nabla S)^2\mathfrak{L}^{-1}(\omega,\nabla\Phi,2p_{\perp})(\varphi_\tau\varphi_{\tau\tau}-\varphi\varphi_{\tau\tau\tau})$$

By substituting these expansions into (34).(35) and omitting the terms cubic in α, we obtain the relation which defines the amplitude α:

$$A=i\,|\alpha|^2\left(\bar{\varphi}^*\left\{3p_{\perp}^2\left(\psi\bar{\varphi}_\tau+\frac{1}{2}\psi_\tau\bar{\varphi}\right)+(\nabla S)^2\left[\frac{1}{2}(\psi_\tau\bar{\varphi}_{\tau\tau}-\bar{\varphi}\psi_{\tau\tau\tau})+\psi\bar{\varphi}_{\tau\tau\tau}-\bar{\varphi}_\tau\psi_{\tau\tau}\right]\right\}\right)_{av}\times$$

$$\times\left\{\left(\psi\left[(\nabla S)^2\partial^2/\partial\tau^2-p_{\perp}^2\right]\bar{\varphi}^*\right)_{av}\right\}^{-1}.$$

The projections of the vector w (in the principal term) on the directions ∇S and $\nabla\Phi$ can be defined from the equations (27), (28), and the projection $w_n = \langle w(\tau,\ \eta+G\tau,x,t),n\rangle$ on the direction n (in the principal term also) satisfy the following equation

$$(\Phi_t+\langle v,\nabla\Phi\rangle+\langle a,\nabla\Phi\rangle)\frac{\partial w_n}{\partial\eta}+\chi\frac{\partial}{\partial\tau}(w_n+\langle a,n\rangle)+(\theta-G\chi)\frac{\partial w_n}{\partial\eta}=0. \qquad (36)$$

In this case the terms $\partial\overline{(w_i w_k)}_{av}/\partial x_k$ will be added to the Reynolds tensions in (3)

$$\frac{dv_i}{dt}=-\frac{\partial(\bar{p})_{av}}{\partial x_i}-\frac{\partial}{\partial x_k}\left[(a_i a_k)_{av}+\left(\overline{w_i w_k}\right)_{av}\right].$$

In equations (4),(5) some additional summands will also appear

$$\frac{da_i}{dt}+a_k\frac{\partial}{\partial x_k}(u_i+a_i)+\lambda\frac{\partial a_i}{\partial\tau}=(\nabla S)^2\frac{\partial^2 a_i}{\partial\tau^2}-\frac{\partial S}{\partial x_k}\frac{\partial\pi^{(1)}}{\partial\tau}-\frac{\partial}{\partial x_i}\pi^{(0)}+$$

$$+\frac{\partial}{\partial x_k}\left[(a_i a_k)_{av}+(\overline{w_i^{(0)}w_k^{(0)}})_{av}\right]-\overline{\langle w^{(0)},\nabla S\rangle\frac{\partial}{\partial\tau}w_i^{(1)}}-\overline{\langle w^{(1)},\nabla S\rangle\frac{\partial}{\partial\tau}w_i^{(0)}}-$$

$$-\overline{\langle w^{(0)},\nabla\Phi\rangle\frac{\partial}{\partial\eta}w_i^{(1)}}+\overline{\langle w^{(1)},\nabla\Phi\rangle\frac{\partial}{\partial\eta}w_i^{(0)}}, \qquad \frac{\partial\lambda}{\partial\tau}+\frac{\partial a_k}{\partial x_k}=0.$$

Here

$$w^{(0)} = w\big|_{h=0},\; w^{(1)} = \frac{\partial w}{\partial h}\Big|_{h=0},\; \pi^{(0)} = \bar{p} - (\bar{p})_{av}\big|_{h=0},\; \pi^{(1)} = \frac{\partial}{\partial h}(\bar{p} - (\bar{p})_{av})\big|_{h=0}.$$

However, there is a situation when this problem cannot be regularized. In partricular, this situation occurs when we have such a symmetric picture that the variables in the Navier-Stokes equation can be separated, so that the main flow is two-dimensional. A more general condition for such a situation is given by the equalities

$$g^{11}\Gamma^3_{22} - g^{13}\Gamma^1_{22} = 0,\; g^{11}\bar{u}^3_{,2} - g^{13}\bar{u}^1_{,2} = 0 \tag{37}$$

Here g^{il}, Γ^l_{ki} are the metric tensor coefficients and the Christoffel symbols of the Lagrangian coordinate system $x^j = x^j(x,t)$, where x^j are constant along the mean flow v, $x^1\big|_{t=0} = S_0(x)$, $x^2\big|_{t=0} = f(x)$, $x^3\big|_{t=0} = \Phi_0(x)$, $\nabla f \,||\, n$, $v^l_{,k}$ is thecovariantderivative with respect to x^k of the l-th component of the vector v in the Lagrangian system of coordinates.

Under conditions (37) one can set in (4) $\langle a, \nabla\Phi\rangle = 0$, and the equations (4),(5) for the remaining components will have the form

$$\frac{\partial\Xi}{\partial t} + \Xi\,\Xi_{,2} + \lambda\frac{\partial\Xi}{\partial\tau} + \alpha\Xi + \beta\Xi^2 = g^{11}\frac{\partial^2\Xi}{\partial\tau^2} + R^2$$

$$\frac{\partial\lambda}{\partial\tau} + \frac{1}{\sqrt{g}}\frac{\partial}{\partial x^2}(\sqrt{g}\;\Xi) = 0$$

where

$$\Xi = \langle a, \nabla x^2\rangle,\; g = \det(g^{ij}),\; R^2 = \frac{1}{\sqrt{g}}\frac{\partial}{\partial x^2}\left(\sqrt{g}\;(a^2)_{av}\right) +$$

$$+\frac{1}{g^{11}}(g^{11}\Gamma^2_{22} - g^{12}\Gamma^1_{22})(a^2)_{av},\; \alpha = \frac{2}{g^{11}}(g^{11}u^2_{,2} - g^{12}u^1_{,2}),$$

$$\beta = -g^{12}\Gamma^1_{22}/g^{11}.$$

The expression $\varepsilon\frac{1}{2\pi}\int_0^{2\pi}\langle V^1(\tau,\eta,x),\nabla\Phi_0\rangle d\eta$ is the initial condition for the function $\langle a,\nabla\Phi\rangle$ in (4),(5), thus the function $\langle a,\nabla\Phi\rangle$ is a value of order ε. So, one can set $\langle a,\nabla\Phi\rangle=\varepsilon\, g(\tau,\eta,x,t)$, $\omega=\varepsilon\tilde{\omega}$, $\chi=\varepsilon\tilde{\chi}$, $\theta=\varepsilon\tilde{\theta}$. In this case the equations (27),(28) do not change, and the component w_n satisfies the equation

$$[\varepsilon^{-1}(\Phi_t+\langle v,\nabla\Phi\rangle)+g]\frac{\partial w_n}{\partial\eta}+\tilde{\chi}\frac{\partial}{\partial\tau}(w_n+\langle a,n\rangle)+(\tilde{\theta}-G\tilde{\chi})\frac{\partial w_n}{\partial\eta}=0 \quad (38)$$

Under our assumptions we have

$$\varepsilon^{-1}(\Phi_t+\langle v,\nabla\Phi\rangle)=0(\varepsilon),$$

(due to (34)), and (38) yields $w_n = \mathcal{O}(1)$. Since χ depends essentially on initial perturbations (25), we can consider in this situation a stochastic picture only and we must apply the probability methods, so, the turbulence in this situation differs essentially from the turbulence in the general position.

3.REFERENCES

1. Maslov,V.P. (1987) Resonance vortexes and coherent structures in turbulent flow , MIEM, Moscow.
2. Maslov,V.P. (1986) "Coherent structures, resonances and asymptotical non-uniqueness for the Navier-Stokes equations with large Reynolds numbers", Uspekhi Mathem. Nauk 41, N6, 19-35.
3. Maslov,V.P. (1986) "Violation of the causality principle for unsteady equations of two- and three-dimensional gas dynamics with large numbers", Theor. and Mathem Physics 69, N3, 361-378.
4. Maslov,V.P. (1988) "Lin-Lees equations for boundary layers in domains with curvilinear boundary",Physica D 33, 266-280.
5. Hirshell,E.H. and Kordulla,W. (1986) "Sheer flow in surface-oriented coordinates", Notes on Numerical Fluid Mech. 4, Vieweg,Braunschweig/Wiesbaden.

MARKOV RANDOM FIELDS, STOCHASTIC QUANTIZATION AND IMAGE ANALYSIS[1]

Sanjoy K. Mitter
Massachusetts Institute of Technology
Electrical Engineering and Computer Science
and
Laboratory for Information and Decision Systems
Cambridge, MA 02139

1 Introduction

Markov random fields based on the lattice Z^2 have been extensively used in image analysis in a Bayesian framework as a-priori models for the intensity field and on the dual lattice $(Z^2)*$ as models for boundaries. The choice of these models has usually been based on algorithmic considerations in order to exploit the local structure inherent in Markov fields. No fundamental justification has been offered for the use of Markov random fields (see, for example, GEMAN-GEMAN [1984], MARROQUIN-MITTER-POGGIO [1987]). It is well known that there is a one-one correspondence between Markov fields and Gibbs fields on a lattice and the Markov Field is simulated by creating a Markov chain whose invariant measure is precisely the Gibbs measure. There are many ways to perform this simulation and one such way is the celebrated Metropolis Algorithm. This is also the basic idea behind Stochastic Quantization. We thus see that if the use of Markov Random fields in the context of Image Analysis can be given some fundamental justification then there is a remarkable connection between Probabilistic Image Analysis, Statistical Mechanics and Lattice-based Euclidean Quantum Field Theory. We may thus expect ideas of Statistical Mechanics and Euclidean Quantum Field Theory to have a bearing on Image Analysis and in the other direction we may hope that problems in image analysis (especially problems of inference on geometrical structures) may have some influence on statistical physics.

This paper deals with the issues just described and above all it suggests a program of research for dealing with the fundamental issues of probabilistic image analysis.

What is the fundamental problem of Image Analysis? It may be stated as follows: Given noisy images (possibly stereo) of the visual world in motion, represent and recognize "structures" in some invariant manner. For example, if the structures of concern are three-dimensional rigid objects then we want this recognition to be invariant under the action of the euclidean group. Thus symmetries play an important role in the whole process.

[1]This research has been supported by the Air Force Office of Scientific Research under grant AFOSR 89-0276 and by the Army Research Office under grant ARO DAAL03-86-K-0171 (Center for Intelligent Control Systems)

R. Spigler (ed.), Applied and Industrial Mathematics, 101–109.

Ultimately, we want to identify the mechanisms (circuits) which perform the full recognition task. There are three aspects to this problem: a) the identification and representation of a priori knowledge of the visual world without which the recognition cannot take place; b) the extraction of knowledge from the imprecise images available about the visual world and c) understanding the correct interaction between a priori knowledge and the imprecise data available in the form of images. It is this formulation of the problematique that suggests a Bayesian framework. In this paper we deal only with some partial aspects of b) and c) and it is here that the connection with statistical physics enters.

The visual world is a continuous world and if we believe that it is important to capture symmetries then mathematical formulations should not be on the lattice $\mathcal{L}$ but should be based on $\mathbf{R}^2$. Even if we are interested in algorithms on a digital machine which necessarily involve discretization, then in order to capture the symmetries in the limit of lattice spacing going to zero attention must be paid to the discrete formulations of the problem (see for example, KULKARNI-MITTER-RICHARSON [1990]).

This paper is organized as follows. In section 2 we discuss Markov fields and Euclidean fields and state the conjecture regarding Osterwalder-Schrader fields. Section 3 is concerned with a variational problem in Image Analysis and its probabilistic interpretation and it shows how these ideas are related to those of Section 2. Finally in Section 4 we discuss Stochastic Quantization.

2 Markov Fields and Euclidean Fields. (NELSON [1973])

Let $\mathbf{R}^d$ denote Euchidean d-dimensional space ($d \geq 2$) and let $\mathcal{S}(\mathbf{R}^d)$ denote Schwartz space. If V is a topological vector space, a linear process over V is a stochastic process φ indexed by V which is linear and such that if $f_\alpha \to f$ in V then $\varphi(f_\alpha) \to \varphi(f)$ in measure. This implies that $\varphi(f)$ for $f \in V$ is a random variable over $(\Omega, \mathcal{F}, \mu)$.

Let φ be a linear process over $\mathcal{S}(\mathbf{R}^d)$. If $\Lambda \subset \mathbf{R}^d$ is open, $\mathcal{A}(\Lambda)$ be the σ-algebra generated by $\varphi(f)$ with support of $f \subseteq \Lambda$ and if Λ is any subset of $\mathbf{R}^d$ let $\mathcal{A}(\Lambda) = \bigcap_{\Lambda' \supset \Lambda} \mathcal{A}(\Lambda')$ where Λ' ranges over all open sets containing Λ. $\mathcal{A}(\Lambda)$ will also denote the set of all random variables which are measurable with respect to $\mathcal{A}(\Lambda)$. Let Λ^c denote the complement of Λ and $\partial\Lambda$ denote the boundary of Λ.

A Markov field on $\mathbf{R}^\mathbf{d}$ is a linear process φ over $\mathcal{S}(\mathbf{R}^d)$ such that whenever Λ is open in $\mathbf{R}^d$ and α is an integrable random variable in $\mathcal{A}(\Lambda)$ then $E(\alpha|\mathcal{A}(\Lambda^c)) = E(\alpha|\mathcal{A}(\partial\Lambda))$. Here E denotes conditional expectation.

Let $O(d)$ be the Euclidean group of $\mathbf{R}^d$. By a representation T of $O(d)$ on some probability space $(\Omega, \mathcal{F}, \mu)$ we mean a homomorphism $\eta \to T(\eta)$ of $O(d)$ into the group of automorphisms of the measure algebra and this group acts in a natural way on random variables.

A Euclidean field is a Markov field φ over $\mathcal{S}(\mathbf{R}^d)$ together with a representation T of $O(d)$ on $(\Omega, \mathcal{F}, \mu)$ such that $\forall f \in \mathcal{S}(\mathbf{R}^d)$ and $\eta \in O(d)$

$$\text{(Covariance)} \quad T(\eta)\varphi(f) = \varphi(f \circ \eta^{-1}), \tag{2.1}$$

and if ρ in the reflection in the hyperplane $\mathbf{R}^{d-1}$

$$T(\rho)\alpha = \alpha, \ \alpha \in \mathcal{A}(\mathbf{R}^{d-l}). \tag{2.2}$$

Consider the mapping

$$\begin{aligned} S_n \ &: \ (\mathbf{R}^d)^n \longrightarrow C \\ &: \ S_n(f_1,\cdots,f_n) = E(\varphi(f_l)\cdots\varphi(f_n)) \end{aligned}$$

and assume it is continuous. By the Schwartz Kernel Theorem, there exists a distribution $\mathcal{S}_n \in \mathcal{S}'(\mathbf{R}^{dn})$ such that $S_n(f_1,\cdots,f_n) = \mathcal{S}_n(f_l \otimes \cdots \otimes f_n)$.

Nelson proves that the sequence of distributions $\mathcal{S}_n$ satisfy euclidean invariance, symmetry and Osterwalder-Schrader positivity, namely, if Λ is the half-space $x^d > 0$, so that $\partial\Lambda$ is the hyperplane $\mathbf{R}^{d-1}$ and Λ^c is the half space $x^d \leq 0$ then $E[(T(p)\bar{\alpha})\alpha] \geq 0$ where $\alpha \in \mathcal{A}(\Lambda)$ and $T(p)\bar{\alpha} \in \mathcal{A}(\Lambda^c)$.

An example of such a euclidean field is obtained as follows. Let $\mathcal{S}_{\mathbf{R}}(\mathbf{R}^d)$ be the real Schwartz space, let $m > 0$ and let H be the real Hilbert space completion of $\mathcal{S}_{\mathbf{R}}(\mathbf{R}^d)$ with respect to the scaler product $< g, (-\Delta + m^2)^{-1} f >$ where Δ is the Laplace operator. Let φ be the unit Gaussian on H, i.e. φ is a real Gaussian process indexed by H with mean zero and covariance given by the scalar product on H. Extend φ to the complexification of H by linearity. Restricted to $\mathcal{S}(\mathbf{R}^d)$, φ is a linear process over it. Nelson proves that φ is a Euclidean field, that is it is Markov and satisfies euclidean invariance, symmetry and Osterwalder-Schrader positivity. Indeed the Markov property, covariance property and reflection implies Osterwalder-Schrader positivity.

Non-gaussian random fields can be constructed using Multiplicative functionals or measure transformations. Let φ be a Markov field over $\mathcal{S}(\mathbf{R}^d)$ with the underlying probability space $(\Omega, \mathcal{F}, \mu)$. We say, that a random variable β is multiplicative if for every open cover $\{\Lambda_i\}$ of $\mathbf{R}^d$, there exists strictly positive β_i in $\mathcal{A}(\Lambda_i)$ with $\beta = \prod_i \beta_i$. We shall see later how we construct $P(\varphi)_2$ fields using these ideas.

To see the connections with Gibbs density let us proceed formally. Let

$$H_0(x) = \frac{1}{2}((\nabla\varphi(x))^2 + m^2(\varphi(x))^2)$$

and consider the formal expression

$$\exp(-\int H_0(x)dx) \prod_{x\in\mathbf{R}^2} d\varphi(x).$$

The rigorous interpretation of this expression is as a Gaussian measure μ_c on $\mathcal{S}'(\mathbf{R}^2)$ with mean zero and covariance $C = (-\Delta + m^2)^{-1}$. This measure corresponds to the free euclidean field. Non-gaussian measures are obtained by considering formal expressions

$$\exp(-\int_{even} [H_0(x) + \lambda P(\varphi(x))]dx) \prod_{x\in\mathbf{R}^2} d\varphi(x)$$

where P is an even polynomial and λ is a coupling constant. The above corresponds to the canonical Gibbs density for transverse vibrations of an elastic membrane subject to the non-linear restoring force $F = -m^2\varphi(x) - \lambda P'(\varphi(x))$, (after integration over momentum variables $\hat{\varphi}(x)$), and $'$ denotes derivative).

Now if φ is the unit Gaussian process over $\mathcal{S}(\mathbf{R}^2)$, then $P(\varphi)$, for an even polynomial does not make sense. To fix ideas let us consider the case where $P(\varphi) = \varphi^4$. It turns out, however that

$$:\varphi^4:(g) \triangleq \int g(x) : \varphi(x) : dx \text{ for } g \in L' \bigcap L^\infty$$

has a well-defined meaning in a limiting sense in $L^2(\Omega, \mathcal{F}, \mu)$, where $:\varphi(x)^4:$ is an element of the 4th Homogeneous class. In a more concrete way, $:\varphi^3 :\triangleq \varphi^3 - 3E(\varphi^2)\Phi$ and this is well-defined. Now if we consider

$$\beta = \frac{\exp(-\int g(x) : \varphi^4(x) : dx)}{E(\exp(-\int g(x) : \varphi^4(x) : dx)}$$

then β is a multiplicative random variable and we are able to construct the non-gaussian measure

$$d\mu = \frac{\exp(-\int : \varphi^4 : dx)}{E \exp(-\int : \varphi^4 : dx)} d\mu_C.$$

This measure is the so-called $(\varphi^4)_2$-measure.

The proof that $(\varphi^4)_2$-measure defines a Markov field is surprisingly difficult and has been accomplished only recently (see ALBEVERIO, HOEGH-KROHN and ZEGARLINSKI [1989]). It depends on the complicated theory of local specifications, related Gibbs states and cluster expansions. On the other hand the property of Osterwalder-Schrader positivity is much easier to verify. With a view to proving the Global Markov property and for problems in Image Analysis we advance the following conjecture.

We define an Osterwalder-Schrader field (O-S) on $\mathbf{R}^2$ to be a linear process φ over $\mathcal{S}(\mathbf{R}^2)$ which satisfies the Euclidean covariance property (2.1) and Osterwalder-Schrader positivity. For example an Osterwalder-Schrader field may be obtained by considering a function of a Markov field. We conjecture that every Osterwalder-Schrader field can be obtained as a function of a Markov Field. We call such a Markov field a Hidden Markov Field for the O-S field.

Let us see this conjecture in the familiar context of stochastic processes. Let $(\varphi_t | t \in \mathbf{R})$, $\varphi : \Omega \to V$ be a stochastic process, which is continuous in probability, stationary, symmetric (i.e. φ_t and φ_{-t} are stochastically equivalent). It is O-S positive if $\forall\ 0 \le t_1 \le \cdots \le t_n$, $\forall$ $f : V^n \to C$, bounded Borel (V is a topological vector space) we have

$$< f(\varphi_{t_1}, \cdots, \varphi_{t_n}), \bar{f}(\varphi_{-t_1}, \cdots, \varphi_{-t_n}) >_{L^2(\Omega,\mathcal{F},P)} \ge 0.$$

Now, let $\tilde{\varphi}_t : \Omega \to \tilde{V}$, be a stationary, symmetric Markov process and let $\psi : \tilde{V} \to V$ be bounded and Borel.

Define a new process by

$$\varphi_t(\omega) = \psi(\tilde{\varphi}_t(\omega)).$$

$\tilde{\varphi}_t$ is called a Markov extension of φ_t. φ_t is symmetric and stationary and satisfies O-S postivity but is not necessarily Markov. The conjecture would be:

Does every stationary, O-S process have a Markov extension?

This conjecture in discrete-time is not true (cf. ARVESON [1986]) but in the form stated previously may be true.

(It is clear that every bounded Borel function of a symmetric, stationary Markov process is O-S positive).

3 Probabilistic View of Image Segmentation

We think of a noisy image as a function $g : \Omega \to \mathbf{R}$ where $\Omega \subset \mathbf{R}^2$ is a bounded, open set. We assume $g \in L^\infty(\Omega)$. A variational formulation of the image segmentation problem due to Mumford and Shah (cf. MUMFORD-SHAH [1989]) is as follows. Approximate g by a function f and a closed set $\Gamma \subset \bar{\Omega}$ such that the following energy function is minimized:

$$E(f,\Gamma) = \beta \int_\Omega |f-g|^2 dx + \int_{\Omega\backslash\Gamma} |\nabla f|^2 dx + \alpha H^1(\Gamma). \tag{3.1}$$

Here f is required to be in $W^{1,2}(\Omega\backslash\Gamma)$, $H^1(T)$ denotes the 1-dimensional Hausdorff measure of Γ and β, α are positive constants. We would like to give a probabilistic interpretation of $E(f,\Gamma)$ by considering

$$\exp(-\int_{\Omega\backslash\Gamma} |\nabla f|^2 dx - \alpha H^1(\Gamma))$$

as a Gibbs density with respect to a suitable reference measure which is to serve as a prior measure on (f,Γ) and $\exp(-\int |g-f|^2 dx)$ as a likelihood function of g given f. The choice of the energy functional is dictated by the requirement that f should approximate g in the L^2-sense, it should be smooth away from the boundaries Γ and the total length of the boundary should be short. Note that for fixed Γ, $\exp(-\int_{\Omega/\Gamma} |\nabla f|^2 dx) \prod_{x\in\Omega\backslash\Gamma} df(x)$ would have a rigorous interpretation as a Gaussian measure μ_C with mean zero and covariance $C = (-\Delta)^{-1}$ on $\mathcal{S}'(\Omega/\Gamma)$. Using the recent work of Surgailis (cf. ARAK-SURGAILIS [1989]) we can give a rigorous interpretation as a density to the following:

$$\exp(-\int_{\Omega\backslash \bigcup_{i=1}^n \Gamma_i} |\Delta f|^2 dx - 2\sum_{i=1}^n \ell(\Gamma_i) - \varphi(n))$$

which, for an appropriate $\varphi(n)$, is a density over $f \in W_0^{2,2}(\Omega\backslash \bigcup_{i=1}^n \Gamma_i)$, n and a set of straight lines Γ_i which form together with $\partial\Omega$ a polygonal partition of Ω (cf. MITTER-ZEITOUNI [1990]). We outline here the basic result of Surgailis which constructs a measure on closed polygonal partitions of Ω.

Let Ω be a closed convex subset of $\mathbf{R}^2$ with smooth boundary. In $\mathbf{R}^2$, choose coordinates (t,x) such that, for all $y \in \Omega$, $t(y)$, $x(y) > 0$. Let $\mathcal{L}_\Omega$ denote the lines which intersect Ω, each line $\ell \in \mathcal{L}_\Omega$ is parameterized by its distance from the origin ρ_ℓ and the angle it forms with the $t = 0$ axis, α_ℓ.

Let $\mu(d\ell)$ be a uniform measure on the set $\alpha_\ell, \rho_\ell | \ell \in \mathcal{L}_\Omega$. The Poisson point process with intensity $\mu(d\ell)$ will be denoted μ^Ω, and the measure it induces on the boundary $\partial\Omega$ by the hitting points $\{(x_\ell, t_\ell) | t_\ell = \inf\{t | \ell \in \Omega\}\}$ is again a Poisson point process on the triple (x, t, ν) with intensity $\mu^{\partial\Omega}$. Here and in the sequel, ν denotes the velocity of the particle, i.e. the tangent of the angle formed by the trajectory and the t axis.

For any line $\ell \in \mathcal{L}_\Omega$, let ν_ℓ denote the slope of ℓ. Clearly, $\mu(d\ell)$ can be considered as a measure on ν_ℓ and x_ℓ, the intersection of ℓ with the x axis. In the sequel, we consider the measure $\mu(d\nu, dx)$ obtained from the uniform measure $\mu(d\ell)$ on α, ρ, i.e.

$$\mu(d\ell) = \mu(d\nu, dx) = dx \frac{d\nu}{(1+\nu^2)^{\frac{1}{2}}}. \tag{3.2}$$

Inside Ω, construct a point process on the quadruple (t, y, ν', ν'') with intensity

$$\mu^P(dt, dy, d\nu', d\nu'') = |\nu' - \nu''| dy dt \frac{d\nu'}{(1+(\nu')^2)^{\frac{3}{2}}} \frac{d\nu''}{(1+(\nu'')^2)^{\frac{3}{2}}}. \tag{3.3}$$

Finally, construct a random partition of Ω as follows:

Pick up on $\partial\Omega$, n_0 triples (t, x, ν) according to the law $\mu^{\partial\Omega}$, and inside Ω, n_1 quadruples (t, y, ν', ν'') according to the Poisson process with intensity μ^P. At each of those points, start a line of slope ν (two lines of slopes ν', ν'' in the case of interior points) and evolve ν according to the Markov transition law

$$P(\nu_{t+dt} \in du | \nu_t = \nu) = |u - \nu| \frac{du dt}{(1+u^2)^{\frac{3}{2}}}. \tag{3.4}$$

Finally, at each intersection of lines (when viewing it in the direction of growing t) kill the intersected lines. Clearly, such dynamics describe a random partition of Ω by polygons, c.f. fig. 1. The basic result of Arak and Surgailis is:

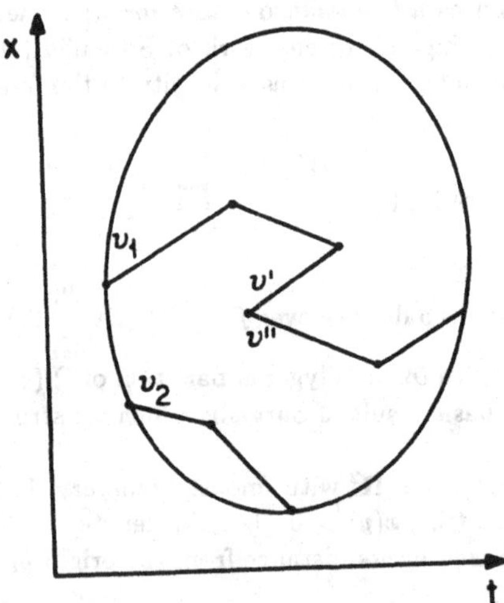

Figure 1: Random Polygonal Parition

Lemma 3.1

$$P(n,\ell \in d\ell_i, i=1,\cdots,n) = \frac{1}{n!}\mu(d\ell_1)\cdots\mu(d\ell_n)\exp(-2\sum_{i=1}^{n}\mathcal{L}(\ell_i)) \tag{3.5}$$

where $\mathcal{L}(\ell_i)$ denotes the length of the i-th segment.

Note that due to the presence of n in (3.5), one can't consider (3.5) directly as a candidate for a density: indeed, if one were to consider $P(n,\ell,\in \underline{\ell}_i \pm \epsilon)$, the required normalization constant (as $\epsilon \to 0$) would have depended on n and therefore, a path with no jumps will be infinitely more likely than a path with one jump.

One way out of this problem is by using an appropriate definition: Let

$$\begin{aligned} Z^n &\triangleq \int_{\mathcal{L}_D}\cdots\int_{\mathcal{L}_D} P(n,\ell_i \in d\ell_i, i=1,\cdots,n) \\ Z &\triangleq \sum_{n=0}^{\infty} Z^n. \end{aligned}$$

$(\frac{Z^n}{Z})$ is the probability of having n lines in a specific partition. Now, one may define:

Definition The prior density of a partition (n,ℓ_i) is given by

$$p(n,\ell_i) = (\frac{Z^n}{Z})\lim_{\epsilon\to 0}\frac{P(n,\ell_i \in \ell_i \pm \epsilon, i=1,\cdots,n)}{2\epsilon^{2n}} \tag{3.6}$$

Combining these ideas with some of the ideas contained in Dembo-Zeitouni (cf. DEMBO-ZEITOUNI [submitted]) the desired result referred to before can be obtained. The more general problem of interpreting $\exp(-\int|\nabla f|^2dx - H^1(\Gamma))$ as a density remains open.

The interpretation of $\beta\int_\Omega|f-g|^2dx + \int_{\Omega\setminus\Gamma}|\nabla f|^2dx + \alpha H^1(\Gamma)$ as a posterior density accomplishes something important. It frees us from obtaining Maximum A Posteriori estimates of (f,Γ) via minimization of the above functional. We can in principle obtain other estimates such as conditional mean estimates. Indeed for closed, convex partitions of Ω it opens up the possibility of doing inference on geometries via Monte-Carlo simulations.

4 Stochastic Quantization (see BORKAR-CHARI-MITTER [1988] for Stochastic Quantization of $(\varphi^4)_2$ fields).

The problem of stochastic quantization for the energy functional (3.1) is to create a Markov process whose invariant measure is $\exp(-E(f,\Gamma))$. In this section we wish to suggest a program for achieving this goal. We concentrate on the prior density $\exp(-\int_\Omega|\nabla f|^2dx - \alpha H^1(\Gamma))$. The first step in the procedure is to replace the above energy functional appearing in the density by an approximating functional

$$\mathcal{E}_n(f,v) = \int_\Omega|\nabla f|^2(1-v^2)^n dx + \alpha\int_\Omega((|\nabla f|^2+|\nabla v|^2)(1-v^2)^n + \frac{n^2v^2}{16})dx. \tag{4.1}$$

In (4.1) the variable $v(x) \in [0,1]$ should be thought of as a control variable, which controls the gradient of f and depends on the discontinuity set Γ, n is a parameter which tends to infinity. The above expression makes sense for $f, v \in W^{1,2}(\Omega)$. $1-(1-v^2)^n$ approximates smoothed neighbourhoods of Γ and the approximate boundaries can be identified with $(1-v^2)^n \simeq 0$. If we denote by $\mu_n(B) = 2(n+1)\int_B v(1-v^2)^n |\nabla v| dx$, then essentially $\mu_n(\Omega) \rightarrow H^4(\Gamma)$ in a weak sense. For details of this approximation scheme see AMBROSIO-TORTORELLI [1990].

The first step in the program would be to identify $\exp(-\mathcal{E}_n(f,v)) \prod_{x \in \mathbf{R}^2} d(f(x), v(x)) \triangleq d\mu_n$ as a measure in a suitable distribution space. The covariances involved will have appropriate boundary conditions. One can then study the weak convergence of this measure $d\mu_n$ as $n \rightarrow \infty$. The functional derivatives of $\mathcal{E}_n$ with respect to f, v can be computed:

$$\frac{\delta \mathcal{E}_n}{\delta f} = -\nabla \cdot (\nabla f \cdot (1-v^2)^n) \tag{4.2}$$

$$\frac{\delta \mathcal{E}_n}{\delta v} = -\alpha \nabla \cdot (\nabla v \cdot (1-v^2)^n) + n(|\nabla f|^2 + \alpha |\nabla v|^2)(1-v^2)^n v + \frac{\alpha n^2}{16} v \tag{4.3}$$

(There are additional terms involving the normal derivatives).

In analogy with the study of stochastic quantization of $(\varphi^4)_2$ fields, the problem of stochastic quantization for fixed n is the study of the coupled pair of stochastic differential equations

$$df(t) = -\frac{\delta \mathcal{E}_n}{\delta f} \cdot dt + dw(t) \tag{4.4}$$

$$dv(t) = -\frac{\delta \mathcal{E}_n}{\delta v} \cdot dt + dv(t) \tag{4.5}$$

where $w(\cdot)$ and $v(\cdot)$ are infinite-dimensional independent Brownian motions.

5 Conclusions

The conceptual program outlined in this paper is quite general and may be applied to other variational problems arising in Image Analysis, for example, those involving curvature terms. These variational problems may be important to obtain non-overlapping segmentations of images with a view to identifying occluded regions.

References

[1] Albeverio, S., Hoegh-Krohn, R. and Zegarlinski, B., Uniqueness and Global Markov Property for Euclidean Fields: The Case of General Polynomial Interactions, *Communications in Mathematical Physics*, Vol. 123, pp. 377-424.

[2] Ambrosio, L., and Tortorelli, V. [1990]: Approximations of functionals depending on jumps by elliptic functionals via G-convergence, to appear in *Communications in Pure and Applied Mathematics*.

[3] Arak, T., and Surgailis, D. [1989]: Markov fields with polygonal realizations, *Prob. Th. Rel. Fields* 80, pp. 543-579.

[4] Arveson, W. [1986]: Markov Operators and O-S Positive Processes, *Journal of Functional Analysis* 66, pp. 173-234.

[5] Borkar, R., Chari, R. and Mitter, S.K. [1988]: Stochastic Quantization of Field Theory in Finite and Infinite Volume, *Journal of Functional Analysis*, Vol. 81, No. 1.

[6] Dembo, A. and Zeitouni, O. [submitted]: Maximum a-posteriori estimation of elliptic Gaussian fields observed via a noisy nonlinear channel.

[7] Geman, S. and Geman, D. [1984]: Stochastic Relaxation, Gibbs Distribution, and the Bayesian Restoration of Images, *IEEE Transactions on Pattern Analysis and Machine Intelligence* 6, pp. 721-741.

[8] Kulkarni, S. , Mitter, S.K. and Richardson, T.J. [1990]: An existence theorem and lattice approximation for a variational problem arising in computer vision, Proceedings, IMA Signal Processing Workshop, summer 1989, published as *Signal Processing, Part I: Signal Processing Theory*, New York: Springer-Verlag.

[9] Marroquin, J.L., Mitter, S.K. and Poggio, T. [1987]: Probabilistic Solution of Ill-posed Problems in Computer Vision, *Journal of the American Statistical Association* 82, No. 397.

[10] Mitter, S.K. and Zeitouni, O. [1990]: An SPDE formulation for image segmentation, *Proceedings, Conference on Stochastic PDEs*, Trento, Italy, January 1990.

[11] Mumford, D. and Shah, J. [1989]: Optimal approximations of piecewise smooth functions and associated variational problems, *Communications in Pure and Applied Mathematics* 42, pp. 577-685.

[12] Nelson, E. [1973]: Probability theory and euclidean field theory, In: Constructive Quantum Field Theory, G. Velo and A.S. Wightman (eds.), it Lecture Notes in Mathematics, Vol. 25, New York: Springer-Verlag.

COMPUTATIONAL METHODS FOR THE BOLTZMANN EQUATION

H. Neunzert, F. Gropengießer, J. Struckmeier
Department of Mathematics
University of Kaiserslautern
P.O. Box 3049
D-6750 Kaiserslautern
F.R.G.

ABSTRACT. This paper contains the basic ideas and practical aspects for numerical methods for solving the Boltzmann Equation. The main field of application considered is the reentry of a Space Shuttle in the transition from free molecular flow to continuum flow. The method used will be called Finite Pointset Method (FPM) approximating the solution by finite sets of particles in a rigorously defined way. Convergence results are cited while practical aspects of the algorithm are emphasized. Ideas for the transition to the Navier Stokes domain are shortly discussed.

1. Introduction

The determination of the flow field around a space shuttle during its reentry is one of the most challenging tasks in computational fluid dynamics. The phase of the reentry where the shuttle moves down from altitudes over 150 km to lets say 70 km is not the most critical one but determines the "initial conditions" for the following critical phase. The paper deals with this first part of the reenty, which belongs to the rarefied gas regime; this means that the gas is not dense enough in order to use the normal continuum flow equations as Navier-Stokes or Euler. The similarity parameter deciding whether this is possible or not is the Knudsen number $Kn = \lambda/L$, where λ is the mean free path and L is the characteristic length of the problem (for example the curvature radius of the shuttle nose). $Kn > 10$ designs the free molecular flow, $0.05 \leq Kn \leq 10$ defines the transitional regime, where Navier-Stokes may and in general will give wrong results.
The first attempt to treat this region was done by Bird [5], using a "Direct Simulation Monte Carlo method" (DSMC). This method developed from 1968 on simulates the microscopic behaviour of the gas, not refering to any equation which has to be solved. DSMC was and is extremely successful, carried by a deep understanding of the physics behind. From a mathematical point of view the situation however is not satisfying, since questions about convergence or the quality of

R. Spigler (ed.), Applied and Industrial Mathematics, 111–140.

approximations cannot even be posed. The equation which is governing the transition regime is the Boltzmann equation, from which the continuum flow equations can be derived in considering certain singular limits $Kn \to 0$ (see for example [7]). This quite complicated equation resisted for a long time all attempts for a rigorous numerical treatment; several steps in this direction were done by Indian, Russian and American authors, but could by far not compete in practical respects with DSMC. Nanbu [24] in 1980 proposed a new method directly related to the Boltzmann equation but again suffering from practical weaknesses. However, starting from Nanbu's idea our group in Kaiserslautern succeeded from 1986 on in treating the following problems:

(i) To clarify the basic principle of particle methods and thereby constructing a whole class of numerical schemes consistent with the Boltzmann equation (including Nanbu's algorithm);
(ii) to give rigorous convergence proofs for these schemes;
(iii) to improve the computational efficiency of these methods such that they are able to treat all the cases handled by DSMC - and hopefully even more.

We want to state clearly that we consider DSMC still to be a practicable method from which we learn a lot especially with respect to physical extensions (as including interior energy, chemical reactions etc.); moreover we believe that minor modification of DSMC can be shown to belong to the class of convergent schemes, since the basic idea could be formulated in an appropriate way.

This paper focusses on problems (i) and (iii), since (ii) was discussed in several papers by Babovsky et al. [2], [4]. We call our method "Finite Pointset Method" FPM, expressing thereby the fact that it is based on the approximation of the solution by mathematical objects depending on finitely many data - as FEM or Finite Differences do. To describe this appoximation ideas, where no Monte Carlo gambling is involved, in a way which is understandable also to the engineer is the first intention; the second is to discuss practical aspects of our FPM and the third to present some ideas for connecting it with the continuum flow domains. Extensions to polyatomic gases and chemical reactions will be presented in another paper [18].

2. The Mathematical Formulation of the Problem

Our main mathematical object we wish to solve is the Boltzmann equation; it describes the evolution of a position-velocity space density $f(t,x,v)$ with $x \in \Omega \subset \mathbb{R}^3$, Ω denoting that part of the space surrounding the shuttle, where neither the free molecular flow description nor the Navier-Stokes equations are correct. Ω may therefore change with time during the reentry but for the time being we keep it fixed and discuss domain decomposition in chapter 6.

We consider here the Boltzmann equation for a one species monatomic gas - generalization to several species of molecules including interior energy of different kind are available but would complicate our presentation unreasonably. Chemical reactions are very important but the research we know is not very much developed in this direction. We use the notation of Cercignani (see [7], pages 57 ff), so that the Boltzmann equation may be written as

$$\frac{\partial f}{\partial t} + v\cdot\frac{\partial f}{\partial x} = \frac{1}{\varepsilon}J(f,f) \tag{1a}$$

with

$$J(f,f)(t,x,v) := \int_{\mathbb{R}^3}\left\{\int_0^{2\pi}\int_0^{\pi/2} B(\theta,\|V\|)\cdot\right. \tag{1b}$$

$$\left.[f(t,x,v')f(t,x,v_1')-f(t,x,v)f(t,x,v_1)]d\theta d\phi\right\}dv_1\ ;$$

here

- $V = v-v_1$
- $v' = v-n\langle n,V\rangle$ and $n = (\sin\theta\cos\phi,\sin\theta\sin\phi,\cos\theta)$
  $v_1' = v_1'+n\langle n,V\rangle$
- $B(\theta,\|V\|)$ characterizes the intermolecular forces and is $\sigma^2\|V\|\cdot\sin\theta\cos\theta$ for a gas of rigid spheres of diameter σ or $\|V\|^{\frac{n-5}{n-1}}\cdot\beta(\theta)$ for an inverse power potential of type $k\|x\|^{1-n}$.
- ε is proportional to the mean free path and therefore also to the Knudsen number.

By integrating with respect to n over the hemisphere $S_+^2 = \left\{n/\|n\| = 1 \text{ and } \langle n,V\rangle > 0\right\}$ instead of $d\theta d\phi$ and by using the differential scattering cross section s one may write the inner integration as

$$\int_{S_+^2} \|V\|s(\theta,\|V\|)[\ldots]d\omega(n)$$

with

$$s(\theta,\|V\|) = \frac{1}{\|V\|\sin\theta}B(\theta,\|V\|)\ .$$

Equation (1) must be supplemented by initial and boundary conditions. Concerning the initial condition the following remark seems to be necessary. There are two time scales:
Time scale 1 refers to the arrival of the shuttle at a certain part of the space, when the unperturbed atmosphere is changed suddenly to the flow field around the shuttle; time scale 2 refers to the change of the unperturbed atmosphere depending on the altitude the shuttle is just

passing. Travelling with the object we may therefore consider stationary solutions of equation (1) with slowly changing boundary conditions of the far field. However, the stationary Boltzmann equation presents a tougher problem both from a theoretical (see for a survey [22] and the papers of Ukai & Asano cited there) as well as from a numerical point of view. What one does is to solve the instationary problem with the unperturbed atmosphere as initial condition, i.e. one operates on time scale 1. During the reentry one has to adapt the initial condition to the altitude the shuttle has arrived. But one should keep in mind that one is interested in different stationary solutions, i.e. in $\lim_{t\to\infty} f(t,x,v)$ but not in the dependence of f on time. t has more the character of an iteration index and may therefore be changed in order to speed up the procedure; to do this in an optimal way is a research problem for its own.

We assume furthermore that the shuttle is at rest, but the gas around moves and we consider (1) together with $f(0,x,v) = \overset{\circ}{f}(x,v)$ given (including the speed of the flight, because the shuttle is at rest) and with boundary onditions.

The choice of boundary conditions is very crucial; there are many indications (see for example [27]) that the solution is much more sensitive with respect to the boundary condition than to a correct model $B(\theta,\|V\|)$ for the intermolecular forces. The boundary $\partial\Omega$ of our computational domain consists of three different parts:

(a) The boundary $\partial\Omega_1$ against the shuttle surface; here we have to model the gas-surface interaction, which creates finally the drag, lift, heat transfer, pressure on the shuttle (see again [7], chapter 3). The general shape of the condition at the surface is

$$|\langle v\cdot n\rangle f(t,x,v) = \int_{\{v':\langle v'\cdot n\rangle<0\}} R(v' \to v;x)f(t,x,v')\cdot |\langle v',n\rangle| dv' \quad \text{for } v\cdot n > 0 , \tag{2}$$

where n is the normal to the surface at x, which points into Ω and R is the scattering kernel representing the information about the surface. It is extremely difficult to find reliable scattering kernels for two reasons:

(i) It is a very complex task to describe what happens to a par ticle entering the surface and interacting with the molecules in the rigid body.

(ii) There are very few experimental data available.

Mathematically simple but unrealistic is specular reflexion with $R_S(v' \to v;x) = \delta(v-v'+2n\langle n,v'\rangle)$;
practically used in the past is diffuse reflexion with $R_D(v' \to v;x) = f_o(x,v)|\langle v,n\rangle|$, where

$$f_o(x,v) = \frac{1}{2\pi R^2 T_o^2(x)} \exp\left(-\frac{\|v-u_o(x)\|^2}{2RT_o(x)}\right) ,$$

$T_o(x), u_o(x)$ denoting the temperature and the velocity of the wall at x.

Maxwell proposed a combination of these kernels, namely

$$R = \alpha R_D + (1-\alpha) R_S$$

and α is called accomodation coefficient. The solution depends very sensitively on α as one can see in figure 1 presenting the lift coefficient of a solution computed numerically with different α.

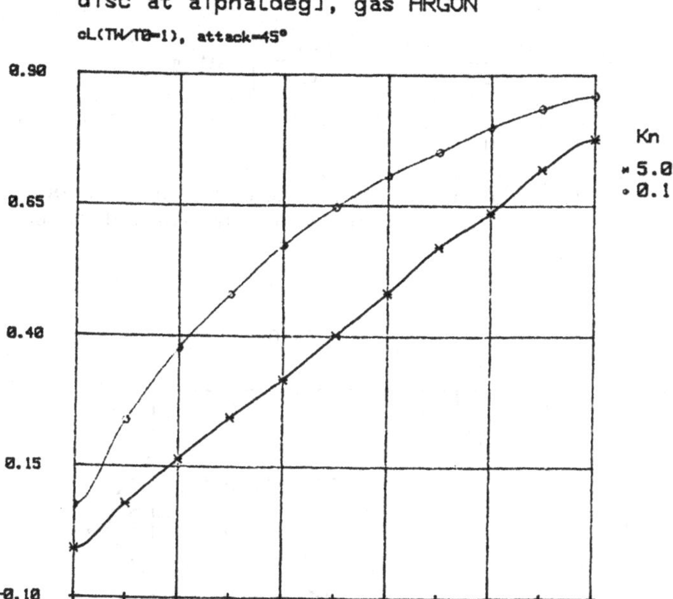

Figure 1

We have no experimental information about the best accomodation coefficient for a given surface. This fact reduces also the practical value of more elaborate models as for example that by Cercignani & Lampis (see [7']), when even two accomodation coefficients are present. However, good benchmark experiments would allow to fit the coefficients and then it is quite likely that the two coefficients in the model mentioned above could be chosen such that there is a good agreement between computation and experiment of lift, drag, heat transfer etc. (Our numerical experiments with all the boundary conditions mentioned above and with different accomodation coefficients point in this direction.)

(b) Our computational domain has to be bounded - we have to introduce an outer boundary $\partial\Omega_\infty$; its distance to the vehicle should be large enough, so that the asumption of thermodynamic equilibrium and spatial homogeneity is justified. This makes no problem in front of the shuttle - $\partial\Omega_\infty^1$ is located just in front of the shock wave. Behind the shuttle - we call this part $\partial\Omega_\infty^2$ - these asumptions are never strictly fulfilled; it is however enough that the mean velocity in the direction of the outer normal n to $\partial\Omega_\infty^2$ is supersonic. This boundary is computed during the calculation and is time dependent; at $\partial\Omega_\infty^2$ the flux into the computation domain vanishes. We therefore have for $\langle v,n\rangle<0$, where n is the outer normal on $\partial\Omega_\infty^1$

$$f(t,x,v) = \frac{1}{2\pi R^2 T^2(t)} \exp\left[-\left(\frac{\|v-u\|^2}{2RT}\right)\right] \quad \text{at } \partial\Omega_\infty^1$$

$$= 0 \quad \text{at } \partial\Omega_\infty^2 \ .$$

(c) But our computational domain Ω has a third kind of boundary, the boundary $\partial\Omega_2$ against the region, where the fluid dynamic equations are valid (figure 2).

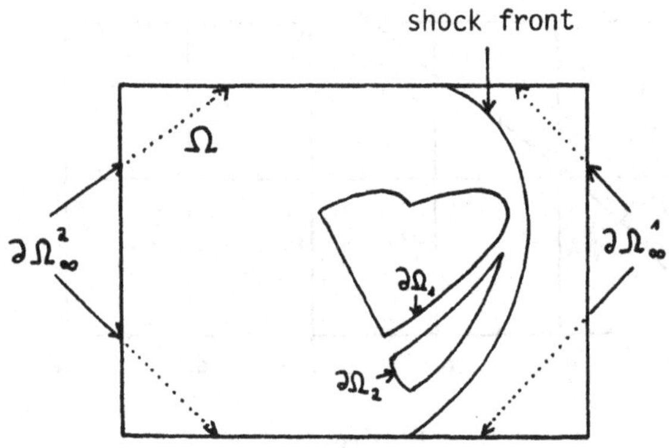

Figure 2

At this boundary the relation between the Boltzmann equation and the Euler or Navier-Stokes equations becomes important. These continuum flow equations are limits of the Boltzmann equation for small Knudsen numbers - the sense in which this limit has to be understood is crucial for our problem. In high altitudes where the Knudsen numbers are everywhere greater than 1 we are sure that the numerical solution of the Boltzmann equation is not more expensive than the solution of the Navier-Stokes or Euler equations even in regions of thermodynamic equilibrium. Further down the computational effort for the Boltzmann equation increases, but there are still domains where the Navier-Stokes equations even with slip boundary conditions are not valid (there is still a small controversy about how "non-valid" the Navier-Stokes equations are in altitudes around 90 km, but there are experiments showing that the errors can be very significant). For Knudsen numbers smaller than 10^{-2} the Boltzmann equation becomes more expensive than Navier-Stokes and the Boltzmann regions become very small (however, the latter are never really empty - there remain kinetic boundary layers and the neighborhoods of small edges and wedges). What is really needed is a mixed flow code discovering automatically where Boltzmann, Euler and NavierStokes are valid - more precisely: Where one can get numerical simplifications from the fact that the solution is near to the Maxwell or Navier-Stokes distribution. We will discuss these questions a bit more in details in chapter 6. In our numerical treatment in chapter 3 and 4 we will neglect $\partial\Omega_2$.

We therefore shall treat the following initial boundary value problem:

$$\frac{\partial f}{\partial t} + v\cdot\frac{\partial f}{\partial x} = \frac{1}{\varepsilon}J(f,f) \tag{1}$$

$$|v\cdot n| f(t,x,v) = \int_{v'\cdot n<0} R(v'\to v;x) f(t,x,v')|v'\cdot n| dv' \quad \text{at } \partial\Omega_1 \tag{2}$$

$$\begin{aligned} f(t,x,v) &= \frac{1}{2\pi R^2T^2(t)} \exp\left(-\frac{\|v-u\|^2}{2RT}\right) \quad \text{for } \langle v,n\rangle < 0 \quad \text{at } \partial\Omega^1_\infty \\ &= 0 \quad \text{for } \langle v,n\rangle < 0 \text{ at } \partial\Omega^2_\infty , \end{aligned} \tag{3}$$

$$f(0,x,v) = f(x,v) \tag{4}$$

(= unperturbed atmosphere depending on the altitude)

3. The basic idea of the Finite Pointset Method (FPM)

The task to solve the problem stated in section 2 numerically poses many difficult practical problems. For a 3-dimensional problem we have at least 7 variables t,x,v with a tendency to more if one includes interior energy; we have the quite complicated integral expression for J; we have finally complicated boundary conditions. Therefore there is little hope for a successful application of standard approximation methods like finite differences or finite elements. We are aware of the fact that there are nice ideas available in the direction of finite differences, ideas developed and used by several authors [29], [1], [12] - but we are also aware that they have only been applied to lower dimensional problems. Practically more successful are particle methods - particles are points with mass, position, velocity (interior energy, charge, etc.). We consider a one species monatomic gas and normalize the total mass to 1; mass is conserved as $\int_{\mathbb{R}^3}\int_{\Omega} f(t,x,v)dxdv$ doesn't depend on t. Having N particles we may assume that each has a mass $\frac{1}{N}$, such that the i-th particle is characterized by $(x_i,v_i) \in \Omega\times\mathbb{R}^3$, i=1,...,N, i.e. by a point in $\Omega\times\mathbb{R}^3$. The collection of all N particles is therefore a finite pointset in $\Omega\times\mathbb{R}^3$

$$\omega_N := \{(x_1,v_1),\dots,(x_N,v_N)\} \ .$$

We want to use finite pointsets for approximating the solution $f(t,\cdot)$ - as we use finite elements to approximate solutions of the Navier-Stokes equations. Refinement of the mesh there means enlargement of the number N here: Convergence would mean that ω_N converges to f with N tending to infinity. The only thing which has to be defined is the distance between ω_N and f - what means that ω_N is near to f?
We explain this in a simple one-dimensional situation - the extension to our situation is straightforward.

Let $\omega_N = \{x_1,\dots,x_n\}$ be a set of points in the interval [0,1] and $f: [0,1] \to \mathbb{R}$ an integrable nonnegative function with $\int_0^1 f(x)dx = 1$. We define the discrepancy of ω_N with respect to f by

$$D(\omega_N;f) := \max_{0\le a<b\le 1} \left|\int_a^b f(x)dx - \frac{1}{N}\,(\text{number of points in } [a,b])\right| \qquad (5)$$

$$= \max_{0\le a<b\le 1} \left|\int_a^b f(x)dx - \frac{1}{N}\sum_{j=1}^{N} \chi_{[a,b]}(x_j)\right|$$

where

$$\chi_M(x) = \begin{cases} 1 & \text{for } x \in M \\ 0 & \text{else .} \end{cases}$$

Discrepancy was defined by Hermann Weyl (see [28]) as a number theoretic concept and was reinvented in probability theory as Kolmogorov-Smirnov distance. It is clear that it compares the relative frequency of points in [a,b] with the "fraction of mass with distribution f in [a,b]" and considers the worst case with respect to all intervals. It is therefore the concretization of the intuitive understanding of f as a continuous particle density, but has nothing to do with any probabilistic interpretation.
The generalization to our situation is simple: If $\omega_N = \left\{(x_1,v_1),\dots,(x_N,v_N)\right\}$ is a pointset in $\Omega\times\mathbb{R}^3$ and $f\ge 0$ integrable on $\Omega\times\mathbb{R}^3$ with integral 1, we define

$$D(\omega_N;f) = \sup_R \left|\int_R f(x,v)dxdv - \frac{1}{N}\sum_{j=1}^{N} \chi_R(x_j,v_j)\right| \qquad (5a)$$

where R denotes an arbitrary axi-parallel 6-dimensional rectangle in $\Omega\times\mathbb{R}^3$

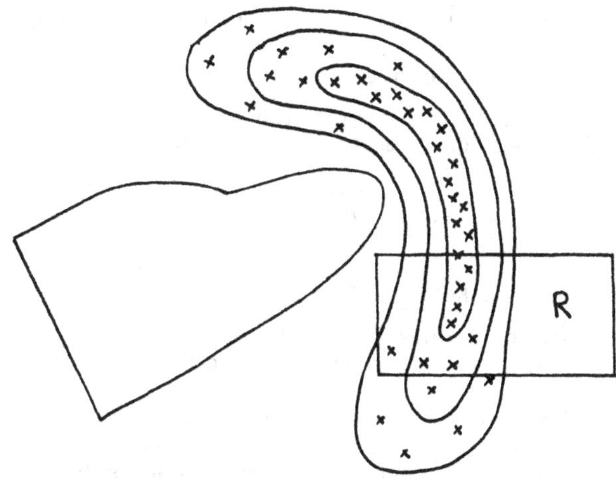

Figure 3

We mention a few consequences of the notion of discrepancy - these and a lot more useful results can be found in [19]; we used this concept for plasmaphysical computation already in 1973, see [25].

a) For every density function f there exists a sequence $(\omega_N)_{N\in\mathbb{N}}$ of finite pointsets such that $\lim\limits_{N\to\infty} D(\omega_N,f) = 0$.

b) The possible speed of convergence is restricted by the following inequalities:

 (i) There exists a constant $A_k[f]$, depending on the dimension k (here in general k=6) and f, such that for all ω_N

$$D(\omega_N,f) \geq A_k[f]\, \frac{(\ln N)^{\frac{k-1}{2}}}{N} . \tag{6a}$$

 (ii) For each f and N prime there exists a pointset ω_N and a constant $B_k[f]$, such that

$$D(\omega_N,f) \leq B_k[f]\, \frac{(\ln N)^k}{N} . \tag{6b}$$

c) For k=2, $\Omega = [0,1]^2$ and $f(x,v) = 1$ in Ω the following construction of ω_N creates sets of "good lattice points", i.e. those, which fulfil the estimate (6b):
 Let α_n be a Fibonacci-number, $\alpha_{n+2} = \alpha_n + \alpha_{n+1}$, $\alpha_1 = \alpha_2 = 1$ and

$$(x_i, v_i) = \left(\frac{(i-1)\alpha_{n-1}}{\alpha_n} \bmod 1\right),\ 1 \le i \le \alpha_n = N;\ \text{then } D(\omega_N, f) = \frac{7}{6N}\ln(6N) + \frac{1}{N},$$

i.e. ω_N is a set of good lattice points.

d) Let $\Phi: \Omega \times \mathbb{R}^3 \to \mathbb{R}$ be of bounded variation $V[\Phi]$ ("in the sense of Hardy and Krause" - see for example [15]), then

$$\left| \int_{\Omega \times \mathbb{R}^3} \Phi \cdot f \, dxdv - \frac{1}{N} \sum_{j=1}^{N} \Phi(x_j, v_j) \right| \le V[\Phi] D(\omega_N, f)$$

(this is the socalled Koksma-Hlavka inequality).

One may conclude from the points (a)-(d) that

- every density f can be approximated arbitrarily well by finite pointsets,
- the quality of approximation is rather slowly increasing with N, but also rather slowly increasing with the dimension k,
- the problem of constructing good approximations asks for quite tricky methods borrowed from number theory,
- the expectation values of functions Φ with respect to the density f are approximated by the averages of Φ over ω_N with an accuracy given by $D(\omega_N, f)$.

This explains, why FPM (or particle methods) are appropriate for problems, where the solution is a density depending on many variables (high k) and where one is mainly interested in functionals (moments, expectation values) of the solution.

We are now able to formulate the general concept of a FPM:

(1) Given the initial value $\overset{o}{f}$, find a good approximation by a finite pointset $\overset{o}{\omega}_N = \left\{(\overset{o}{x}_1, \overset{o}{v}_1), \ldots, (\overset{o}{x}_N, \overset{o}{v}_N)\right\}$.

(2) Find a time evolution of the points

$$t \to \omega_N(t) = \left\{(x_1(t), v_1(t)), \ldots, (x_N(t), v_N(t))\right\}$$

with $\omega_N(0) = \overset{o}{\omega}$ such that $\omega_N(t)$ is a good approximation of $f(t,\cdot)$, the solution of the Boltzmann equation at time t.

One may express this concept also by saying that one has to find an algorithm constructing for each N a $\overset{o}{\omega}_N$ such that $D(\overset{o}{\omega}_N, \overset{o}{f}) \to 0$ and one has to find an evolution $\omega_N(t)$ such that $D(\omega_N(t), f(t)) \to 0$ for $0 \le t \le T$.

4. The time-space discretized Boltzmann equation

As explained above we have to solve two problems, the initialization and the evolution. The first one is relatively easy, the second one quite complicated.

(1) Given $\overset{o}{f}$, construct $\overset{o}{\omega}_N$ such that $D(\overset{o}{f}, \overset{o}{\omega}_N)$ is small. We would like to say: is as small as possible. This defines a min max problem: Given

f and N construct

$$\min_{\omega_N} \max_{R} \left| \int_R \overset{o}{f} \,dxdv - \frac{1}{N} \sum_{i=1}^{N} \chi_R(x_i,v_i) \right| .$$

We are aware of some attempts to attack the "inner" maximization problem, which can be reduced to a problem of combinatorial optimization - but we know until now nothing about the full problem. However we believe that a solution is crucial - we do not need algorithms which converge quickly with N tending to infinity, but give good results for given, relatively small N. What one can do practically today is the following:

One constructs first a pointset $\tilde{\omega}_N$ approximating the equidistribution with f(P) = 1 for P in the k-dimensional (k=6) unit cube. We mentioned one method in 3c for k=2. In the general case we use the following approximation procedure for the initial distribution $\overset{o}{f}$ for the initial distribution f: The spatial domain gets a rectangular grid structure at the beginning of the simulation procedure (this grid structure arises from the space discretization of the Boltzmann equation, which is explained in details later in this chapter). Now we approximate $\overset{o}{f}$ in each cell of the grid separately - one has to realize that $\overset{o}{f}$ is spatially homogeneous. The positions of the particles are simply equidistantly distributed over the whole spatial domain. This leads to a particle number N_c in each cell, which is nearly proportional to the cell size. The velocity distribution in each cell is approximated by using the modified Hammersley sequence in the 3-dimensional unit cube $[0,1]^3$, given by $(\frac{2i-1}{2N_c}, \Phi_2(i), \Phi_3(i))$, where Φ_2 and Φ_3 are the van-der-Corput sequences for the primes 2 and 3 [19]. The transformation from $[0,1]^3$ to the velocity domain $\mathbb{R}^3$ then mainly contains a transformation to spherical coordinates and the calculation of the inverse of the error function [3].

$$\text{erf}(x) = \frac{2}{\sqrt{\pi}} \int_0^x e^{-t^2} dt .$$

The modified Hammersley sequence leads - even for small particle numbers - to a good approximation of the moments of $\overset{o}{f}$, especially of the temperature.

We mention that another solution consists in just using random number generators. However, a good set of random number has to care for more than just for approximating the equidistribution. Since we consider this aspect to be important, we want to explain it shortly for the simple 1-dimensional case. What does it mean that a finite pointset $\omega_N=\{x_1,\dots,x_N\}$ is approximately uniformly distributed in [0,1]? First we want to have the discrepancy $D(\omega_N,\chi_{[0,1]})$ to be small. In this sense the set $\left\{\frac{1}{2N},\frac{3}{2N},\dots,\frac{2N-1}{2N}\right\}$ is best - its discrepancy is optimal $\frac{1}{N}$. This set would not be considered to be a good choice for pseudorandom numbers. Why? As a sequence with this ordering these numbers are too much correlated. A simple expression for this is: Take the set $\omega_N^{(2)}$ of points in the unit square defined as pairs of subsequent members of the sequence, i.e.

$$\omega_N^{(2)} := \left\{ \left(\frac{1}{2N}, \frac{3}{2N}\right), \left(\frac{3}{2N}, \frac{5}{2N}\right), \ldots, \left(\frac{2N-3}{N}, \frac{2N-1}{N}\right) \right\} ,$$

then $D\left(\omega_N^{(2)}, \chi_{[0,1]^2}\right)$ is extremely bad, always larger than $\frac{1}{4}$, since the points are all very near to the diagonal. Therefore a set created by a good pseudorandom generator should have the property that $D\left(\omega_N^{(2)}, \chi_{[0,1]^2}\right)$ is small. We may continue in considering tripels, general k-tupels of subsequent numbers in ω_N to get $\omega_N^{(k)} = \left\{(x_1,\ldots,x_k),(x_2,\ldots,x_{k+1}),\ldots,(x_{N-k+1},\ldots,x_N)\right\} \subset [0,1]^k$ and compare it with the uniform distribution in the unit cube $[0,1]^k$. We say that ω_N is a k-fold approximation of $\chi_{[0,1]}$, if

$$D\left(\omega_N^{(j)}, \chi_{[0,1]^j}\right) \longrightarrow 0 \quad \text{for } N \to \infty,\ 1 \le j \le k \ .$$

A "real" random number generator would produce a sequence $(x_n)_{n\in N}$ such that $\omega_N = \langle x_1,\ldots,x_N \rangle$ would be a k-fold approximation for any k. A pseudorandom number generator has at least a $k \ge 2$. But what do we need for our simulation? It depends on the problem and we have to make a careful analysis to answer this question. What we want is that the final result of the simulation approximates the solution as good as possible - in the sense of discrepancy. In order to achieve this goal we have to find an approximation of some data as the initial value or the collision process - and only a convergence analysis can tell us, whether these approximations have to be simple, double or k-fold with $k \ge 3$. This convergence analysis is given in [4] and it tells us that there is no reason to approximate the initial value better than simple. This means that we would loose approximation quality in using pseudorandom generators without gaining anything. The situation is different for the simulation of collisions as we shall soon see.

A warning might be necessary: In using the same deterministic set of low discrepancy again and again in a simulation procedure, one makes small errors in each step but the errors may have always the same sign and therefore may accumulate, meanwhile using a worse but always different set of pseudorandom numbers may lead to a cancellation of errors. Being aware of such systematic errors makes it easy to avoid them - we shall discuss such a problem in connection with treating the boundary condition ("numerical freezing").

(2) Our main task is to find a time evolution $\omega_N(t)$ of our pointset. We have N particles each of mass $\frac{1}{N}$ at position $x_1,\ldots,x_N$ and with velocities $v_1,\ldots,v_N$. How should we move them? The first idea may be: As nature, i.e. as billiard balls if we have a rigid sphere model. Would this fulfil our condition, would it converge? Would $\omega_N(t)$ corresponding to our billiard game converge to the solution of the Boltzmann equation f(t), if $\omega_N(0)$ tends to ^of? This question is not a numerical one - it is the question whether the Boltzmann equation can be derived from classical mechanics as a limit $N\to\infty$. It is therefore a very old and fundamental question - and its answer today is a weak "yes". It is the result by O.E. Landford [20], improved by Illner and Pulvirenti [16]; it is a "weak" yes, since it holds only with probability one and for a very short time (one may also look at [7], pages 40-57). Since nature cannot be imitated even by the best supercomputer (even if we consider only 10^5 particles, it would be impossible to follow the fate of each individual particle), we have to give up our first idea and must invent an artificial evolution simpler to handle but with the same effect: to remain near to the solution of the Boltzmann equation. This evolution will proceed in discrete time steps and will depend on a grid in position space; it will therefore not approximate the solution of the original Boltzmann equation but one of the time and space discretized Boltzmann equation. Babovsky and Illner showed in [4] that under certain conditions we can come as near as we like to a solution of this discretized equation - by increasing N; and we can make the difference of this solution to the "real one" as little as we like - by making the time step and the grid size smaller.

a) The time discretization of the Boltzmann equation is easy: One discretizes t by considering only $j\Delta t$, j=0,1,2,... and approximates (we consider only the first time step)

$$\frac{\partial f}{\partial t}+v\frac{\partial f}{\partial x}\Big/_{t=0} = \frac{\partial}{\partial t}f(t,x+vt,v)\Big/_{t=0}$$

by

$$\frac{1}{\Delta t}\Big[f(\Delta t,x+v\Delta t,v)-f(0,x,v)\Big] \ .$$

If x+vt reaches the boundary $\partial\Omega$ during $0\leq t\leq\Delta t$, we have to include the boundary condition. We will not write down how this can be done exactly; $x+v\Delta t$ must be substituted by the position of a particle which hit the surface and was reflected according to a law consistent with (2). We will describe this in more details in the next chapter. The step size Δt has to be chosen such that we have in average less than one collision per particle during Δt - we shall see the mathematical reason for this also in the next chapter. We have to reach the stationary limit $t\to\infty$, which we discover by the fact that f is nowhere changing during one timestep. In practical applications it took between 100 and 150 timesteps to reach the stationary state neglecting fluctuations.

Physically the time $T = \Delta t \cdot$ number of timesteps can be interpreted as the relaxation time of an equilibrium gas suddenly perturbed by the reentry object. To obtain useful results we of course have to do some time averaging in order to avoid fluctuations of the numerical scheme. The size of these fluctuations is strongly depending on the choice of the random numbers, especially on those which we use to simulate the boundary condition at $\partial\Omega_\infty$. Low discrepancy sequences allow a quite precise approximation of the outer boundary condition with very small fluctuations. We obtain optimal condition with a time-averaging over only 20 to 30 timesteps. We have to emphasize that we do not average over several independent runs. This of course reduces the computation time drastically.

b) <u>The space discretization</u> means that we divide Ω into cells and approximate $x \to f(t,x,v)$ by a function $\tilde{f}(t,x,v)$, which is constant in each cell

$$\tilde{f}(t,x,v) \approx f_c(t,v) \qquad \text{for } x \in \text{Cell } c\ .$$

This is quite a rough approximation of f as a function of x by step functions; however we have not yet another idea since we need cellwise homogeneity for treating the collision term. It is crucial that the collision operator doesn't "work on the variable x", but only on v. x is involved in $\frac{\partial f}{\partial x}$ on the right hand side which we have already taken into account by the time discretization; otherwise it plays only the role of a parameter. For

$$\tilde{f}(t,x,v) = \sum_{\text{Cell } c} f_c(t,v)\chi_c(x)$$

we now construct the time-space discretized Boltzmann evolution (we again write only the first time step).
We may be tempted to write

$$\tilde{f}(\Delta t,x,v) = \tilde{f}(0,x-v\Delta t,v) + \frac{\Delta t}{\varepsilon}J(\tilde{f},\tilde{f})$$

but we have to realize that $\tilde{f}(0,x-v\Delta t,v)$ is not a step function on the cell grid and we have to decide what we really mean by $J(\tilde{f},\tilde{f})$.
In order to get a correct step function, we should do a coarse graining i.e. smear out the function $\tilde{f}(0,x-v\Delta t,v)$ over a cell. Precisely, we have to substitute $\tilde{f}(0,x-v\Delta t,v)$ by the step function

$$(P\tilde{f})(x,v) := \frac{\int_c \tilde{f}(0,x-v\Delta t,v)dx}{\text{volume of } c} \qquad \text{for } x \in \text{Cell } c\ .$$

The operator $P\tilde{f}$ can be interpreted as the projection of $\tilde{f}(0,x,v)$ on the space of all step functions (or even as conditional expectation of f with respect to the algebra of cells).

$J(\tilde{f},\tilde{f})$ must as well be a step function. We may choose the argument f of J to be just $\tilde{f}(0,x,v)$, which is a step function and - since J doesn't work on x - remains it after application of J. This corresponds to a simple Euler step for the integration of

$$\frac{\partial f(t,x+vt,v)}{\partial t} = J(f,f)(x,vt,v) \ .$$

It seems to be better (but there is no rigorous proof for it) to interprete $\tilde{f}$ in $J(\tilde{f},\tilde{f})$ as $P\tilde{f}$.
We get then, writing Jg instead of J(g,g)

$$\tilde{f}(\Delta t,x,v) = (1+\frac{\Delta t}{\varepsilon}J)(P\tilde{f})(x,v) \ . \tag{7}$$

This is consistent with the convergence proof in [4] and means that we have two fractional steps, one consisting in free flow (including boundary effects) plus coarse graining, the other one treating the collisions.
We should mention that DSMC follows a different strategy in mixing free flow and collisions during a time step Δt by the socalled time counter procedure. It might give better approximations (at least, if this procedure is really consistent with the Boltzmann equation which is not clear), but it makes the algorithm less easy vectorizable - a quite important aspect for supercomputers.
We see, moreover, that this equation combines various cells, since in general $x-v\Delta t$ is not in the same cell as x. This gives rise to a computational restriction with respect to our cell grid: We will have to move particles through the grid and have to refind them after a motion very quickly, i.e. we need fast algorithms to determine in which cell a particle is. This problem is by far easier to solve for a coordinate grid, where the cells are axiparallel "rectangles". We therefore avoid a complicated shape of the cells and prefer to diminuish the cell size in regions of steep gradients. An adaptive grid structure has been developed where the adaption is done during the time evolution of $\omega_N(t)$ with respect to the flow properties. This method runs as follows: Due to the fact that the initial distribution is a global Maxwellian we can start with a rather coarse grid with cell diameters of several mean free paths. During the evolution process the grid has to be refined to cell sizes of about one mean free path in critical regions (shock, boundary layer etc.). In the cells the distribution function is supposed to be homogeneous in space. We can fulfil this requirement by observing the following inequality which arises during the proof of convergence of the algorithm [4]:

$$\sup|f(t,x+\Delta x,v)-f(t,x,v)| \ \exp(v^2) \leq B\cdot\Delta x$$

for some $B>0$ and all "cell sizes" Δx.
Transferring this to our finite pointset

$$\omega_N = \{(x_1,v_1),\ldots,(x_N,v_N)\} \subset \cup\, \omega_{N_c},$$

this requires a refinement of the grid in the following way: Choose B such that the criterion is just fulfilled for cells with a Maxwellian distribution. Then divide each cell c into parts A_i^c, $i=1,\ldots,s$, of equal size and the velocity domain filled by particles (this domain has finite volume) into equally sized subsets W_j, $j=1,\ldots,\ell$.
Then compute

$$\delta_c = \max_{\substack{i,k\in\{1,\ldots,s\}\\ j=\{1,\ldots,\ell\}}} \frac{e^{\bar{v}_j^2}}{N_c \Delta x_{i,k}} \Big| \{\text{number of points with } x \in A_i^c, v \in W_j\} - \{\text{number of points with } x \in A_k^c, v \in W_j\} \Big|$$

where $\bar{v}_j^2$ is the mean velocity square in W_j and $\Delta x_{i,k}$ is the distance of the centers of the subcells A_i^c and A_k^c.

Finally check, whether $\delta_c \leq B$. If yes, no refinement is necessary, otherwise a further subdivision is necessary. This procedure gives a good grid adaption near the boundary and within the shocks. The generation of the refined grid, however, is rather time consuming, because it is not well vectorizable. One therefore should restrict the refinement procedure to every tenth time step and keep the grid fixed in between. Experience shows that this doesn't influence the results of the computation.

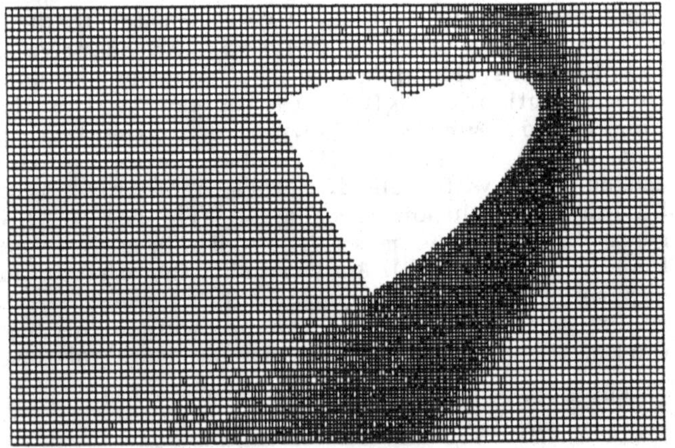

Figure 4

5. THE FINITE POINTSET METHOD FOR TIME-SPACE DISCRETIZED BOLTZMANN EQUATIONS

We know how to initialize, i.e. how to construct

$$\omega_N(0) \longrightarrow \tilde{f}(0) = f(0) = \overset{o}{f} .$$

We have to construct a one-step evolution $T_{\Delta t}$ for $\omega_N(0)$, such that $\omega_N(\Delta t) = T_{\Delta t}\omega_N(0)$ converges with $N\to\infty$ to $\tilde{f}(\Delta t)$, the solution of (7). The time evolution $\tilde{f}(0) \to \tilde{f}(\Delta t)$ given by (7) consists of two components: The free flow $x\to x-v\Delta t$ including boundary conditions and the coarse graining P - and $1+\frac{\Delta t}{\varepsilon}\cdot J$. We do the same for $T_{\Delta t}$: It consists first of free flow and then of collisions.

a) The free flow $\omega_N(0) \longrightarrow \hat{\omega}_N(\Delta t)$. Here we have to approximate $P\tilde{f}(x,v)$ by $\hat{\omega}_N(\Delta t)$, when $\omega_N(0)$ approximates f(0). $\hat{\omega}_N$ is just given by "simulating nature": We move the points of $\omega_N(0)$ freely with their velocities over Δt; if they hit the boundary we simulate again nature. That is

$$\hat{\omega}_N(\Delta t) = \left\{(x_1+v_1\Delta t,v_1),\ldots,(x_N+v_N\Delta t,v_N)\right\}$$

if no particle hits $\partial\Omega$. If particle j hits the boundary, i.e. if there is a τ with $0\le\tau\le\Delta t$, such that $x_j+v_j\tau \in \partial\Omega$, then v_j is changed according to the scattering law we discussed with boundary condition (2). If we have specular reflection, i.e. $R=R_S$, then v_j is changed into $v_j'=v_j-2n\langle n,v_j\rangle$, where n is the normal vector to $\partial\Omega$ at $x_j+v_j\tau$. With this new velocity the free flow continues until Δt is over: We get

$$x_j+v_j\tau+(\Delta t-\tau)v_j' .$$

All deterministic boundary conditions (where R has a δ-behaviour) can be treated similarly. If we have complete accomodation, we have to take a v_j' according to the distribution $f_o(v)$, the half-space Maxwellian with the temperature of the wall; in each boundary cell, quite many particles, say the subset ω_M, will hit the wall and we only have to care that $D(f_o,\omega_M)$ is small. This can be done again in a more systematic way or with help of a pseudorandom generator [3]. However, the systematic error mentioned in chapter III may become significant - we call this effect, which was studied in [23], numerical freezing". This effect arises by using uniform distributed sequences like the van der Corput-Halton or the Hammersley sequence for the generation of the outgoing velocities. It can be shown that the approximation of the energy flux from the boundary into the computational domain is always below the exact value.
This error, indeed small in a single time step, sums up during the computation and leads finally to a significant error of the energy in the whole region. This trend can be avoided by constructing sequences, which, in an oscillating way, approach the true value

from above and below; this can be easily done without changing the discrepancy.

We have also studied the Cercignani-Lampis model and construct a random number generator for the corresponding distribution:
The scattering kernel $R(v' \to v)$ of the Cercignani-Lampis model is given by

$$R(v' \to v) = |v \cdot n| f_o(v) H(v,v').$$

Here $f_o(v) = \frac{1}{2\pi R^2 T_o^2} \exp\left(-\frac{\|v\|^2}{2RT_o}\right)$ with wall temperature T_o,

$$H(v,v') = \frac{1}{\alpha_n \alpha_t (2-\alpha_t)} I_o\left(\frac{2\sqrt{1-\alpha_n}}{\alpha_n} v_n v_n'\right) \cdot$$

$$\cdot \exp\left\{-\frac{1-\alpha_n}{\alpha_n}(v_n^2+v_n'^2) - \frac{(1-\alpha_t)^2}{\alpha_t(2-\alpha_t)}(v_t^2+v_t'^2) + \right.$$

$$\left. + \frac{2(1-\alpha_t)}{\alpha_t(2-\alpha_t)} v_t' \cdot v_t \right\},$$

where $\begin{cases} v_n = (v \cdot n) \cdot n \\ v_t = (v \cdot t) \cdot t \end{cases}$ (c normal vector, t tangential vector with respect to the wall) and I_o denotes the modified Bessel function of first kind and zeroth order.
This scattering kernel can be separated into the normal and tangential components of v

$$R(v' \to v) = R_1(v_n' \to v_n) R_2(v_t' \to v_t),$$

with the corresponding accomodation coefficients α_n and α_t. The outgoing velocity v_t can be calculated by the simple formula

$$v_t^{(1)} = (1-\alpha_t)v_t'^{(1)} - \left\{-\alpha_t(2-\alpha_t)\ln(1-v_1)\right\}^{1/2} \cos(2\pi v_2)$$

$$v_t^{(2)} = (1-\alpha_t)v_t'^{(2)} - \left\{-\alpha_t(2-\alpha_t)\ln(1-v_1)\right\}^{1/2} \sin(2\pi v_2)$$

where (v_1,v_2) is a random point in the square $[0,1]^2$.

The calculation of v_n is much more complicated: The distribution of v_n is given by a socalled Polya-Aeppli-distribution. The algorithm for the generation of v_n we use can be found in [19].

After having moved all particles with their velocities we have to "refind" them, precisely: We have to know the particles contained in each cell after the motion; this is a simple task for a coordinate grid but costs much computing time for a usual finite volume grid - a little problem if we want to connect our code with a usual Navier-Stokes code.

b) The collisions.

We were left with N_c particles in cell c - for each cell. This is an approximation for Pf(x,v), a step function constant in each cell. It remains to approximate $(1+\frac{\Delta t}{\varepsilon}\tilde{J})P\tilde{f}$. Again, x is only a parameter for J, i.e. the cells can be treated independently. Take an arbitrary cell c, $P\tilde{f}$ restricted to c is a function f_c of v alone - approximated by the pointset $\{v_1,\ldots,v_{N_c}\}$ of the velocities of the N_c particles in c. The positions of these particles may be forgotten until a new time step with a new free flow. (7) is now

$$f_c(\Delta t,v) = (1+\frac{\Delta t}{\varepsilon})J(f_c,f_c)(0,v)$$

i.e. nothing else but the time discretized <u>spatially homogeneous</u> Boltzmann equation.

Since particles are now only "velocity particles" without any position, a "natural game" simulating collisions is not possible. But still there must be some interaction between the particles - the quadratic term J makes it necessary. A collision is an interaction of two particles, which results in new velocities of both collision partners. But which pair of particles will interact?

The solution can be found in considering matchmaking procedures in some societies: Meanwhile in our society everybody finds his partner for a marriage by really meeting him/her somewhere (i.e. by moving around and "colliding" with him/her), there are some other societies with organized matchmaking - a marriage brooker selects the couples (there are no local effects, at least if our cells are not too big).

The two methods are completely different on a microscopic level but may give the same result from an economical or demographical point of view. This is what we need: A matchmaker which gives (with much less particles) the same macroscopic results as nature. In order to understand this matchmaker condition we rewrite the spatially homogeneous Boltzmann equation in a way, which was found by H. Babovsky [2]. For theoretical reasons we have to assume that

$$\int_{S^2_+} \|v\| s(\theta,\|v\|)\,d\omega(n) := \alpha(\|v\|) \leq A\,. \tag{8}$$

Without this assumption things could go wrong since it would be possible that $f_c\,(\Delta t,v)$ becomes negative - remind that (7) is with $V=v-v_1$

$$f_c(\Delta t,v) = \left(1 - \frac{\Delta t}{\varepsilon}\int_{\mathbb{R}^3} \alpha(\|V\|)f_c(v_1)dv_1\right)f_c(v)$$

$$+ \frac{\Delta t}{\varepsilon}\int_{\mathbb{R}^3}\int_{S^2_+} \|V\|s(\theta,\|V\|)f_c(v')f_c(v_1')\,d\omega(n)\,dv_1$$

so that (8) guarantees posivitity of $f_c(\Delta t)$ if $\Delta t < \frac{\varepsilon}{A}$ (coming down during the reenty forces Δt has to become smaller). This condition (8) is not a trivial one - for rigid spheres, $\alpha(\|V\|)$ is proportional to $\|V\|$ and therefore not at all bounded. Practically the situation is not as bad: A is proportional to $\|V\|$ and if there is a bound for the relative velocity of two colliding particles, A remains bounded. But one has to be careful even practically: If the relative speed of many pairs of particles is rather high, Δt must be very small.

We really need $\frac{\Delta t}{\varepsilon}\cdot\alpha(\|V\|) < 1$ for all particles in our cell c. Now we define a function $\Phi_{v,v_1}(y)$ on a circle κ of radius $\frac{1}{\sqrt{\pi}}$ in the plane, i.e. for $0 \leq \|y\| \leq \frac{1}{\sqrt{\pi}}$, $0 \leq \alpha \leq 2\pi$, which depends on v and v_1 and which decides whether the particles with velocities v and v_1 collide at all and if, what will be the impact parameter n of the collision. One has to realize that these impact parameter vectors n have a certain distribution (depending on $\|V\|$, if one has not the special case of rigid spheres) given by the differential cross section s; s is the density of deflection n on the unit hemisphere S_+^2 after a collision of v with v_1. Our function Φ_{v,v_1} has essentially the task to transform the uniform distribution on our circle κ into this distribution of n on S_+^2.

$\Phi_{v,v_1}(y)$ is constructed in the following way: Let us define a radius r_o by

$$\pi r_o^2 = \frac{\Delta t}{\varepsilon}\,\alpha(\|V\|) < 1$$

i.e. $r_o < \frac{1}{\sqrt{\pi}}$ such that the circle $r \leq r_o$ lies in κ. r_o depends on $\|V\|$. For $r_o < r < \frac{1}{\sqrt{\pi}}$, $\Phi_{v,v_1}(r\cos\alpha, r\sin\alpha)$ is defined to be that n which means no (or better a grazing) collision $\theta = \frac{\pi}{2}$, $\phi = \alpha$. The probability that a random number generator for uniform distribution in κ produces a point in the ring $r_o < r < \frac{1}{\sqrt{\pi}}$ is equal to the probability that the two particles with velocity v and v_1 respectively do not collide at all during Δt.

For $0 \leq r \leq r_o$ a collision will happen and we have to define $\Phi_{v,v_1}(r\cos\alpha, r\sin\alpha) = n = (\theta,\phi)$. Again $\phi = \alpha$, meanwhile $\theta = \theta(r)$ (not depending on ϕ) is given as the inverse of the function $r(\theta)$ defined by

$$r^2(\theta) = \frac{2\Delta t}{\varepsilon}\int_0^{\theta} B(\phi, \|V\|)\,d\phi$$

(since $B \geq 0$, r is strictly monotone and the inverse exists).

One realizes that

$$r^2(\tfrac{\pi}{2}) = \frac{\Delta t}{\varepsilon}\int_0^{\pi/2} B(\Phi,\|V\|)d\Phi$$

$$= \frac{2\Delta t}{2\Pi\varepsilon}\int_{S_+^2} \|V\| s(\theta,\|V\|)dn$$

$$= \frac{\Delta t}{\varepsilon\pi}\alpha(\|V\|) = r_o^2$$

It is shown in [2] that $n=\Phi_{v,v_1}(y)$ has the correct distribution given by B if y is uniformly distributed on κ. With $n=\Phi_{v,v_1}(y)$ we have immediately the result of the collision of v,v_1:

$$v' = v-n\langle n,V\rangle,\quad v_1' = v_1+n\langle n,V\rangle .$$

One realizes that $v'=v$, $v_1'=v_1$ in case of $r \geq r_o$ (then $\theta=\frac{\pi}{2}$ and since θ is the polar angle against V, n is orthogonal to V). In general, v' and v' are now functions of v, v_1 and y through $n = \Phi_{v,v_1}(y)$.

Φ depends on $\|V\| = \|v-v_1\|$ and naturally on B - its values as a function of $\|V\|$ can be computed and stored for each gas at the beginning of a run.

Our equation (7) can now be written as follows:

For every set in the velocity space, for example for every "rectangle" R in $\mathbb{R}^3$ the equation

$$\int_R f_c(\Delta t,v)dv = \int_{(v,v_1,y):v'\in R} f_c(v)f_c(v_1)dvdv_1dy \tag{7a}$$

must hold. The left hand side just gives the mass of the gas in cell c with velocity in R; the right hand side means: integrate over those v, v_1, y which guarantee that

$$v' = v-\Phi_{v,v_1}(y)\cdot\langle\Phi_{v,v_1}(y),v-v_1\rangle$$

is in R. This integration set is a subset of $\mathbb{R}^3\times\mathbb{R}^3\times\kappa$. (7a) is a weak formulation of (7), and it leads immediately to the rule for our brooker.

What he has to do is to choose out of N_c^2 possible collision pairs (we accept that our "daily life" picture goes wrong now: We allow a particle to become a partner of itself - and we don't have two classes as males and females!) precisely N_c pairs and to assign to each an impact parameter $y \in \kappa$ (if $|y| \geq r_o$, the collision partners do not really collide).

The crucial thing is now that in order to get convergence this assignment $v_i \to v_{j(i)}$, $y_i \in \kappa$, $i=1,\ldots,N_c$ has to be made in such a way that the 8-dimensional finite pointset

$$\Omega_{N_c} := \left\{(v_1, v_{j(1)}, y_1), \dots, (v_{N_c}, v_{j(N_c)}, y_{N_c})\right\}$$

is as near as possible (in the sense of discrepancy) to $f_c(v)f_c(v_1)\chi_\kappa(y)$, where $\chi_\kappa(y)$ is again the uniform distribution in κ; more precisely we have to make sure that

$$D(\Omega_{N_c}, f_c(v)f_c(v_1)\chi_\kappa(y)) \to 0 \quad \text{for} \quad N_c \to \infty.$$

One must be a bit careful here: $f_c(v)f_c(v_1)\chi_\kappa(y)$ is not normalized to 1 but has a total mass of $(\int f\,(v)dv)^2 \cdot 1$ - one has to use the normalized version of this function in the definition of D. Theoretically everything is clear: <u>Each</u> procedure selecting Ω_{N_c} according to this condition will converge (see again [2]).

But how do we select $v_{j(i)}$ and y_i practically? y is just uniformly distributed on κ - we may therefore use a random number generator for getting a sample of N_C points in κ; we may also use a deterministic method to construct $y_1, \dots, y_{N_c}$ such that $D\left(\left\{y_1, \dots, y_N\right\}, \chi_{\kappa_c}(y)\right)$ is as small as possible - but should permutate this selection in every time step to avoid systematic errors. The problem of selecting $v_{j(i)}$ may become more clear by just looking at a one-dimensional case (figure 5).

Selection of N points
from N^2 points
a) Monte Carlo

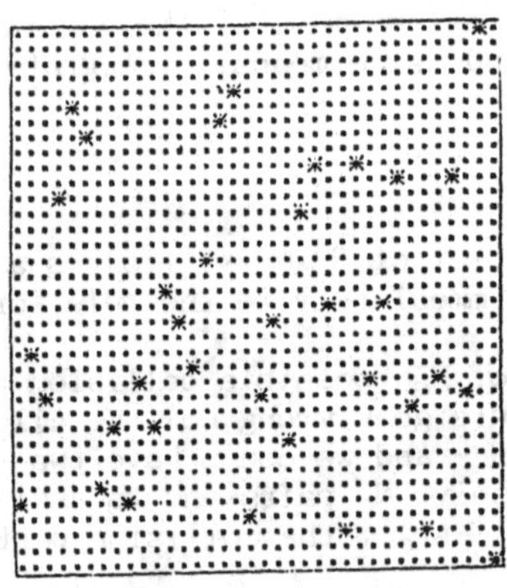

Selection of N points
from N^2 points
b) Low Discrepancy

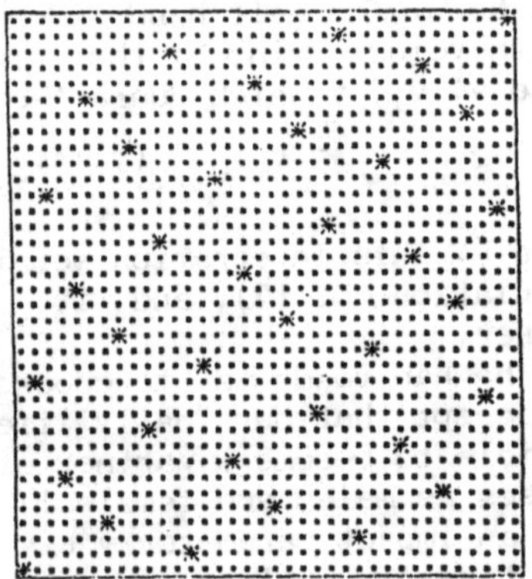

Figure 5

We consider the (v,w)-plane, have a set $\omega_N = \{v_1,\dots,v_N\}$ on the v-axis near to f(v) and the same set on the w-axis near to f(w). The dots in the plane indicate possible collision pairs. Out of them we have to select N "crosses", such that the set of N crosses is near to the product f(v)f(w).

In the terminology of chapter 4 this means: Can we give to ω_N such an order, that $\omega_N^{(2)}$, the set of pairs is near to $f(v)\cdot f(\omega)$, i.e. that ω_N with this order is a double approximation of f. Therefore in treating the collisions, which are quadratic in nature, one needs a sampling, or better, an approximation of f by point sets, which is of second order.

The simplest method is to play with the index: Take a sample $r_1,\dots,r_N$ for the uniform distribution in [0,1] from a random number generator and let

$$j(i) = [Nr_i] + 1 ,$$

where $[Nr_i]$ is the Gauss bracket. This fulfils our condition at least with probability one as is shown in [2]. Different i do not lead to different j(i) - the symmetry of the collision partners is lost, some particles may play the role of a collision partner several times. Clearly energy and momentum is not conserved in such a collision process, but these conservation laws are satisfied in the mean. This "Monte Carlo method" has therefore some disadvantages but is simple and therefore mostly used.

A bit more complicated is the idea to choose in every raw and column only one cross, i.e. to choose randomly a permutation π of $\{1,2,\dots,N\}$ and put the crosses at $(v_{\pi(1)},v_{(\pi(2)}),(v_{\pi(3)},v_{\pi(4)}),\dots,(v_{\pi(N)},v_{\pi(N)})$ - one has to consider only even numbers N, but this is not a real restriction. This methods conserves energy and momentum.

There are certainly better solutions for the selection procedure but all we found until now are too expensive. But since our condition is rather weak, there are many possibilities; it is for example not necessary that crosses and dots coincide.

With which method fulfilling the condition for Ω_{N_C} ever, we are at the end of the time step: We have collision pairs $(v_i,v_{j(i)})$ and collision parameters y_i - we determine v_i' for all $i=1,\dots,N_c$ and end up with all particles having new positions and new velocities. This set is near to $f(\Delta t,x,v)$, the solution of the time-space discretized Boltzmann equation - and we can begin a new time step.

We have not treated interior energy, chemical reactions in the interior or at the wall and ionisation effects.

To model these effects and to evaluate the reliability of the models, is a research field in its own; there are ideas by Bird, Nanbu and others in use, but there is still no proof that they give correct results for realistic reentry problems.

6. A FEW REMARKS ON THE TRANSITION TO NAVIER-STOKES

When the shuttle comes down, the atmosphere becomes denser, the mean free path λ_∞ smaller, the computational effort for FPM (and any other Boltzmann solver) larger. Between 85 and 70 km λ_∞ drops by a factor of 10 and the number of cells in a 3-dimensional calculation has to be increased by a factor of 10^3. There is only one way out: The distriution f approaches a local Maxwellian when $\lambda_\infty \to 0$ - and one has to use this information numerically. The procedures making this approach more precise are the socalled Hilbert and Chapman-Enskog expansions of the solution f with respect to the parameter ε (proportional to λ_∞). A first order approximation is - in both expansions - given by the Euler equation, while Chapman-Enskog gives in second order Navier-Stokes. (Rigorous results on Hilbert expansion may be found for example in [6] and there are many books on the formal aspects, for example [7], [8]). In short, the distribution function

$$F_{NS}(t,x,v) = \text{Maxwell}[\rho(t,x),u(t,x),T(t,x)](v)$$
$$- \varepsilon\ \text{Correction}[\rho(t,x),u(t,x),T(t,x),\eta(t,x),\lambda(t,x)](v)$$

is a solution of the Boltzmann equation with a residuum of order ε^2, if ρ, u, T are solutions of the Navier-Stokes equations with viscosity η and thermal conductivity λ depending on the differential cross section (and the temperature) and if as usual

$$\text{Maxwell}[\rho,u,T](v) = \frac{\rho}{(2\pi RT)^{3/2}} \exp\left[-\frac{1}{2RT}|v-u|^2\right] .$$

The correction term is quite lengthy and includes some coefficients given as solutions of an integral equation involving the linearized collision operator and therefore again the differential cross section - we do not give details about it.

F_{NS} defines - for incoming molecules - the boundary conditions for the Boltzmann region coming from the Navier-Stokes domain. More complicated is the other direction: Given f in a transition layer one has to approximate it by a function of type F_{NS} in order to get good ρ, u, T as boundary conditions for the Navier-Stokes. The idea to do this is to compare the fluxes of mass, momentum and energy for f and F_{NS} across the boundary and determine ρ, u, T such that these fluxes are equal:

$$\int_{v\cdot n\geq 0} v\cdot n\ \phi_k(v) f(t,x,v)dv = \int_{v\cdot n\geq 0} v\cdot n\ \phi_k(v) F_{NS}[\rho,u,T](t,x,v)dv$$

with

$$\phi_k(v) = \begin{cases} 1 & \text{for } k=0 \\ v_k & \text{for } k=1,2,3 \\ |v|^2 & \text{for } k=4 \end{cases}$$

gives 5 conditions for the five unknowns ρ, u and T. We remark that one gets in this way also slip boundary conditions for the Navier-Stokes equation. There is a good survey in [9]; we refer also to the Thèse de Doctorat de l'Université Paris Nord of F. Coron [10], where many of related equations are discussed.

We believe that theoretically the transition from Boltzmann to Navier-Stokes and vice versa may be settled - if we know, where the transition layer is. Numerically there are still many problems in the details, for example the different grid structures for Finite Volume techniques for Navier-Stokes and Finite Pointset Methods. It seems to be a tempting idea to use the socalled Boltzmann schemes for solving Euler or Navier-Stokes, since they are derived from a kinetic basis (see for example [17] or [26]), or even use the fact that f_{NS} solves the Boltzmann up to ε^2 [11]. However, these schemes are not yet enough developed to compete with the schemes normally used for planes etc.

Another problem seems to be very important with respect to the transition: Where is it allowed to pass from Boltzmann to Navier-Stokes or - where is it necessary to leave Navier-Stokes and to use Boltzmann? The answer to the first question seems to be easier: If f is sufficiently near to F_{NS}, one may use Navier-Stokes. Near with respect to which metric? Discrepancy may be a choice, relative entropy

$$\int f \ln \frac{f}{F_{NS}} \, dxdv$$

another one; all of them may create to many numerical difficulties and a simpler "solution" can be to consider the local Knudsen number Kn(x), where locality stems from the fact that the mean free path λ_n depends on x and the typical length depends on the typical length of the shuttle in vicinity (nose curvature, chord length etc.). This "solution" has some draw backs - mainly since Kn(x) is not so rigorously defined and since Kn(x) small (how small?) is certainly not sufficient. These problems are even more serious, if one is faced by the second question: Assume that we solve only Navier-Stokes, even in regions where it is physically not justified - how do we recognize this fact?

One proposal by Bird and others is to consider $\frac{\nabla\rho}{\rho}$, which should not become to large. We believe (and C. Bardos and his group are working in the same direction) that a sensitive feature is the entropy flux, which in the Boltzmann picture is defined by

$$Q_E^B[f] := -k \int_{\mathbb{R}^3} J(f,f) \ln f \, dv$$

and which for Navier-Stokes may be gained from inserting F_{NS} instead of f into Q_E^B, getting an expression depending only on ρ, u, T, λ, η, ∇T and

$$\langle\nabla u\rangle = \left[\frac{1}{2}\left(\frac{\partial u_i}{\partial x_j} - \frac{\partial u_j}{\partial x_i}\right) - \frac{1}{3}\,\mathrm{div}(u)\cdot f_{ij},\ 1\leq i,j\leq 3\right],$$

namely

$$Q_E^{NS}[f_{NS}] := -\frac{\lambda}{T^2}\nabla T\cdot\nabla T - \frac{2\eta}{T}\langle\nabla u\rangle : \langle\nabla u\rangle .$$

It is a consequence of the Boltzmann equation that in the interior of Ω (not at the boundary)

$$Q_E^B[f] \leq 0$$

for any exact solution f of the Boltzmann equation with equality only for f=Maxwell.
In Navier-Stokes calculations however, especially in shock r egions, Q_E^{NS} may become positive. Positivity of Q_E^{NS} is therefore a sign that F_{NS} is not near to the exact solution of the Boltzmann equation - too "much positivity" force us to leave Navier-Stokes. What means "too much" is still to be investigated, maybe by comparing Q_E^B and Q_E^{NS}, if both solutions are available.

A complete flow code, which is the final goal, should be able to discover automatically, where Euler, Navier-Stokes and Boltzmann regions are located and to shift automatically the "free boundary" between these regions during the reentry. The Boltzmann region, in high altitudes covering almost everything behind the shock will shrink and end up just to form a kinetic layer around the shuttle. But there is still some way to go.

7. SOME COMPUTATIONAL RESULTS

Let us finally demonstrate by a few examples that FPM is a successful method to compute complicated flow fields. In order to do so we will only show some pictures and refer for details to our papers [13] and [14].

Figure 6 gives an impression of the Mach field (the space dependent sound speed) around a 2-dimensional "double" ellipse (a shape considered to be near to a shuttle), figure 7 shows the corresponding grid refinement. Figure 8 presents a result from 3-dimensional computation, a section of the temperature field around a flat disc at 75° angle of attack ([13]). Far away from the disc the Knudsen number is 0.1, the gas temperature is 190 K and the speed of the gas (remember that we consider objects at rest but the gas moving) is Mach 15.6. Wall and stagnation temperature are considered to be equal. Figure 9 turns to a more realistic 3-dimensional object, a delta wing, and gives the density field around it [14]. Here we have Kn=0.01, T_∞=13.5 K, T_{wall}=620 K and V_∞=Mach 20.2. All the calculations presented here were done in the frame of the European Space Agency programme called "HERMES development programme" and were sponsored by a "contrat d'études et de recherches" between La Société Avions Marcel Dassault-Breguet Aviation and the University of Kaiserslautern. The calculations were done on a Fujitsu VP100 (in parts VP400) in Kaiserslautern (and Karlsruhe).

Figure 6

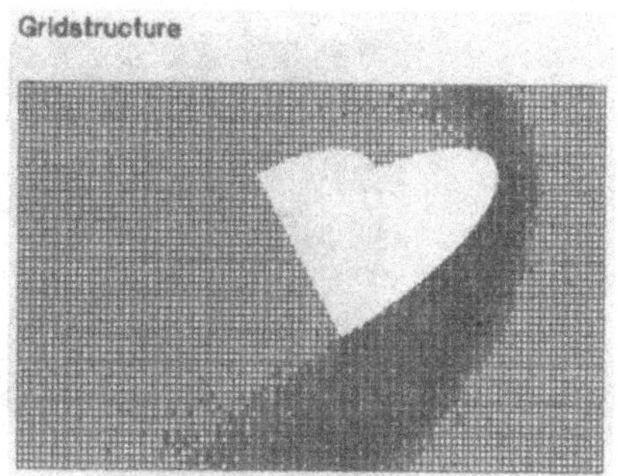

Figure 7

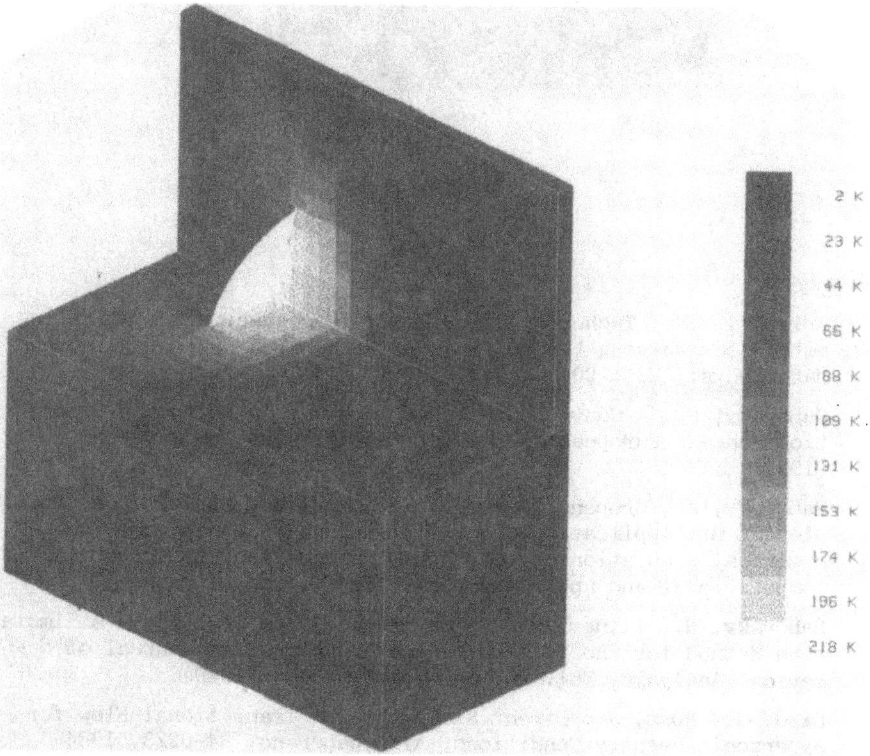

Figure 8

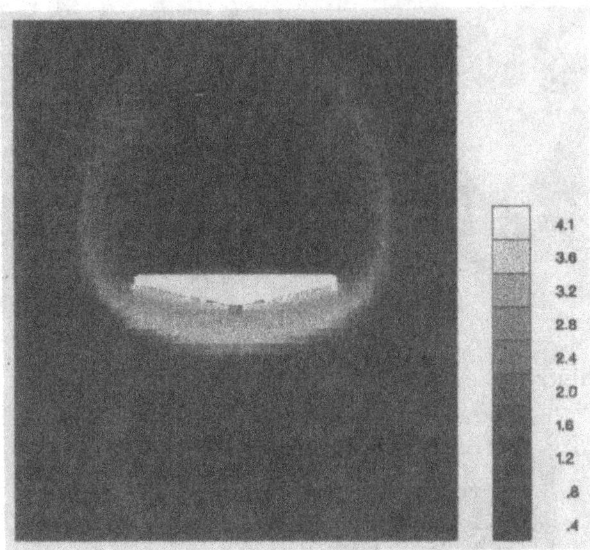

Figure 9

References

[1] **Aristov, V.V., Tscheremissine, F.G.:** The conservative splitting method for solving the Boltzmann equation, USSR Comp. Math. and Math. Phys., Vol. 20, p. 208 (1980)

[2] **Babovsky, H.:** A Convergence Proof for Nanbu's Boltzmann Simulation Scheme, European Journal of Mechanics B/Fluids, 8, no. 1 (1989)

[3] **Babovsky, H., Gropengießer, F., Neunzert, H., Struckmeier, J., Wiesen, B.:** Application of well-distributed sequences to the numerical simulation of the Boltzmann equation, to appear in Computational and Applied Mathematics

[4] **Babovsky, H., Illner, R.:** A Convergence Proof for Nanbu's Simulation Method for the full Boltzmann equation,SIAM Journal of Numerical Analysis, Vol. 26, no. 1, pp. 45-65 (1989)

[5] **Bird, G., Moss, J.:** Direct Simulation of Transitional Flow for Hypersonic Reentry Conditions, AIAA paper no. 84-0223, 1984

[6] **Caflish, R.:** The fluid dynamic limit of the nonlinear Boltzmann equation, Comm. Pure & Appl. Math., Vol. 33, pp. 651-666 (1980)

[7] **Cercignani, C.:** The Boltzmann Equation and its Applications, Springer 198

[7'] **Cercignani, C.:** Scattering Kernels for Gas-Surface Interaction, Workshop on Hypersonic Flows for Reentry Problems, Antibes Tome 1, 1990

[8] **Chapman, S., Cowlings, T.G.:** The mathematical theory of non-uniform gases, Cambridge University Press, 1970

[9] **Coron, F.:** Derivation of slip boundary conditions for the Navier Stokes System from the Boltzmann Equation, To appear in Journal of Statistical Physics

[10] **Coron, F.:** Applications de la théorie cinétique à l'aérodynamique hypersonique: une approche mathématique, Thèse de Doctorat de l'Université Paris Nord

[11] **Elizarova, T.G., Chetverushkin, B.N.:** Kinetical consistent finite difference gasdynamic schemes, to appear in: Japanese Journal of Aerospace Science

[12] **Frezzotti, A., Parani, R.:** Direct Numerical Solution of the Boltzmann Equation for a Binary Mixture of Hard Sphere Gases, to appear in Meccanica

[13] **Gropengießer, F., Neunzert, H., Struckmeier, J., Wiesen, B.:** Rarefied gas flow around a disc with different angles of attack, in preparation

[14] **Gropengießer, F., Neunzert, H., Struckmeier, J., Wiesen, B.:** Rarefied gas flow around a 3d-deltawing at low Knudsen number, in preparation

[15] **Hlavka, E.:** Funktionen von beschränkter Variation in der Theorie der Gleichverteilung, Ann. Math. pura et appl. IV, 54 (1961)

[16] **Illner, R., Pulvirenti, M.:** Global validity of the Boltzmann Equation for two- and three-dimensional rare gas in vacuum: Comm. Math. Phys. 121, 143-146 (1989)

[17] **Kaniel, S.:** A kinetic model for the compressible flow equations, Indiana University Mathematics Journal, Vol. 37, No. 3, 1988

[18] **Körber, S., Wiesen, B.:** A comparison of a Microscopic and a Phenomenological Model for a Polyatomic Gas, in preparation

[19] **Kuipers, L., Niederreiter, H.:** Uniform distribution of sequences, John Wiley & Sons (1974)

[20] **Landford, O.E.:** The evolution of large classical systems, Proceedings of the 1974 Battelle Rencontres on Dynamical Systems, J. Moser ed., Springer Lecture Notes in Physics no. 35, pp. 1-111 (1975)

[21] **Lécot, C.:** Low Discrepancy sequences for solving the Boltzmann Equation, to appear

[22] **van der Mee, C.V.M.:** Stationary Solutions of the Nonlinear Boltzmann Equation in a Bounded Spatial Domain, Math. Meth. in the Appl. Sciences, Vol. 11, pp. 471-481 (1989)

[23] **Mißmahl, G.:** Randwertprobleme bei der Boltzmann Simulation, Diplomarbeit, University of Kaiserslautern (1990)

[24] **Nanbu, K.:** Direct Simulation Schemes derived from the Boltzmann Equation, J.Phys. Japan 49, p. 2042 (1980)

[25] **Neunzert, H., Wick, J.:** Die Theorie der asymptotischen Verteilung und die numerische Lösung von Integrodifferentialgleichungen, Num. Math. 21, pp. 234-243 (1973)

[26] **Perthame, B.:** Boltzmann type schemes and the entropy property, to appear in SIAM Journal on Numerical Analysis

[27] **Platkowski, T.:** private communication

[28] **Weyl, H.:** Über die Gleichverteilung von Zahlen mod 1, Math. Ann. 77 (1916)

[29] **Yen, S.M.:** Numerical Solution of the Nonlinear Boltzmann Equation for Nonequilibrium Gas Flow Problems, Ann. Rev. Fluid Mech., Vol. 16, pp. 67-97 (1984)

A CLASS OF MOVING BOUNDARY PROBLEMS ARISING IN INDUSTRY

J.R. OCKENDON
Mathematical Institute
24-29 St Giles
Oxford OX1 3LB
U.K.

ABSTRACT. There is a large class of moving boundary problems of industrial relevance in which, as in the Signorini problem, the moving boundary has spatial codimension two. Examples include saturated-unsaturated porous medium flows when the unsaturated region is narrow, various types of supercooled solidification problems, impact problems in ship hydrodynamics, models for electrical painting, and contact problems. Providing that time is increasing, a surprising number of these problems are amenable to a smoothing similar to the Biaocchi transformation. This means they can be reduced to well-posed families of variational inequalities, parameterised by time, for which simple numerical algorithms are available. Even though a formal time reversal is possible in several of the problems, the failure of the smoothing transformations suggests that such time-reversed problems may be ill-posed. Corroboration of this conjecture might be provided by a formal linear stability analysis but this requires an approach different from the usual stability analysis of moving boundaries with codimension unity.

1. Introduction

This paper is motivated by some industrial processes which give rise to a special class of mathematical models. These take the form of free boundary problems in which the free boundary has two dimensions fewer than the space in which the field equations have to be solved. Because of the consequent ease of solving the field equations, it seems possible that such free boundary problems may be classifiable, at least in some simple cases where the field equation is Laplacian.

The problems will only be described schematically here, but fuller details of the physical background can be found in the references. The common feature is that the geometry of the free boundary is as in Fig. 1 and is such that it lies near a segment of a prescribed boundary Γ_0.

R. Spigler (ed.), Applied and Industrial Mathematics, 141–150.

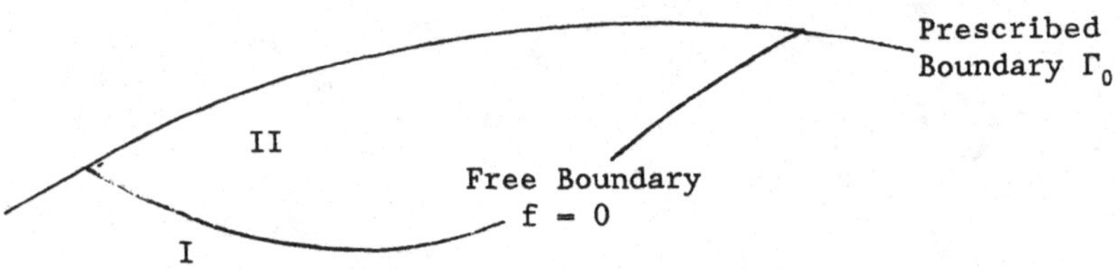

Figure 1.

The free boundary will be denoted by $f(x,y,z,t) = 0$ and its normal velocity $-|\nabla f|^{-1}\partial f/\partial t$ by v_n. In general the field equation must be solved both in region I and in the "thin" region II but we will mostly be concerned with one-phase problems, for which the solution in region II is trivial. No explicit mention will be made until later of other fixed boundary conditions, or initial conditions, although the latter must be compatible with the fact that $f = 0$ is near Γ_0.

We first list the three principal problems as in Fig. 1. In all cases the model has been made dimensionless, the field equation is Laplacian, and $\partial/\partial n$ denotes the derivative along the normal from region I to region II.

TABLE 1.

Problem [reference]	Field Equation	Condition on f=0	Condition on Γ_0
P_1 Shallow dam with seepage face. [1,2]	$\Delta p = 0$ in I; p = pressure	$p = 0$; $-\frac{\partial}{\partial n}(p+y) = v_n$ (y is vertical coordinate)	Seepage face: $p = 0$; $\frac{\partial}{\partial n}(p+y) \leq 0$
P_2 Ship slamming in zero gravity. [3,4]	$\Delta\phi = 0$ in I; ϕ = velocity potential	$\frac{\partial\phi}{\partial n} = v_n$; $\frac{\partial\phi}{\partial t} + \frac{1}{2}\|\nabla\phi\|^2 = 0$	Ship: $\frac{\partial\phi}{\partial n} = V_n$, ($V_n$=normal velocity of Γ_0); $\frac{\partial\phi}{\partial t} + \frac{1}{2}\|\nabla\phi\|^2 \leq 0$
P_3 Hele-Shaw flow with impermeable boundary [5,6]	$\Delta p = 0$ in I; p = pressure	$p = 0$; $\Rightarrow \frac{\partial p}{\partial t} - \|\nabla p\|^2 = 0$	Impermeable boundary: $\frac{\partial p}{\partial n} = 0$; $p \geq 0$

The inequality conditions on Γ_0 are those which are conventionally imposed, sometimes tactictly, and we will subsequently consider the effect of relaxing some of these constraints. P_3 is not of direct industrial relevance but it is related both to P_2 and to the Stefan model for phase changes, which has many industrial applications, [7].

Except in cases of steady flow ($\partial/\partial t = 0$) and for certain simplified geometries, very little is known theoretically about any of the problems P_1-P_3. P_1 has the largest literature and much of what is known about its solution comes from the theory of variational inequalities and generalisations thereof [8]. P_3 is also amenable to variational techniques in the case where $p \geq 0$ everwhere in region I, [9], but not even the weak solution of P_2 is well understood because, despite many efforts to analyse the simplest case of the so-called "wedge-entry" problem [10], the singularities where $f = 0$ meets Γ_0 remain unidentified.

Concerning the stability of possible classical solutions of these problems, only a formal local linear analysis is possible. P_1 is stable or unstable according as the component of v_n in the positive y-direction is greater or less than - 1, P_2 is neutrally stable and P_3 is stable or unstable according as to whether p is greater or less than zero respectively in the vicinity of $f = 0$.

For each of these problems we now take Γ_0 to be $y = \varepsilon g(x,z,t)$ $0 < \varepsilon \ll 1$, so that its normal is nearly the y-axis; we than write $f = y - \varepsilon h$, rescale time with ε, and take the formal limit as $\varepsilon \to 0$ to retrieve the following "codimension-two" free boundary problems for Laplace's equation in $y < 0$. For simplicity we still denote the rescaled time by t and also we have taken the y-axis to be an axis of symmetry in each case; the "codimension-one" free boundary is just denoted by $x = D(t)$.

TABLE 2.

Problem [reference]	Conditions on $y=0,\ 0 < x < D$	Conditions at $y = 0,\ x = D$	Conditions on $y = 0,\ x > D$
P_1' $p + y = \varepsilon\phi$ [1]	$\phi = h$ $\dfrac{\partial\phi}{\partial y} = -\dfrac{\partial h}{\partial t}$	$h = g$ $\lvert\nabla\phi\rvert < \infty$	$\phi = g$ $\dfrac{\partial\phi}{\partial y} \leq 0$
P_2' "Ship" is $g = S(x) - Vt$ [4]	$\phi = 0$ $\dfrac{\partial\phi}{\partial y} = \dfrac{\partial h}{\partial t}$	$h = S(D) - Vt$ $(\lvert\nabla\phi\rvert = \infty)$	$\dfrac{\partial\phi}{\partial y} = -V$ $\dfrac{\partial\phi}{\partial t} \leq 0$
P_3' $g = 0$ [11]	$p = 0$ $\dfrac{\partial p}{\partial y} = -\dfrac{\partial h}{\partial t}$	$h = 0$ $(\lvert\nabla p\rvert = \infty)$	$\dfrac{\partial p}{\partial y} = 0$ $p \geq 0$

Problem P_3' can be related to P_2' when $S = 0$ by writing $\phi = - Vy - p$ and redefining h. Also it is easiest to interpret P_2' as a model for the slamming of a ship of small "deadrise angle" when the conditions on $y = 0$ for x greater and less than D are reversed and we will indeed use these reversed coordinates in the next section.

It would be a simple matter to extend any of these models to the case of more general Γ_0 as long as region II remains thin.

The rigorous justification of any of these limiting models poses a difficult mathematical challenge and will not be discussed here. We will simply record some comparable limiting models which have appeared in the literature in a variety of different physical situations.

P_4. Electrical painting, [12,13]. This is a two-phase problem with nonlinear conditions on $f = 0$, but, in the steady state codimension-two limit, it is closely analogous to P_1'.

P_5. Inviscid flow down a step, [14]. Here a "Prandtl-Batchelor" model is assumed for the vorticity within the circulating region. The stream function is harmonic and, on $y = 0$, it satisfies

$$\psi = h, \quad \frac{\partial}{\partial x}\left(\frac{\partial \psi}{\partial y} + \frac{1}{2h^2}\right) = 0, \quad 0 < x < D$$

$$\psi = 0, \quad x > D$$

with $|\nabla\psi| < \infty$ at $x = D$. An integro-differential equation can be derived for h, [15].

P_6. Screw Dislocations, [16]. The adoption of a simple mobility law means that in the limiting codimension-two free boundary problem, the stress function ϕ, which is harmonic, satisfies, on $y = 0$

$$\frac{\partial}{\partial t}\left(\frac{\partial \phi}{\partial y}\right) + \frac{\partial}{\partial x}\left[\left(1 + \frac{\partial \phi}{\partial x}\right)\frac{\partial \phi}{\partial y}\right] = 0, \quad 0 < x < D$$

$$\frac{\partial \phi}{\partial y} = 0, \quad D < x$$

with $|\nabla\phi| < \infty$ at $x = D$, [17].

There are many other, more famous, examples of codimension-two free boundary problems such as contact and crack problems in solid mechanics, the sintering of solids or liquids, the spread of thin films, etc. These often involve more complicated field equations than the Laplacian but it is interesting to note that they have usually only been considered in the case where dD/dt has one sign; for example there are very few references to "unloading" in contact problems [18]. The question of time-reversibility is closely related to that of well-posedness and we will refer to it again later. P_2' and P_3' are formally time-reversible when we replace ϕ or p by their negatives, and also change the sign of V in P_2'.

2. Smoothing Transformations

The problem P_1' does not need any preliminary transformations to enable the existence and uniqueness of its weak solution to be proved. Indeed, as shown in [2], it is an evolutionary variational inequality even though P_1 is only a variational inequality in steady flow with suitable geometry and after the application of a so-called "Baiocchi" transformation [8].

The situation is more interesting for P_2' and P_3'. In order to interpret P_2' in terms of the slamming of a finite ship, as mentioned earlier it is convenient to impose the free and prescribed boundary data on $y = 0$, for x greater and less than D respectively. Thus, when the fluid occupies the entire half space $y < 0$, and suitable assumptions are made about the behaviour of ϕ at infinity, it has been pointed out in [3] that if we define

$$\Phi(\underset{\sim}{x},t) = \int_0^t \phi(\underset{\sim}{x},\tau)d\tau \tag{1}$$

then $$\Delta\Phi = 0 \tag{1a}$$

with $$\Phi = 0,\ \frac{\partial\Phi}{\partial y} = h - h_0 \text{ on } y = 0,\ x > D(t), \tag{1b}$$

where $h(x,0) = h_0(x)$ is the prescribed initial elevation of the fluid. Now if we make the crucial assumption that V and dD/dt are both positive and define $w(D(t)) \equiv t$, we find that on $y = 0$, $x < D$,

$$\frac{\partial\Phi}{\partial y} = \left[\int_0^w + \int_w^t\right]\frac{\partial\phi}{\partial y}\, d\tau$$

$$= h(x,w) - h_0(x) - V(t-w)$$

$$= S(x) - V\tau - h_0(x). \tag{1c}$$

Hence we have that $|\nabla\Phi| < \infty$ at $x = D(t)$ and, on $y = 0$,

$$\Phi \le 0,\ \frac{\partial\Phi}{\partial y} + Vt + h_0(x) - S(x) \le 0 \tag{2a}$$

$$\Phi\left[\frac{\partial\Phi}{\partial y} + Vt + h_0(x) - S(x)\right] = 0. \tag{2b}$$

Thus it is easy to prove that we have variational inequality for Φ and well-posedness of a weak solution is assured when $V > 0$, even for the slamming of a three-dimensional ship. Practical experience suggests that despite the formal time-reversibility of P_2', the solution of the initial value problem when $V < 0$ has characteristics different from that when $V > 0$.

The situation is similar for P_3', where again the possibility of making a transformation such as (1) depends crucially on the sign of dD/dt.

3. Stability and Well-Posedness for Problems P_2', P_3', P_1'

The advantage of working with codimension-two rather than codimension-one free boundary problems is well illustrated for the problems P′ because explicit solutions of Laplace's equation can be exploited. We begin by considering P_2' which can be interpreted physically in a less ambiguous way than P_3'. In a two-dimensional flow whose configuration is as for the analysis in (1,2), the solution for which $|\nabla\Phi| \to 0$ at infinity is

$$\Phi = \mathrm{Rl}\,\{-V(y+\sqrt{D^2-(x+iy)^2}\}. \tag{3}$$

The condition at x = D gives that

$$S(D) - Vt - h_0(D) = \int_0^t \left[-V + \frac{VD(t)}{\sqrt{D^2(t)-D^2(\tau)}}\right] d\tau \tag{4a}$$

and hence that assuming $V > 0$, $w = D^{-1}$ is given by

$$w(x) = 2/\pi V \int_0^x \frac{[S(\xi)-h_0(\xi)]d\xi}{\sqrt{x^2-\xi^2}} . \tag{4b}$$

This solution can be readily extended to cases where V is non-constant. Also it can be shown to give a realistic force distribution on the surface of a two-dimensional ship [4].

We are now in a postion to attempt a perturbation solution about (4) in which the free boundary x = D varies in the z-direction with a small amplitude δ. It is easiest to perform this analysis locally in space and time, so we only consider the "semi-infinite" problem in which, on y = 0

$$\frac{\partial\phi}{\partial y} = 0 \ , \ x < D \tag{5a}$$

$$\phi = 0, \ \frac{\partial\phi}{\partial y} = \frac{\partial h}{\partial t} \ , \ x > D, \tag{5b}$$

with $h \sim O(x^{-\frac{1}{2}})$, $\phi \sim A\,\mathrm{Rl}\sqrt{-x-iy}$ at infinity, where A is prescribed. The term in V in (5b) has been neglected because the local velocities near x = D, y = 0 in the solution (4) are so large. The solution of (5) is now

$$\phi = A\, \mathrm{Rl}\sqrt{D-x-iy} \tag{6a}$$

$$h = -AV_0^{-1}\sqrt{x-D}\quad, x > D \tag{6b}$$

where

$$D = V_0 t. \tag{6c}$$

Like A, V_0 is also a constant to be prescribed (say, from matching with an outer solution such as (4)) but A and V_0 will have the same sign and only when they are both positive will we have a slamming problem. We now change to moving axes $\xi = x - V_0 t$ in which case (5b) becomes

$$\frac{\partial h}{\partial t} - V_0\frac{\partial h}{\partial \xi} = \frac{\partial \phi}{\partial y}\ ; \tag{7}$$

we seek the free boundary in the form

$$\xi = \delta \cos nz . e^{\sigma t} + O(\delta^2),\ \delta \ll 1, \tag{8}$$

where n is a prescribed positive wavenumber and σ is a growth rate to be determined if possible.

The relevant solution of Laplace's equation in polar coordinates centred at $\xi = 0$ is

$$\phi = Ar^{\frac{1}{2}}\sin\theta/2 + \delta B r^{-\frac{1}{2}} e^{\sigma t - nr}\cos nz.\sin\theta/2 + O(\delta^2) \tag{9}$$

for some arbitrary constant B which implies that

$$h = -AV_0^{-1}\sqrt{\xi} + \delta h_1(\xi)e^{\sigma t}\cos nz + O(\delta^2), \tag{10}$$

where, from (7), h_1 satisfies

$$-V_0\frac{dh_1}{d\xi} + \sigma h_1 = Be^{-n\xi}/2\xi^{3/2}. \tag{11}$$

In order to satisfy the growth condition $h \sim O(x^{\frac{1}{2}})$ as $x \to \infty$, we need $h_1 \to 0$ as $\xi \to \infty$ and this means $h_1 = O(\xi^{-\frac{1}{2}})$ as $\xi \to 0$. This in turn means we need an " inner" asymptotic expansion when ξ and y are of $O(\delta)$. The details of the construction of this expansion are given in [4], but the conclusion is that $h_1 \sim A\xi^{-\frac{1}{2}}(1+\sum_1^\infty \alpha_n \xi^n)/2$ as $\xi \to 0$ and, combined with (11). this implies that if $\sigma/V_0 > 0$, then $\sigma/V_0 = -n$. This contradiction shows that σ and V_0 have opposite signs which gives us some analytical evidence that the slamming problem is stable but the time-reversed "exit" problem is unstable. The stability analysis does not, unfortunately, yield a dispersion relation between σ and n, even though such a relation is readily deducible for the codimension-one problem P_2, with the result quoted after Table 1. The reason that it is not available here is that the assertion that the codimension-two free boundary has the form (8) contains less information than a comparable assertion about the free boundary perturbation for P_2; we need more information to be fed into (11), such as the initial displacement h_0, to determine a dispersion relation.

This suggestion that the stability of P_2' or P_3' depends crucially on the sign of dD/dt is corroborated by the fact that when we attempt an explicit solution of the two-dimensional problem as in (4b) when $V < 0$, we need to perform an analytic continuation of the initial free boundary $h_0(x)$, which itself must be assumed analytic.

Some quite different conjectures can be made concerning the stability and well-posedness of P_1'. Indeed, the analysis in [2] discusses the variational inequality in situations where dD/dt can be positive or negative, although this formulation cannot give much explicit information about D(t). Moreover, there are some situations in which P_1' is explicitly solvable; for example, when $g = 0$ and $h(x,0) = h_0(x)$ is negative and has compact support, it is easy to verify that

$$\frac{\partial h}{\partial t} - \frac{1}{\pi}\int_{-\infty}^{\infty} \frac{\partial h/\partial\xi \, d\xi}{\xi - x}$$

because the solution to this equation, namely

$$h(x,t) = \frac{t}{\pi}\int_{-D(0)}^{D(0)} \frac{h_0(\xi)\,d\xi}{t^2+(x-\xi)^2}$$

is negative everywhere† and tends to h_0 as $t \downarrow 0$. In this case there is no free boundary because the dry region extends instaneously to infinity.

Such a linear analysis is probably not possible when the support of h_0 is in a region where $g > 0$, because then there is numerical evidence [1] that D moves with finite speed.

A stability analysis as in (5-11) can still be attempted. The local problem (5) now satisfies, on $y = 0$,

$$\phi = 0 \ , \ x < D$$

$$\phi = h, \ \frac{\partial\phi}{\partial y} = -\frac{\partial h}{\partial t} \ , \ x > D.$$

Thus (6) is replaced by

---

†When $g \neq 0$, the term $\int_{-D(0)}^{D(0)} g'(\xi)\log[1+t^2/(x-\xi)^2]\,d\xi$ must be added to the h, and this may result in the violation of the constraint $h \leq g$.

$$\phi = A \, \mathrm{Im}\, [i(D-x-iy)]^{\alpha}$$

$$h = -A \sin \alpha\pi.(x-D)^{\alpha} \quad , \; x > D$$

where $D = V_0 t$ and $\tan \alpha\pi = - V_0^{-1}$, $\pi < \alpha < 2\pi$. Again A and V_0 are to be prescribed but now A is always positive. Unfortunately a linear perturbation about this solution now leads to a Wiener-Hopf problem which cannot be solved as easily as in (9), and we will not pursue this here.

4. Conclusions and Conjectures

For the codimension-one problems we have considered, we recall that, as far as local linear stability is concerned, P_1 is stable or unstable depending on how large is the downward velocity of the free boundary; P_2 is neutrally stable, independent of the direction of motion of the free boundary; P_3 is stable or unstable depending whether the free boundary is expanding or contracting region 1.

It is clearly dangerous to carry such statements over to the codimension-two specialisation because, for example, the codimension-one free boundary may "turn over" near the points where it meets Γ_0, in which case the ideas of expanding and contracting become ambiguous. However, the analysis in Section 3 suggests that

(i) if P_i is stable for both an expanding and a contracting free boundary, so is P_i',

(ii) if P_i is stable for expanding and unstable for contarcting free boundaries, so is P_i',

(iii) if P_i is neutrally stable, then P_i' can switch from stable to unstable depending on the direction in which the boundary moves.

Acknowledgement

The author is grateful to C.M. Elliott, D.A. Spence, P. Wilmott, W.W. Wood and especially S.D. Howison for helpful discussions during the preparation of this work.

References

1. Aitchison, J.M., Elliott, C.M. and Ockendon, J.R. (1983) 'Percolation in Gently Sloping Beaches', IMA J. App. Math., 30, 269-287.

2. Elliott, C.M. and Friedman, A. (1985) 'Analysis of a model of percolation in a gently sloping sandbank', SIAM J. Math. Anal., 16, 941-954.

3. Korobkin, A.A. (1982)'Formulation of penetration problem as a variational inequality'. Din. Sploshnoi Sredy, 58, 73-79.

4. Howison, S.D., Ockendon, J.R. and Wilson, S.K. (1990) 'A note on incompressible water entry problems at small deadrise angles', (Preprint).

5. Elliott, C.M. and Ockendon, J.R. (1982) 'Weak and Variational Methods for Free and Moving Boundry Problems', 59, Pitman Research Notes in Math.

6. Lacey, A.A., Howison, S.D. and Ockendon, J.R. (1988) 'Hele-Shaw free boundary problems with suction', Q. Jl. Mech. App. Math., 41, 183-193.

7. Ockendon, J.R. and Hodgkins, W.R. (1975) 'Moving Boundary Problems in Heat Flow and Diffusion', Oxford.

8. Chipot, M. (1984) 'Variational Inequalities and Flow in Porous Media', Applied Mathematical Sciencies, 52, Springer.

9. Elliott, C.M. and Janovsky, V. (1981) 'A variational approach to Hele-Shaw flow with a moving boundary', Proc. Roy. Soc. Edin., 93-107.

10. Greenhow, M. (1987) 'Wedge entry into initially calm water', Appl. Ocean Res., 9, 214-223.

11. Lacey, A.A., Howison, S.D., Ockendon, J.R. and Wilmott, P. (1990) 'Irregular Morphologies in Unstable Hele-Shaw Free Boundary Problems', Q.Jl. Mech. App. Maths, (to appear).

12. Aitchison, J.M., Lacey, A.A. and Shillor, M. (1984) 'A model for an electropaint process', IMA J. App. Math. 33, 17-31.

13. Marquez, V. and Shillor, M. (1987) 'The electropainting problem with overpotentials, SIAM J. Math. Anal., 18, 788-811.

14. Childress, S. (1966) 'Solution of Euler's equation continaing finite eddies', Phys. Fluids, 9, 860-872.

15. O'Malley, K., Fitt, A.D., Jones, T.V., Ockendon, J.R. and Wilmott, P. (1990) 'Models for high Reynolds-number flow down a step', J. Fluid Mech., (to appear).

16. Head, A.K., Howison S.D., Ockendon, J.R., Titchener, J.B. and Wilmott, P. (1987) 'A continuum model for two-dimensional dislocation distributions, Phil. Mag. A15, 617-629.

17. Ockendon, H. and Ockendon, J.R. (1983) 'Dynamic dislocation pile-ups, A47, 707.

18. Turner, J.R. (1979) 'The frictional unloading problem on a linear elastic half-space, IMA J. Appl. Math, 24, 439-470.

SYSTEMS WITH NON-FADING MEMORY ENCOUNTERED IN THE MODELLIZATION OF INDUSTRIAL PROBLEMS

M. PRIMICERIO
Dipartimento di Matematica "U. Dini"
Università di Firenze
Viale Morgagni 67/A, 50134 Firenze
Italy

ABSTRACT. Two examples of industrial problems originating mathematical problems of non-local type are illustrated: the degradation of the rheological properties of coal-water slurries and the crystallization of polypropylene. In both cases the model is described, the parameters identification is discussed, and some analytical and numerical results are presented.

0. Introduction

The aim of this talk is to give an account for two examples of industrial problems that have been studied by our group in Florence (A. Fasano, E. Comparini, D. Andreucci and myself) with the occasional cooperation of other colleagues (V. Capasso, S. Paveri-Fontana, R. Ricci) with whom we had, from time to time, fruitful discussions.

The two problems exhibit a common feature, since they originate mathematical models incorporating a "non-local" law, in the sense that the process is essentially history-dependent. Another common feature is that in both cases .... we were successful: the model proved to fit the experimental data and to be able to predict the results of other experiments, one of the keys of the success being the cooperative attitude of the companies involved and in particular the excellent partners we found as responsible of the project from the industrial side.

The talk will be divided in two parts: in the first part, I will describe the problem of degradation of coal-water slurries. In the second part the crystallization of polypropylene will be dealt with.

1. The degradation of coal-water slurries

1.1. THE PROBLEM

A coal-water slurry is a dense suspension of coal

R. Spigler (ed.), Applied and Industrial Mathematics, 151–172.

particles in water. Suitable optimization of the particle size distribution and the use of an appropriate chemical additive acting as a fluidizer allow to produce highly concentrated suspensions (up to 60-70% in weight) having good rheological properties and thus easily pumpable in long pipelines and burnable without previous dewatering.

The technology of slurries has received a great deal of attention, expecially starting from the oil crisis of the Seventies, and the main directions of the mathematical research are the study of the rheology of the slurries at low, moderate and high shear rates, as well as the production of sprays.

The italian-based company Snamprogetti is currently completing the construction of the 250 km long pipeline which will carry coal slurry from the mines of Belovo to Novosibirsk in the USSR.

The model we were asked to set up deals with the degradation of the rheological properties of slurries. As it is shown in Fig.1, when a slurry is circulating for a sufficiently long time in a test loop or stirred in a vessel, its apparent viscosity increases with time. This increase is different at different nominal shear rates and can become dramatic so that the suspension is no longer pumpable.

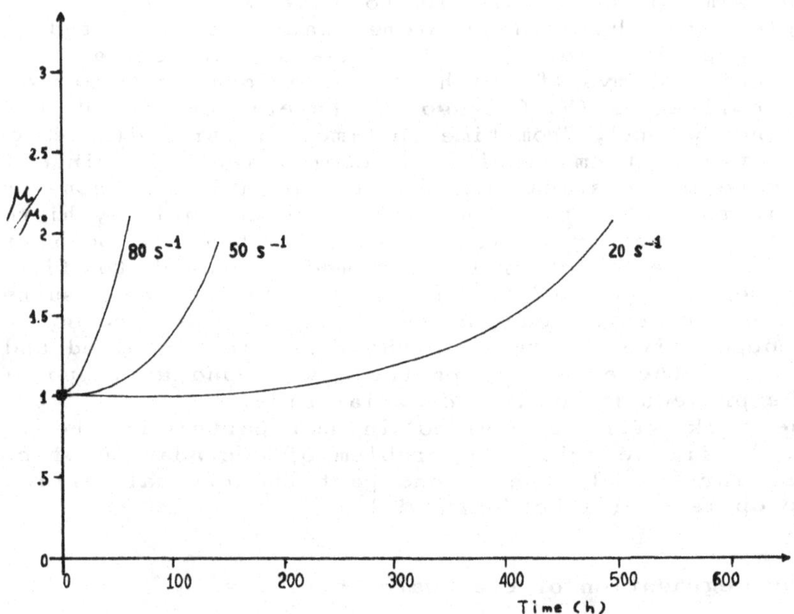

Figure 1

Our starting point was the identification of the mechanical quantities which are relevant. The second step was to give a microscopic interpretation aimed at finding possible ways of controlling the phenomenon.

1.2. A MACROSCOPIC MODEL

The basic assumption of our model is that (in the moderate range of shear rates from 20 to 100 sec^{-1}) the degradation depends on the energy which is dissipated in the system; see [7], [17] for a better explanation of the origin and the reasons of this idea. Thus, if μ represents a measure of the viscosity of the fluid, the governing law is assumed to be

$$\frac{d\mu}{dt} = F(W) \tag{1.2.1}$$

where W is the power locally dissipated during the motion.

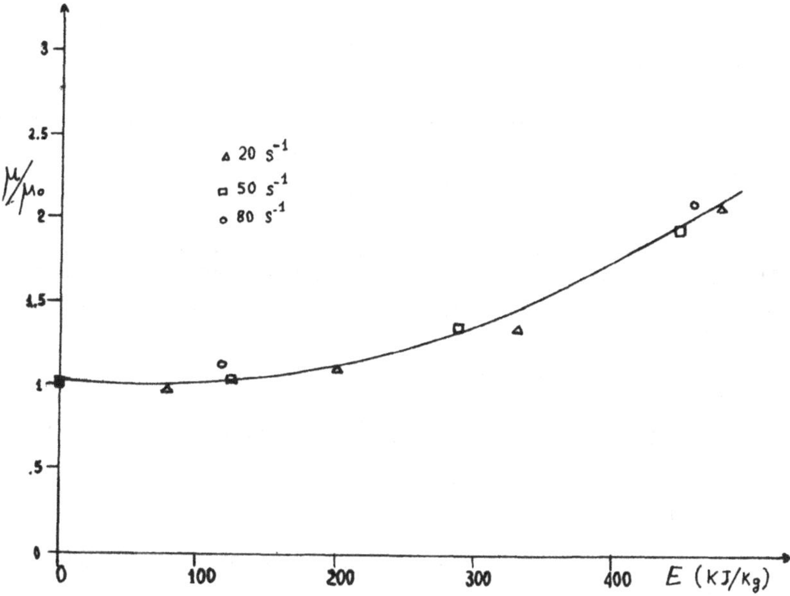

Figure 2

The exact meaning of μ in (1.2.1) depends of course on the rheological constitutive law we assume for the slurry. If the fluid could be thought as being newtonian at each time t, μ would obviously coincide with the viscosity η and this is the model problem we will consider first, because of its simplicity. Otherwise, as we will see, one is led to identify μ with the apparent viscosity at a fixed shear rate.

In any case, as far as an overall behaviour is concerned, experiments confirm the basic assumption. Actually, Fig. 2 displays the same experimental results of Fig. 1, but this time the apparent viscosity is plotted versus the total dissipated energy: the experimental data correspond basically to the same curve for the three different values of the nominal shear rate.

The success of the whole research has been possible also because of the very good work performed at the laboratories of Snamprogetti and of Eniricerche [9], [10]; discussions and cooperation with Drs. D. Ercolani and E. Carniani and Dr. S. Meli have been always stimulating and effective.

Let us come to the mathematical formulation of a model problem.

We consider a one-dimensional laminar flow (actually, vorticity seems to be irrelevant also in the industrial situation) between two parallel planes y=0 and y=d. We assume

$$\underline{v} = v(y,t)\, \underline{e}_1, \tag{1.2.2}$$

and we consider the case in which the boundary conditions are

$$v(0,t) = 0 \ , \qquad v(d,t) = F(t) \ , \tag{1.2.3}$$

and no pressure gradient nor body forces are present. If the fluid is thought to be newtonian, the Navier-Stokes equation is simply (Lagrange and Euler derivatives coincide):

$$\rho_0 \frac{\partial v}{\partial t} = \frac{\partial}{\partial y}\left(\eta \frac{\partial v}{\partial y}\right) . \tag{1.2.4}$$

Equation (1.2.4) has to be supplemented with an initial condition

$$v(y,0) = v_0(y) \ , \tag{1.2.5}$$

and with the specification of the law (1.2.1). Assuming the linear law

$$\dot{\eta} = \lambda W \ , \qquad \eta(y,0) = \eta_0(y) \ , \tag{1.2.6}$$

i.e.

$$\dot{\eta} = \lambda \eta \left(\frac{\partial v}{\partial y}\right)^2 , \qquad \eta(y,0) = \eta_0(y) \tag{1.2.7}$$

and solving (1.2.7) formally, we obtain

$$(1.2.8) \qquad v_t = a[v]\, v_{yy} + b[v]\, v_y \ ,$$

with

$$a[v] = \rho_0^{-1} \eta_0(y) \exp\left\{\lambda \int_0^t v_y^2(y,\tau)\, d\tau\right\} \ ,$$

$$b[v] = \rho_0^{-1} \left\{\frac{\partial \eta_0}{\partial y} + 2\lambda\eta_0(y) \int_0^t v_y(y,\tau) v_{yy}(y,\tau)\, d\tau\right\} \exp\left\{\lambda \int_0^t v_y^2(y,\tau)\, d\tau\right\}.$$

Problem (1.2.8), (1.2.3), (1.2.5) is clearly of non local character, in the sense that the solution at time t depends on the whole history of the process (as it is actually the case in the motion of slurries). Existence of a classical solution for sufficiently short times has been proved in [17] using a fixed point argument in a closed subset of the Banach space $C_{2+\delta}$ of the functions having Hölder continuos derivatives v_{yy} and v_t.

In [7] elementary solutions have been considered, among which we quote the elementary solution corresponding to a stationary velocity field

$$(1.2.9) \qquad \eta = \eta_0 \exp[\lambda v_0^2 t/d^2] \ , \qquad\qquad v = v_0\, y/d \ .$$

The corresponding solution with separate variables for the geometry of the rotational viscometer is

$$(1.2.10) \qquad \eta = \eta_0 \frac{R_1^2}{r^2} \exp\left(\frac{\lambda\, v_0^2 t}{R_1^2}\right) \ , \qquad\qquad v = v_0 \frac{r}{R_1} \frac{\ln(r/R_2)}{\ln(R_1/R_2)}.$$

Both (1.2.9) and (1.2.10) have been tested experimentally.

Concerning the axial motion in a pipe of radius R we have to point out that the experiment are carried out so that the pressure gradient is adjusted to keep the average discharge constant; moreover, since at each passage in the pump the slurry is mixed up, it is reasonable to look for a solution in which the viscosity is independent of the radius, $\eta=\eta(t)$ and that the law of evolution is

$$(1.2.11) \qquad \frac{d\eta}{dt} = \lambda \eta \langle v_r^2 \rangle \ , \qquad\qquad \eta(0) = \eta_0 \ ,$$

where

$$(1.2.12) \qquad \langle v_r^2 \rangle = \frac{2}{R^2} \int_0^R r v_r^2\, dr \ .$$

In this case we have a stationary profile of the velocity

(1.2.13) $$v = \frac{G_0}{4\,\eta_0}\,(R^2 - r^2)\ ,$$

(where G_0 is the initial value of the gradient produced by the pump), and a viscosity increasing with time according to the law

(1.2.14) $$\eta = \eta_0 \exp\Big(\frac{\lambda\, G_0^2\, R\, t}{24\, \eta_0^2}\Big).$$

This means that, in order to keep the discharge constant, G has to increase as η:

(1.2.15) $$G(t) = G_0 \eta / \eta_0\ ,$$

i.e. following an exponential law.

While this law is satisfied within a good approximation in early stages of pumping, some experiments do not fit this scheme. The reason is that, as it can be seen from the diagram of the stress vs. shear rate of a slurry taken at different times, in some cases the fluid can be newtonian only for a short time interval, after which a Bingham-plastic behaviour clearly appears [2]. In other cases the fluid is of Bingham type from the very beginning i. e.

(1.2.16) $$\dot\gamma = [\tau - \tau_0]^+ / \eta_B$$

and τ_0 is history dependent as it is shown in Fig. 3.

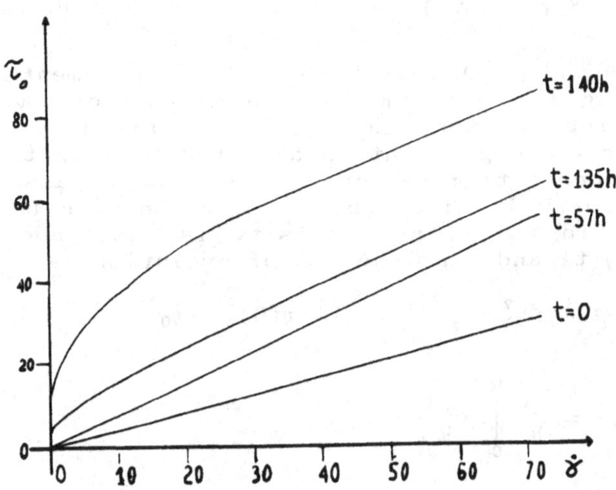

Figure 3

Once again the experimental results show that there are different curves τ_O vs. time for different values of the velocity but when τ_O is plotted versus the total energy dissipated in the flow (Fig. 4) the points corresponding to different velocities lie on the same curve. The same experiments show that η_B is essentially constant during the motion. Some aspects of this problem are currently under additional investigation.

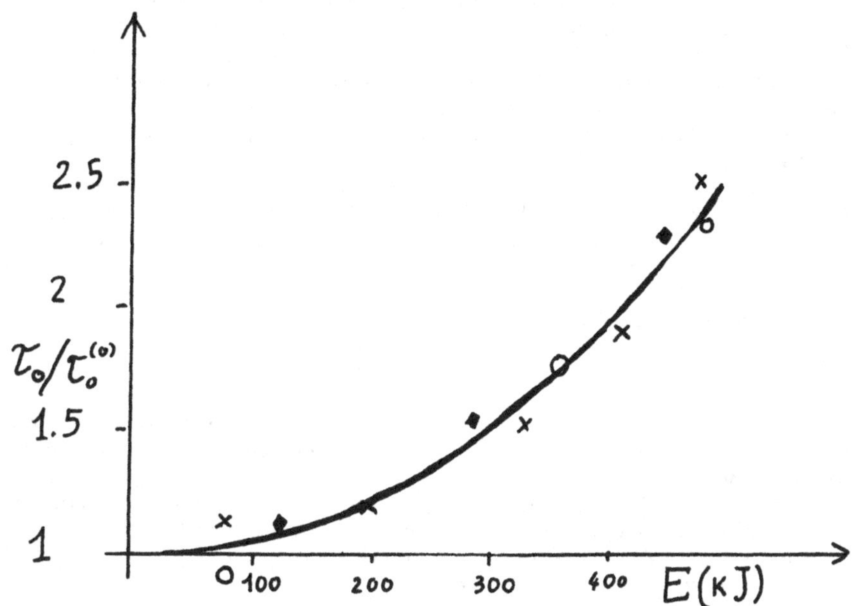

Figure 4

1.3. A MICROSCOPIC PICTURE

The crucial point has been to model how the fluidizing agent works. According to the results obtained by Prof. Ferroni and Gabrielli of the Department of Chemistry of the University of Florence, we may assume that the molecules of the additive can be found in the slurry in three different states:

A) dissolved in water;
B) adsorbed on the coal particles in an “efficient” way, i.e. with their hydro-repellent side facing the liquid;
C) adsorbed on the coal particles in a “non-efficient” way.

According to this picture, a slurry has its minimum viscosity when all the "loci" on the coal particles are occupied by molecules of the group B.

Apparently, a velocity gradient can induce the transition of some molecules from state B to state C. The molecules removed from state B can be replaced by molecules of group A as long as they are available. We assume that the rate of transition from B to C is determined by the power dissipated in the slurry and we can remark that this corresponds to relate this transition to the number of collisions the particles undergo during the motion. We will also assume that this transition is irreversible and that the dynamics of the adsorption A→B follows the rules of a first order reaction. Moreover the transition B→C is considered irreversible. In some cases, a transition A→C is also possible. In the slurries we are considering and at room temperature its influence seems to be irrelevant.

Thus we are led to the following system of ordinary differential equations where A(t), B(t) and C(t) represent the relative concentration of the molecule in the corresponding state.

$$\text{(1.3.1)}\quad \begin{aligned} \dot{A} &= -\alpha A(B_0-B), & A(0)&=A_0, \\ \dot{B} &= \alpha A(B_0-B)-f(W)B, & B(0)&=B_0, \\ \dot{C} &= f(W)B, & C(0)&=C_0. \end{aligned}$$

The evolution of the rheological properties of the slurry, in this picture, are determined by the value of B at each time.

As we pointed out in [11], the ideal case $A_0=0$ is particularly interesting, since it gives information on the function f(W). Indeed, in this case, the system (1.3.1) reduces to the single equation

$$\text{(1.3.2)}\quad \dot{B} = -f(W)B, \qquad B(0)=B_0.$$

Since we known from the experiments that the rheological properties of the slurry depend on time only through the energy dissipated, this means that

$$\int_0^t f(W(\tau))\,d\tau = F\left(\int_0^t W(\tau)\,d\tau\right),$$

for arbitrary t, and it can be seen immediately that f and F have the same form (take e. g. t=1 and $W(\tau)=c$) and have to be linear.

Therefore

$$\text{(1.3.3)}\quad f(W) = \lambda W,$$

$$\text{(1.3.4)}\quad B = B_0\exp[-\lambda_0 E], \qquad E(t) = \int_0^t W(\tau)\,d\tau,$$

independently of the rheological law assumed for the fluid.

This can be interpreted by saying that the intrinsic time scale for the evolution of B is the energy dissipated (the age of population B is determined by the average number of collisions among the particles).

Another useful approximate argument consists in thinking that the replacement reaction A→B takes place at infinite speed, so that, setting A+B=D one has

(1.3.5) $$\dot{D} = -\lambda W \min(D,B_0) , \qquad D(0) = A_0+B_0 .$$

As far as $D \geq B_0$ (1.3.5) simply gives

(1.3.6) $$D-D_0 = -\lambda B_0 E$$

(describing the adsorption of A) so that one can have an idea of the values of λ and B_0 making experiments on different samples of the same slurry with different excess of additive. In Fig. 5 apparent viscosity is plotted against the dissipated energy in slurries with 0.5%, 0.75% and 1% additive.

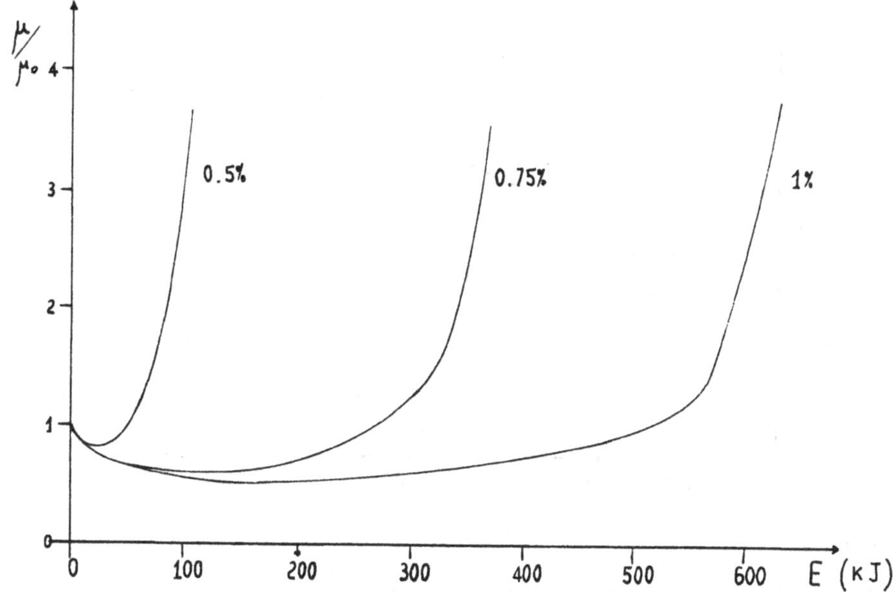

Figure 5

According to the model above we could expect that the rheological properties remain constant as long as $D \geq B_0$; after the dissipation of an energy $\bar{E}$, proportional to the initial excess of additive, the viscosity begins to increase.

The curves in Fig. 5 show that there is a period in which the viscosity decreases; but if we disregard, for the time being, this fact which will be analized later and we look for the values $\bar{E}$ after which the apparent viscosity exceeds the initial value and we identify this situation as the saturation of all the loci (i.e. $D(\bar{E})=B_0$), we find that the curves fit surprisingly well the values $B_0=0.45\%$, $\lambda B_0=10^{-3}(kJ)^{-1}$.

Starting from $\bar{E}$ the three curves have the same slope and

$$D = B_0 \exp[-\lambda(E-\bar{E})] \quad , \tag{1.3.7}$$

where $\bar{E}$ is of course different for the three experiments.

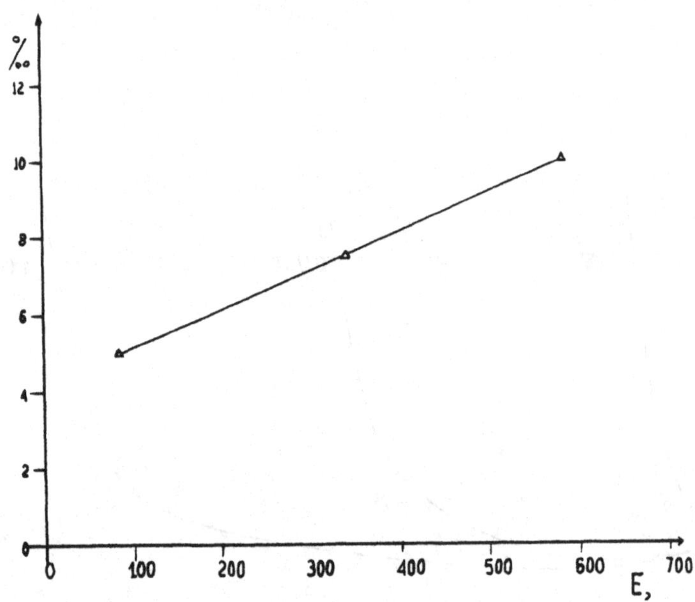

Figure 6

Fig. 6 shows that the energy to be dissipated for doubling the apparent viscosity is linearly related to the initial concentration of the additive, a result which is very promising for the validity of our model. Indeed, according to (1.3.6), (1.3.7), the validity of the model requires that λ and B_0 do not depend on the initial quantity of additive and that

$$E_2 = \bar{E} + (\ln 2)/\lambda = A_0/\lambda B_0 + (\ln 2)/\lambda \ , \tag{1.3.8}$$

E_2 being the energy necessary for doubling the viscosity.

To interpret the initial decreasing of the apparent viscosity, we have to take into account that a slurry is a complex fluid whose rheology cannot be described by just one parameter. Moreover, besides the mechanism of adsorption of the additive described by the system (1.3.1), we have to incorporate in our model the fact that motion produces a better dispersion of the mixture, i. e. the separation of aggregates of coal particles. This mechanical effect is similar to the phenomenon of fluidization induced by the motion in thixotropic media.

The simplest way of taking into account the two mechanisms separately is to assume a Bingham constitutive relationship; of the two parameters of this law - τ_0 and η_B - the first is linked to the quantity of efficiently adsorbed additive, while the latter is related to the degree of dispersion caused by the motion.

In this spirit we shall write

$$\tau_0 = \tau_0(B) \tag{1.3.9}$$

where τ_0 increases for B decreasing.

Moreover, arguing as above, we will write for η_B the differential equation

$$\begin{cases} \dot{\eta}_B = -\beta W(\eta_B - \eta_B^{\infty}) + r(\eta_B^0 - \eta_B) \ , \\ \eta_B(0) = \eta_B^0 \end{cases} \tag{1.3.10}$$

where η_B^{∞} is the Bingham viscosity corresponding to the maximum dispersion and r is the rate of recombination tending to reproduce the initial situation.

A preliminary study of the complete model (1.3.1), (1.3.9), (1.3.10) has been carried out in [11] and the agreement with the experimental date is fairly good.

2. The crystallization of polypropylene

2.1. THE PROBLEM

The process of injection moulding is very well know and largely used by plastic manufacturing industries: the melted polymer is injected in a cold mould so that solification starts, accompanied by shrinking of the material. Then an overpressure is applied to fill the mould completely and temperature is decreased to produce a rapid solidification.

Unlikely pure liquid metals, polymers solidify over a wide range of temperatures; moreover during solidification only a fraction of the polymer crystallizes, undergoing a real change of phase accompained by "ordering" of the molecules and release of latent heat. The remaining part undergoes a continuous transition to an amorphous glassy state; this is not a change of phase but simply a large increase of viscosity. For each polymer two temperatures θ_g and θ_m can be defined such that for $\theta>\theta_m$ the polymer is liquid and for $\theta<\theta_g$ crystallization can not take place and the polymer is solid.

It is evident that the fraction of the crystalline part as well as the dimensions and structure of the crystals which are formed depend on the evolution of the pressure and temperature fields during the whole process. It is also evident that the prediction (and, possibly, control) of these features is a major goal, since the mechanical properties of the solid plastic are strongly dependent on them.

The problem was brought to our attention by Dr. S. Mazzullo (Himont Italia), who also joined our investigation at each step cooperating with us not only providing the relevant experimental data but also contributing to the discussions with stimulating ideas.

We will refer to the case of polypropylene, a polymer made of chains of 1000÷2000 monomeric units of the form

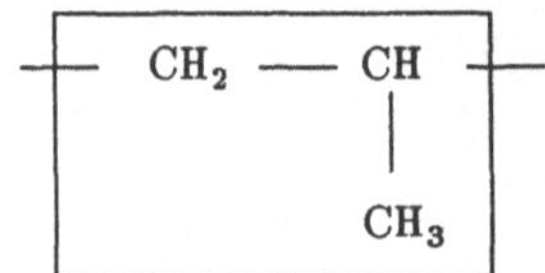

In Section 2.2 and 2.3 we will deal with an ideal isothermal solidification, while in Sec. 2.4 we will sketch the treatment of the complete model.

In this first approach we will neglect motion (which also affects the crystallization process), mechanical effects, polymorphism of crystals, etc.

2.2. ISOTHERMAL SOLIDIFICATION: THE MODEL

Consider a sample of liquid polymer at a temperature $\theta\in(\theta_g,\theta_m)$ and assume that it is possible to keep this temperature constant throughout the process. It can be observed that crystallization is a result of two phenomena: nucleation, i. e. the appearance of elementary crystals (nuclei) in the melt, and growth of almost spherical crystalline aggregations (spherulites) around the original nucleus. Moreover, it can be seen that there is a fraction of the melt which does not crystallize even after a long time. We always refer to a unit volume and define the crystallinity $w(t)$ as the volume of the crystallized part at time t. The above statement means that there is a maximum value w_m of the crystallinity which cannot be exceeded.

Summing up, we have to characterize at each temperature four quantities describing the process: the volume of the nuclei, the maximum crystallinity, the nucleation rate, the growth rate.

Below, we discuss briefly these four quantities, referring the reader wishing a comprehensive exposition of the matter to the recent work by Janeschitz-Kriegl and coworkers [8].

Volume of the nuclei. From thermodynamical considerations it is possible to characterize the critical radius below which a nucleus would be unstable. We will denote it by r_0. In non-isothermal processes its dependence on the temperature should be taken into account.

Maximum crystallinity. This value $w_m \leq 1$ is a characteristic of the polymer and will also depend on temperature. The existence of a fraction $1-w_m$ of non-crystalline solid is related to two main effects: the existence of oligomers, i.e. of shorter chains of molecules, which do not crystallize, and geometric effects connected to the lamellar structure of the spherulites.

Nucleation. At a given temperature nuclei appear everywhere in the melt. The rate of birth of new nuclei per unit volume at time t is a function of crystallinity $w(t)$ - since no nuclei can appear in the already crystallized part - and possibly of t. An explicit dependence on time is assumed e. g. in the case of the so called heterogeneous nucleation: one postulates that the melt contains from the beginning potential "loci" where nucleation can take place provided that the loci are "activated". Supposing that the probability of activation follows the law of radioactive decay with a time of activation τ_a and denoting by $N(t)$ the number of nuclei formed at time t, one has

$$\frac{dN}{dt} = \dot{N}_o\,(1-w)\,e^{-t/\tau_a} \tag{2.2.1}$$

where the term $1-w$ is introduced (assuming $w_m=1$) to express the fact that new nuclei can appear just in the

uncrystallized fraction 1-w.
Following [8] a spectrum of activation times for different kinds of "loci" is introduced thus writing

$$\frac{dN(t)}{dt} = \sum \dot{N}_i (1-w) e^{-t/\tau_i} .$$

Growth. It is generally assumed that, at a given temperature, the radial growth rate of the spherulites is constant as far as they do not impinge. Then a virtual crystallinity W is introduced as the value which would be obtained keeping this growth rate constant throughout the process (and thus neglecting impingement). A correction is introduced to take impingement into account. Following complicated geometrical arguments, Avrami [3] writes

$$\frac{dw}{dt} = (1-w)\frac{dW}{dt} . \tag{2.2.2}$$

If one assumes $\dot{N}=\dot{N}_1$ constant and $r_o=0$, for a three dimensional case it is clearly $W=\frac{4}{3}\pi\dot{N}_1\dot{R}_0^3\frac{t^4}{4}$ where $\dot{R}_o$ is the constant radial growth rate. According to (2.2.2) the so-called Avrami's rule is obtained

$$w(t) = 1-e^{-kt^{n+1}} \tag{2.2.3}$$

where n is the number of dimensions; (2.2.3) has been widely used mainly because of its simplicity (some "generalizations" of (2.2.3) have also been proposed, just on experimental basis, substituting n by an empirically found constant).
But Avrami's approach does not seem to be satisfactory at high crystallinity; thus different arguments have been used to derive a-posteriori corrections to pass from W to w (see [13], [20]). In any case the information of the distribution of the radii of the spherulites is lost, so that other Authors preferred to by-pass the microscopic approach and introduced directly a macroscopic evolution law of the form $\dot{w}=f(w)$ (see [4], [16], [19]).

In the model we propose, the philosophy is in some sense reversed: an a-priory correction is introduced on the growth rate of all the spherulites, so that their radial growth proceeds with a speed $\dot{R}_0$ $f(w,t)$. We assume that $f(0,t)=1$, $f(w_m,t)=0$ and that the function is monotonically decreasing in w.
Similarly, we will assume that the rate of nucleation is given by $\dot{N}_0 g(w,t)$ where g has the same behaviour of f. Of course both $\dot{R}_0$ and $\dot{N}_0$ will be temperature-dependent.
Accordingly, the following equation is easily obtained:

(2.2.4) $$\dot{w}(t) = \beta_n r_0^n \dot{N}(t) + n\beta_n \dot{R}(t)\int_0^t \dot{N}(\tau)\, r^{n-1}(t,\tau)\, d\tau,$$

where n=1, 2 or 3 is the number of dimensions and β_n is a dimensional constant ($\beta_1=2$, $\beta_2=2\pi$, $\beta_3=4\pi$) while $r(t,\tau)$ is the radius at time t of the spherulite which has born at time τ.

Since in our picture the growth speed of all the equivalent spherulites is the same at a given time, the function $\frac{\partial r(t,\tau)}{\partial t}$ is independent of τ and has been called $\dot{R}(t)$. Note that $\frac{\partial R(t,\tau)}{\partial \tau}=-\dot{R}(\tau)$, since $r(t,\tau)=r_0+\int_\tau^t \dot{R}(s)\, ds$.

Introducing the assumption $\dot{R}=\dot{R}_0 f(w,t)$, $\dot{N}=\dot{N}_0 g(w,t)$ we have the integro-differential equation for w(t)

(2.2.5) $$\dot{w} = \beta_n r_0^n \dot{N}_0 g(w,t) + n\beta_n \dot{N}_0 \dot{R}_0 f(w,t)\int_0^t g(w(\tau),\tau)\Big(r_0 + \dot{R}_0\int_\tau^t f(w(s),s)\, ds\Big)^{n-1} d\tau$$

(2.2.6) $$w(0) = 0.$$

Again, (2.2.5) is a model whose solution at time t depends on the values taken by f and g on the whole interval (0,t).

2.3. ISOTHERMAL SOLIDIFICATION: IDENTIFICATION OF THE PARAMETERS

Let us consider the special case in which

(2.3.1) $$\dot{R}(t) = \dot{R}_0 f(w) , \quad f(0) = 1 , \quad f(w_m) = 0 ,$$

(2.3.2) $$\dot{N}(t) = \dot{N}_0 g(w) , \quad g(0) = 1 , \quad g(w_m) = 0 .$$

Moreover, let us assume $g(w)=cf(w)$, so that

(2.3.3) $$\dot{N}(t) = c\dot{R}(t) ,$$

and

(2.3.4) $$\dot{w} = \beta_n r_0^n c\dot{R} - cn\beta_n \dot{R}(t)\int_0^t \dot{R}(\tau)\,(r(t,\tau))^{n-1} d\tau = \beta_n c\, r^n(0,t)\,\dot{R}(t) .$$

It is easily found that

(2.3.5) $$w(t) = \frac{\beta_n c}{n+1}\left(r^{n+1}(t,0) - r_0^{n+1}\right) ,$$

(2.3.6) $$w(t) = \frac{\beta_n c}{n+1}\left\{\left(\frac{N(t)}{c}+r_0\right)^{n+1} - r_0^{n+1}\right\} .$$

According to (2.3.4) and (2.3.5) the crystallinity is simply related to the radius of the largest spherulites or to their number.

Furthermore, if one assumes for f(w) the form

(2.3.7) $$f(w) = (1-w/w_m)^q ,$$

one has from (2.3.4) and (2.3.5)

(2.3.8) $$\dot{w} = c\beta_n(1-\tfrac{w}{w_m})^q(k_n w + r_0^{n+1})^{n/(n+1)} ,$$

where $k_n=(n+1)/\beta_n c$.

It has been pointed out by Mazzullo [15] that for the case $r_0=0$ this corresponds to a law similar to the empirical equation used in [4]

(2.3.9) $$\dot{w} = A w^{1-\gamma}(1-w/w_m)^{1+\gamma} ,$$

if $q=(n+2)/(n+1)$.

Let us come back to (2.2.5) in the special case

(2.3.10) $$f(w) = g(w) = 1-w/w_m$$

and denote for simplicity

$$\delta = 1-w/w_m .$$

We have

(2.3.11) $$-w_m\dot{\delta} = AB^n\delta(t) + nA\delta(t)\int_0^t \delta(\tau)\left(B+\int_\tau^t \delta(s)\,ds\right)^{n-1} d\tau ,$$

(2.3.12) $$\delta(0) = 1 .$$

Here, $A=\beta_n N_0 \dot{R}_0^n$ is essentially the crystallized volume after one second and $B=r_0/\dot{R}_0$ is the ratio between the radius of the elementary nuclei and their increase after 1 sec. These are the two quantities which have to be identified by experiments.

In figures 7 and 8 the squares denote the experimental values obtained in various isothermal solidification processes, while the solid curve is found fitting the points with expression (2.3.11) (for n=3). The experiments consist in cooling the melted polymer down to the desired temperature and to measure the evolution of the crystallinity. It can be seen that the curves fit the experiments quite well for temperature $\theta=134°C$ while the fit is worse for $\theta=124°C$. In our opinion, this is due to the fact that some crystals are formed in the first stage of the experiment (the "quench") so that at the begining of the isothermal process we have a non vanishing crystallinity.

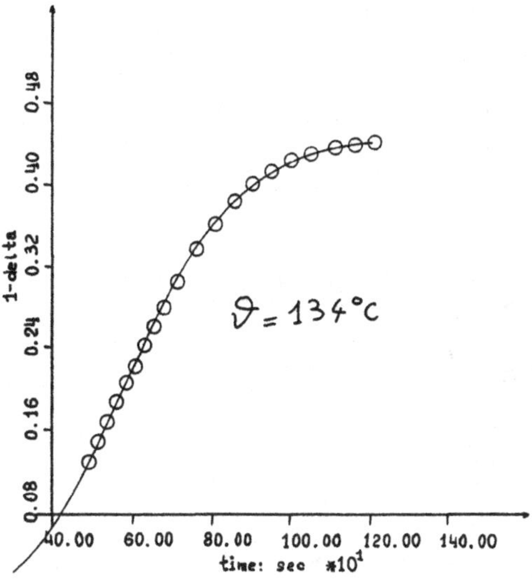

Figure 7

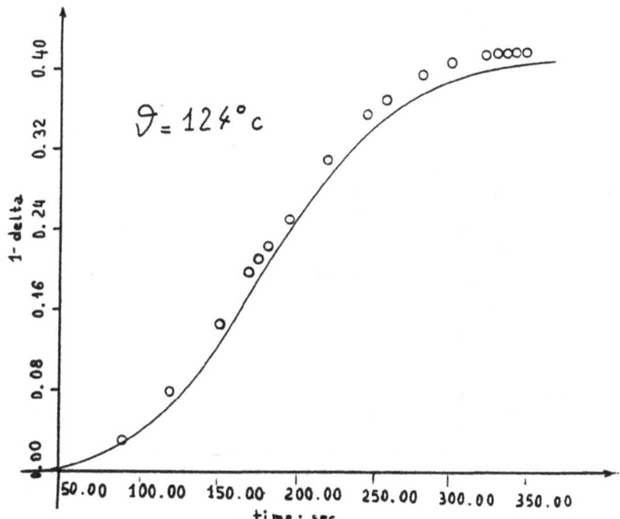

Figure 8

In Figures 9 and 10 we show how our model can also give an information on the distribution of the radii of the equivalent spherulites, an information which is lost in the classical models using a-posteriori corrections to take impingement into account.

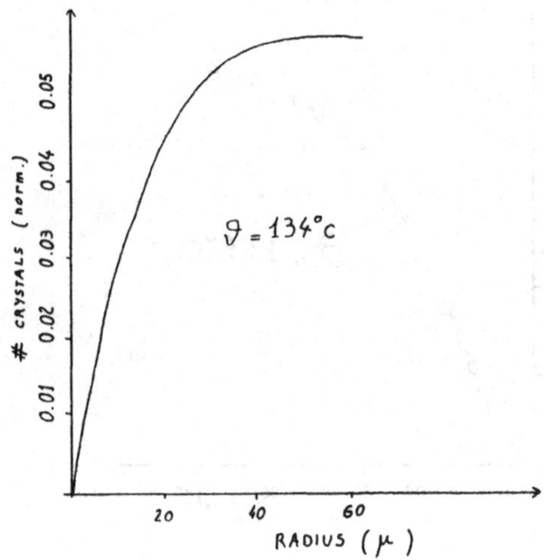

Figure 9

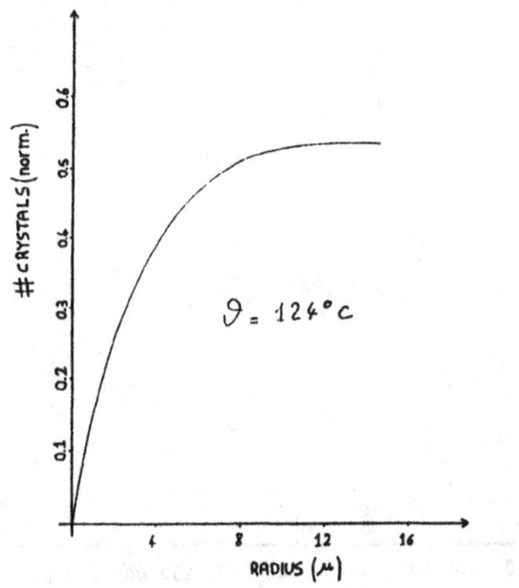

Figure 10

Let us conclude this Section with a short remark concerning other phenomena for which a similar model could be appropriate.

First, the solidification of magmas studied in vulcanology (see. e. g. [5]). In this case just a small fraction of the melt actually crystallizes and thus the influence of the impingement corrections is less relevant. On the other hand the dependence of the parameters on the temperature seems to be more complicated, but the mathematical form of the model should be the same as in the case discussed above.

Second, a large class of solid-solid transitions (as for instance the austenite-pearlite transition in steel); the mathematical models currently used are based on the Avrami's rule (2.2.3) or on its generalizations. The main problem is that, even in cases in which (2.2.3) holds for isothermal situations, it can be extended to non-isothermal processes only if the ratio between the nucleation rate and the virtual radial growth rate is constant (the so-called isokinetic assumption).

Since most solid-solid trasition do not seem to obey the isokinetic assumption, suitable approximations have been proposed to generalize the concept of isokinetic processes and the applicability of Avrami's rule (see [6] and [12] for a concise review). Thus, the so-called additivity rule [18] has been introduced giving rise to an interesting mathematical problem (see [21] for the analysis of its well- posedness).
We believe that a more appropriate modellization of this class of phenomena could be based on the general approach we described above, and we expect that a better understanding of the crystallization of polymers and the verification (or falsification) of our model will be helpful also in the description of austenite-pearlite transitions.

By the way, the transparency of polymers allows to carry out experiments based on optical methods and thus to follow the microscopic evolution of the phenomenon and not just of quantities averaged over a macroscopic volume.

2.4. NON ISOTHERMAL CRYSTALLIZATION

We assume that the model described in the previous two sections holds over the whole range of temperatures between θ_g and θ_m. Thus our first goal is to identify the two parameters A and B over the whole range of temperatures.

Unfortunately, reliable isothermal experiments can be performed over a rather small interval of temperatures; indeed, approaching θ_m makes the experiment extremely slow and not easily reproducible, while approaching θ_g gives rise to the difficulty we pointed out in the discussion of fig.9.

Following classical thermodynamic arguments (see [14], [22]), we postulated for A and for B in (2.3.11) a temperature dependence of the form

$$\alpha \exp[-a/(\theta-\theta_g)-b/(\theta_m-\theta)] \quad ,$$

where a and b are constants related to the enthalpy of crystallization of the pure crystal and to the free energy of activation governing the short distance diffusion of the crystallizing molecules across the phase boundary; and we extrapolated from the values of A and B obtained by fitting the curves of isothermal crystallization.

The values $A(\theta)$ and $B(\theta)$ obtained will be used to predict the crystallization as a function of time which will be obtained in experiments in which the temperature is controlled (but no longer constant), e. g. with a constant cooling rate. The work is still in progress, but the preliminary results are good.

The final step is to consider a complete problem in which the crystallization is coupled with the evolution of the thermal field in the polymer. Actually the temperature of a crystallizing sample can be controlled in small scale experiments using thermostats which can absorb almost istantaneously the latent heat released during the crystallization.

In practical cases these variations are by no means negligible and the problem to be solved is

$$c\theta_t - \mathrm{div}(k \,\mathrm{grad}\theta) = \lambda w_t \tag{2.4.1}$$

where c and k are the heat capacity and the conductivity (which, in principle, could depend on w as well) and λ is the latent heat of solidification.

Equation (2.4.1) has to be coupled with (2.2.5) where r_0, w_m, $\dot{N}_0$, $\dot{R}_0$ are prescribed functions of θ and with (2.2.6) and the appropriate initial-boundary conditions for θ.

Of course the model is history-dependent and non standard. We have investigated its well-posedness in the special case w_m=constant and when f and g are proportional to δ. Since the constants of proportionality are allowed to depend on the temperature in different ways, we are not in a case in which the so-called isokinetic approach is justified and thus, even in the simpler case $r_0=0$, we cannot use a macrokinetic equation of the form e. g. of (2.3.9).

We write (2.2.5) in the form (n=3)

$$-\frac{1}{w_m}\frac{\partial\delta}{\partial t} = n_1\delta(t) + n_2\delta(t)\int_0^t n_3\delta(\tau)\left(n_4 + \int_\tau^t n_5\delta(s)\,ds\right)^2 d\tau$$

where the n_i are given functions of position, time and temperature. We assumed

$$n_i \geq 0\ , \qquad i=1,\ 2,\ \ldots.\ 5$$

$$n_i(x,t,.) \in W_1^\infty(\mathbf{R})\ ,$$

$$n_i(.,.,\theta) \in H^{\alpha,\frac{\alpha}{2}}(Q_T)\ ,$$

where $Q_T=\Omega\times(0,T)$, $T>0$ and Ω is the domain where the problem is considered.

Under suitable regularity assumptions on k, c, λ and on the initial and boundary data we proved a well-posedness result for arbitrary T. The solution is essentially classical: $\theta\in H^{2+\beta,1+\beta/2}(\bar{Q}_T)$, $w\in H^{\beta,\beta/2}(\bar{Q}_T)$.

The proof of the existence theorem can be found in [1] and it will not be duplicated here.

REFERENCES

[1] Andreucci, D., Fasano, A. and Primicerio, M. (1990) 'On a mathematical model for the crystallization of polymers', Proocedings of the 4th ECMI Meeting, Kluwer Acad. Publ., Dordrecht.

[2] Astarita, G. and Marrucci, G. (1974) 'Principles of Non-Newtonian Fluid Mechanics', Mac. Graw-Hill, London.

[3] Avrami, M. (1939) 'Kinetics of Phase Change I', J. Chem. Phys. 7, 1103-1112 (1939); see also the same journal 8, 212-224 (1940); 9, 117-184 (1941).

[4] Berger, J. and Schneider, W., 'A zone model of rate controlled solidification', Plastics and Rubber Processing and Applications 6, 127-133 (1986).

[5] Brandeis, G. (1984) 'Nucleation, Crystal Growth and the Thermal Regime of Cooling Magmas', J. of Geophysical Res. 89, 10161-10177 (1984).

[6] Christian, J. W. (1965) 'The Theory of Transformations in Metals and Alloys', Pergamon Press, Oxford.

[7] Comparini, E. et al. (1987) 'Un modello matematico per la reologia di CWS.', Report Dept. Math. University of Florence, 1987.

[8] Eder, G., Janeschitz-Kriegl, H. and Liedauer, S. (1989) 'Crystallization processes in quiescent and moving polymer melts under heat transfer conditions', Report Institute of Chemistry, Univ. of Linz, (1989).

[9] Ercolani, D. et al. (1988) 'Shear degradation of concentrated coal-water slurries in pipeline flow', Proceedings of the 13th. Conference on Slurry Technology, Denver 1988.

[10] Ercolani, D. et al. (1988) 'Effect of mechanical and thermal energy on CWS rheology', Proceedings of the 11th Conference Hydrotransport of Solids, Stratford 1988.

[11] Fasano, A. and Primicerio, M. (1990) 'Modelling the rheology of coal-water slurries', Proceedings of the 4th ECMI Meeting, Kluwer Acad. Publ., Dordrecht.

[12] Hayes, W. J. (1985) 'Mathematical models in material sciences', M. Sc. Thesis, Oxford 1985.

[13] Malkin, A. YA. et al. (1984) 'General treatment of polymer crystallization kinetics', Parts 1 and 2, Polym. Eng. Sci., 24, 1396-1401, 1402-1408 (1984).

[14] Mandelkern, L. (1964) 'Crystallization of Polymers', Mac Graw-Hill, New York.

[15] Mazzullo, S. (1990) 'Modello di granulazione, accrescimento ed arresto della cristallizzazione nei polimeri semicristallini', Report Himont Italia, 1990.

[16] Mazzullo, S., Paolini, M. and Verdi C. (1989) 'Polymer crystallization and processing: free boundary problems and their numerical approximation', Proceedings of the 3rd ECMI Meeting, Kluwer Acad. Publ., Dordrecht.

[17] Primicerio, M. (1988) 'Dynamics of slurries', Proceedings of the 2nd ECMI Meeting, Kluwer Acad. Publ., Dordrecht.

[18] Scheil, E. (1935) 'Anlaufzeit den Austenitimwandung', Arch. fur Eisenhuttenwesen 8 (1935) 565-579.

[19] Schneider, W. et al. (1988) 'Non-isothermal crystallization of polymers', Intern. Polymer Proceessing II 314, 151-154 (1988).

[20] Tobin, M. C.,'Theory of phase transition kinetics with growth site impingement', I and II, J. Polym. Sci., 12, 394-406 (1974) and 14, 2253-2257 (1976).

[21] Visintin, A., 'Mathematical models of solid-solid phase transition in steel', IMA J. Appl. Math. 39 (1987), 143-157.

[22] Wunderlich, B. (1976) 'Macromolecular Physics' volume 2 Academic Press, New York.

A STOCHASTIC PARTICLE SYSTEM MODELLING THE BROADWELL EQUATON

Mario PULVIRENTI

Dipartimento di Matematica,Università di L'Aquila,L'Aquila (Italy)

A discussion on the Boltzmann-Grad limit for a stochastic particle system yelding the Broadwell equation is presented.

The dynamics of the molecules of a rarefied gas obeying the Newton laws is usually,more conveniently,described by the Boltzmann equation.A rigorous connection between the Newton dynamics and the Boltzmann equation is an outstanding problem not yet fully proven.

By an inspection of the equation governing the motion of a system of N billiard balls (see Ref.1 for a detailed analysis) it follows that the Boltzmann equation is the formal limit,when $N\to\infty, d\to 0, Nd^2\to$const (here d denotes the diameter of the spheres) of such equation,provided that the factorization property for the correlation functions is fulfilled.To give a rigorous proof of the validity of the Boltzmann equation,one has to prove that the Newton dynamics converge,in the above limit (usually called the Boltzmann-Grad limit) to the solutions of the Boltzmann equation.

The solution of the above problem is a very difficult task and,up to now,very few results are known.The first,due to Lanford,shows the validity of the Boltzmann equation for short times.Successively,R.Illner and the author of this note,proved the same result,globally in time,for a cloud of rare gas in the vacuum [3].In both cases the average number of collisions is small and this allows the control of the nonlinear term in the Boltzmann equation.Thus most of the interesting cases are not covered by the above results so that the problem of the validity of the Boltzmann equation seems far to be understood.

In this situation may be of some interest the study of simpler models yielding in (the analogue of) the Boltzmann-Grad limit other kinetic equations.I am going to present a result due to S.Caprino,A.De Masi,E.Presutti and myself [4] concerning a rigorous derivation of the Broadvell equation starting from a stochastic system of particles in a lattice.The model is the following.

In a box of side one $\Lambda=[0,1]^2$ consider the lattice:

$$\Lambda_\varepsilon=\{q=(q_x,q_y)\in\Lambda \mid q_x=n_1\varepsilon, q_y=n_2\varepsilon \quad n_i=0......\varepsilon^{-1}\} \qquad 1)$$

($\varepsilon^{-1}=K$, K integer to be fixed later) and a system of N (identical) particles whose configurations are denoted by:

$$X=\{x_1,...,x_N\}$$

$$x_i=(q_i,v_i) \qquad q_i\in\Lambda_\varepsilon, v_i\in\Omega \qquad 2)$$

where q_i and v_i are the position and the velocity of the i-th particle and

$$\Omega=\{(0,1);(1,0);(0,-1);(-1,0)\} \qquad 3)$$

is the velocity space,consisting of four elements.

We denote by $\Gamma_\varepsilon=\Lambda_\varepsilon\times\Omega$ the phase space of a single particle and,accordingly,Γ_ε^N denotes the phase space of our system.To avoid boundary conditions, we assume periodicity.

R. Spigler (ed.), Applied and Industrial Mathematics, 173–177.

The dynamics of the system is a stochastic process,valued in Γ_ε^N,described in the following way .The free motion of each particle is a Poisson process of intensity ε^{-1} associated to the transition:

$$(q_i,v_i) \rightarrow \qquad (q_i+v_i\varepsilon,v_i) \tag{4}$$

All these Poisson processes are independent.

The interaction among the particles is described by N(N-1)/2 independent Poisson processes of intensity one (each for any pair of particles) associated to the transition:

$$(q_i,v_i,q_j,v_j) \rightarrow \qquad (q_i,v_i^\perp,q_j,v_j^\perp) \tag{5}$$

where:

$$v^\perp=(-v_2,v_1) \text{ if } v=(v_1,v_2) \tag{6}$$

and the transition 5) takes place only if the particles involved are in the same site oif the lattice.

We want to investigate the behavior of the system in the limit:

$$\varepsilon\rightarrow 0, N\rightarrow\infty, N\varepsilon^2\rightarrow 1.$$

In such a limit the number of particles per site is expected to be finite so that each particle undergoes a finite number of collisions per unit time.On the other hand the displacement per unit time of each particle is finite since the generator of the process 4) converges,at least formally,to the generator of the free stream.

For a probability measure μ_ε which is symmetric in the exchange of particles,we introduce the rescaled correlation functions:

$$\rho_j^\varepsilon(x_1....x_j) =$$

$$(N\,\varepsilon^2)^{-j}\, N(N-1).......(N-j+1) \sum_{x_{j+1}....x_N} \mu_\varepsilon(x_1....x_j x_{j+1}....x_N) \tag{7}$$

Proceding as in the derivation of the BBGKY hierarchy for Hamiltonian systems,we obtain:

$$\frac{d}{dt}\rho^\varepsilon{}_j=(\varepsilon^{-1}L_0+L_I)\rho^\varepsilon{}_j+\varepsilon^2 N C_{j,j+1}\rho^\varepsilon{}_{j+1} \tag{8}$$

where

$$C_{j,j+1}\rho^\varepsilon{}_{j+1}(x_1....x_j)= \sum_{i=1}^{j}\{\rho^\varepsilon{}_{j+1}(x_1,....q_i\, v_i^\perp,....x_j\, ,q_i\, -v_i^\perp)-$$

$$\rho^\varepsilon{}_{j+1}(x_1,.q_i\, v_i,...,x_j\, ,q_i\, -v_i\,)\} \tag{9}$$

L_0 and L_I are the generators of the processes 4) and 5) respectively:

$$L_0f(X)=\sum_{i=1}^{|X|} \{f(X_i)-f(X)\} \tag{10}$$

$$L_I f(X)=\frac{1}{2}\sum_{i;j=1;i\neq j}^{|X|} \{f(X_{ij}^{\perp})-f(X)\}\chi_{ij}(X). \tag{11}$$

where $|X|$ is the cardinality of X and:

$$X_i=\{x_1,..\ (q_i-v_i\varepsilon,v_i).....x_{|X|}\} \tag{12}$$

$$X_{ij}^{\perp}=\{x_1,..\ (q_i,v_i^{\perp})...(q_jv_j^{\perp}).....x_{|X|}\} \tag{13}$$

Finally χ_{ij} denotes the characteristic function of the event in which the particles i and j are in the same site with opposite velocities.

The formal limit $\varepsilon\to 0, N\to\infty, N\varepsilon^2\to 1$ of the Eq.s 8) yields the following infinite hierarchy:

$$\frac{\partial}{\partial t}g_j=-(\sum_{i=1}^{j} v_i\cdot\partial_{q_i})g_j+C_{j,j+1}g_{j+1} \tag{14}$$

which we call the Broadwell hierarchy.The reason why the term $L_I\rho^{\varepsilon}_j$ disappears in 14) is due to the fact that the flow generated by $\varepsilon^{-1}L_0$ gives vanishing probability to configurations in which two tagged particles sit in the same site. We have called the set of Eq.s 14) the Broadwell hierarchy because,denoting by g any solution of the two-dimensional Broadwell equation,

$$\frac{\partial}{\partial t}g(q,v) =- v\cdot\partial q\ g(q,v) +[g(q,-v^{\perp})g(q,v^{\perp})-g(q,v)g(q,-v)], \tag{15}$$

The products:

$$g_j(x_1....x_j;t)=\prod_{i=1}^{j}g(x_i;t) \tag{16}$$

solve the Broadwell hierarchy 14) .

The target is to prove rigorously the above limit.We make the following hypothesis at time zero.Consider $g_0\in C^1(\Gamma)$ a positive initial value for the Broadwell equation 15) satisfying the normalization condition $\sum_v\int g_0(q,v)dq=1$.Let μ^{ε}_0 be the probability measure on Γ_{ε}^N defined as follows:

$$\mu^{\varepsilon}_0(x_1,...,x_N)=\prod_{i=1}^{N}\ \int_{\Delta(q_i)} g_0(q,v_i)dq \qquad x_i=(q_i,v_i) \tag{17}$$

where $\Delta(q_i)$ is the atom of the partition in squares induced by the lattice Λ_{ε} ,whose left low corner is q_i.

Let μ^{ε}_t be the distribution of the process $\{x_1,......,x_N\}$ at time t .Consider the associated rescaled correlation functions ρ^{ε}_t .Then:

Theorem (see Ref.4)

Let $g \in C^1([0,T],\Gamma)$ be a solution of 15) with initial datum g_0 and $g_{j,t}$ given by 16). Then:

$$\lim \| \rho^{\varepsilon}_{j,t} - g_{j,t} \|_{\infty} = 0 \quad \text{as} \quad \varepsilon \to 0, N \to \infty, N\varepsilon 2 \to 1 \qquad 18)$$

uniformly in $t \in [0,T]$.

In the above Theorem the existence of a classical solution to the Broadwell equation is assumed.This is known only for very special cases,being the existence problem for such an equation even more involved than that for the Boltzmann equation.Such an assumption,however,does not make trivial the validity problem,as I shall try to explain.Following Lanford [2],it is not hard to prove the above Theorem for short times.The knowledge of a global in time solution for the limit equation suggest to try to iterate the convergence argument.Here we find a deep difficulty.Actually writing down an equation for the differences:

$$\Delta_j(t) = g_{j,t} - \rho^{\varepsilon}_{j,t} \qquad 19)$$

one finds the estimate,valid for short times $t<t^*$:

$$\| \Delta_j(t) \|_{\infty} \leq C^j \sqrt{t\varepsilon} \qquad 20)$$

where C increases with $\sup_{j,t} (\| \rho^{\varepsilon}_{j,t} \|_{\infty})^{1/j}$ .The convergence expressed by estimate 20) does not give us a uniform control on j (and it cannot be elsewhere) so that we have not a sufficiently good control on $\rho^{\varepsilon}_{j,t^*}$ to iterate the argument.

To overcome this difficulty we introduce the folloving family of functions:

$$u_t(X) = \sum (-1)^{|X/Y|} g_t(X/Y)\, \rho^{\varepsilon}_t(Y) \qquad 21)$$

where $X=\{x_1.....x_j\}$ and the sum is performed over all subset Y of X.Obviously this definition reduces to:

$$u_t(X) = \prod_{i=1}^{j} (g_t - \rho^{\varepsilon}_{1,t})(x_i) \qquad 22)$$

if $\rho^{\varepsilon}_{j,t}$ would strictly factorize.Therefore,at fixed time, the u's are expected to be of order $c^j \omega(\varepsilon)^j$,where $\omega(\varepsilon)$ is infinitesimal in ε.The control on the size of the functions u,gives a measure of the tendency of the system,not only to converge to the limiting kinetic behavior,but also to gain the factorization property.Therefore a short time control on the u-functions could,at least in priciple,have the chance to be iterated to reach arbitrary times.However a good control of the u-functions,even for short times,is problematic.Following the definition one arrives to an evolution equation for the u-functions which is very difficult to handle with.However,taking in mind the probabilistic meaning of the u-functions (which are essentially the expectation of the deviation of the normalized occupation number of the process from from the solution g),one can estimate,by means of an iterative procedure,the probability that a sample of the process is far from the right kinetic behaviour and find that this is small.

The techniques exploited in 4 cannot be applied to deterministic systems or even to continuous systems for both of which new ideas are needed.

For an up to date review concerning kinetic and hydrodynamical limits for stochastic particle systems the reader is addressed to ref 5.

REFERENCES

1.C.Cercignani,Theory and Application of the Boltzmann Equation.Scottish Ac.Press (1975)

2.O.Lanford,Time Evolution of Large Classical Systems,Moser E.J. ed.Lecture Notes in Physics.Vol.38 pp.1-111 Springer (1975)

3.R.Illner and M.Pulvirenti Comm.Math Phys. 121 pp. 143-146 (1989)

4.S.Caprino,A.De Masi,E.Presutti and M.Pulvirenti,A Derivation of the Broadwell Equation. CARR report in Math Phys n.22/89 (1989)

5.A.De Masi and E.Presutti,Lecture on Collective Behavior of Particle Systems CARR report in Math Phys n.5/89 (1989)

THEORY AND APPLICATION OF STEKLOV-POINCARÉ OPERATORS FOR BOUNDARY-VALUE PROBLEMS

A. QUARTERONI* and A. VALLI**

* Dipartimento di Matematica del Politecnico, Milano, and Istituto di Analisi Numerica del C.N.R., Pavia, Italy

** Dipartimento di Matematica, Università di Trento, Povo, Italy

1. Introduction.

This paper deals with interface operators in boundary value problems: how to define them, which is their meaning in both mathematical and physical sense, how to use them to derive numerical approximations based on domain decomposition approaches.

When matching partial differential equations set in adjacent subregions of a domain Ω of $\mathbb{R}^n$, the interface operators ensure the fulfillement of transmission conditions between the different solutions. From the mathematical side, they make it possible to reduce the overall boundary value problem into a subproblem depending solely on the trace of the solution upon the interface. Once the solution of such a problem is available, the original solution can be reconstructed through the solution of *independent* boundary value problems within each subregion. The above independency feature is very likely behind the increasing interest for the use in the recent years of interface operators in scientific computing. Indeed, the subdomain approach yields a problem to be solved for the interface gridvalues only, then a family of reduced problems are left to solve simultaneously, hopefully by a multiprocessor architecture.

Following such an approach, the keystep is how to solve effectively the interface problem. The latter, for a linear boundary value problem, is nothing but a linear system for the so called capacitance matrix (the finite dimensional counterpart of the interface operator).

One can introduce interface operators in a variety of situations. In some of them, interface is merely a mathematical artifice. This is, e.g., the case of constant coefficient equations (such as the Poisson equation, or the Stokes equations with constant viscosity) in a domain with regular shape. In other situations, on the contrary, interfaces do really exist and they are due to very significant variations either in the geometry of the domain or in the physics of the problem. The latter is the case of interfaces separating regions with different flow regime (viscous and inviscid), or with different thermal conductivity (yielding advection-diffusion or simply advection equations). There are many examples of this type in fluidmechanics, aerodynamics and thermodynamics.

We want here to introduce the interface operators (also known as Steklov-Poincaré ope-

R. Spigler (ed.), Applied and Industrial Mathematics, 179–203.

rators, see, e.g., [Ag]) for several type of problems; to characterize them in terms of global differential operators through suitable prolungation operators, and to study their most relevant properties.

From the theoretical point of view, this will enable us to prove for instance existence results for boundary value problems arising from the coupling of equations of different character.

From the numerical point of view, any finite dimensional approximation to the Steklov-Poincaré problem based on a particular numerical method (such as finite elements, finite differences, or spectral methods) yields the so called capacitance matrix system. Since the Steklov-Poincaré operators are generally unbounded, the associated capacitance matrix is ill conditioned. Iterative procedures are henceforth in order, and preconditioning techniques owe to be adopted, otherwise the convergence rate would deteriorate as far as the dimension of the algebraic system goes up. Preconditioners are usually obtained after the Steklov-Poincaré operator is split into terms attributable to the different subregions.

The choice of a convenient iterative method is a matter of which properties are enjoyed by the Steklov-Poincaré operator at hand (symmetry, positivity, etc.). We will focus from case to case to fixed point iterations, preconditioned Richardson iterations, preconditioned conjugate gradient iterations.

Furthermore, we will provide interpretation of the above iterative processes as iterative methods among subdomains, yielding problems of the same form as the one originally prescribed. In the limit one gets a numerical solution which most often coincides with the genuine global single-domain solution. In some other cases, however, the limit solution achieved via the Steklov-Poincaré approach is a novel one. This approach is therefore useful in such cases to devise new numerical methods to approximate the prescribed boundary-value problem.

An outline of the paper is as follows. Section 2 traces the guidelines of our investigation. Sections 3, 4 and 5 address respectively to elliptic equations, Stokes equations for viscous flows, and generalized Stokes equations for both viscous and inviscid compressible flows. Finally, in Sections 6 and 7 we introduce numerical approximations and effective iteration-by-subdomains solvers which are well suited for parallel computation.

2. Position of the problem and notations.

The geometric situation we will consider is like the one depicted in Fig. 2.1:

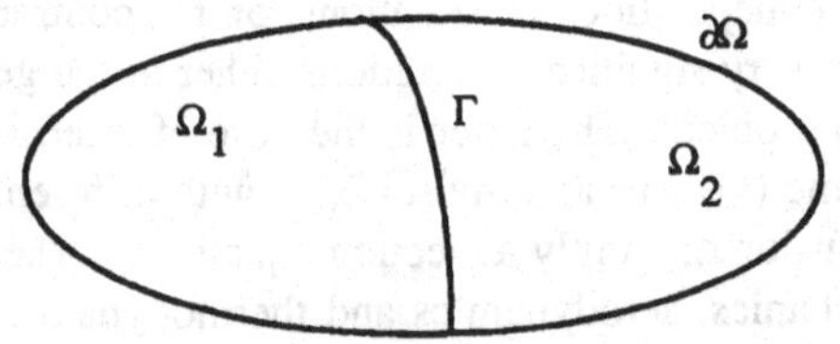

Fig. 2.1. Decomposition of $\Omega : \overline{\Omega} = \overline{\Omega}_1 \cup \overline{\Omega}_2$

We are given a differential problem in Ω

$$Lu = f \quad in \quad \Omega \ , \tag{2.1}$$

supplemented with suitable boundary conditions on the boundary $\partial\Omega$. To the problem (2.1) we associate two boundary value problems

$$Lu_1 = f \qquad in \ \ \Omega_1 \ , \qquad Lu_2 = f \qquad in \ \ \Omega_2 \tag{2.2}$$

and we require that u_k satisfies upon $\partial\Omega_k \cap \partial\Omega$ the same boundary condition as u, for $k = 1, 2$.
We are concerned with the problem of finding proper matching conditions between u_1 and u_2 to be fulfilld at the interface $\Gamma = \partial\Omega_1 \cap \partial\Omega_2$ in order that the split problem (2.2) be equivalent to the original problem (2.1).
When the above problems are set in a correct mathematical framework, some of the transmission conditions arise a priori from the fact that u is the solution to the overall problem (hence u is sought in a precise function space), and that u_k must be the restriction of u to $\Omega_k, k = 1, 2$. The remaining transmission conditions need to be enforced directly through the so-called Steklov-Poincaré operator. The latter, denoted by S, depends on the differential problem at hand, and it allows the determination of λ (the unknown solution u at the interface) via a problem of the form

$$S\,\lambda = \chi \ . \tag{2.3}$$

Since χ is a given function, the above problem can be solved autonomously, prior to (2.2). (Let us notice that, once the solution λ to the Steklov-Poincaré problem is available, the two reduced problems (2.2) supplemented with this boundary condition on Γ can be solved in an independent fashion).
In general the Steklov-Poincaré operator can be split into two terms, say S_1 and S_2, pertaining to the subdomains Ω_1 and Ω_2, respectively. In all cases we consider hereafter, one has precisely

$$S = S_1 + S_2 \tag{2.4}$$

Owing to (2.4), the analysis of S can be carried out directly on the operators L (actually, on its associated bilinear form) within Ω_1 and Ω_2.
On all cases considered hereafter, S satisfies the following property (P):

(P) *Both S_1 and S_2 (and therefore S) are positive and self-adjoint operators. Moreover, they induce two equivalent inner products.*

In view of finite dimensional approaches to the problem (2.1) (or equivalently, (2.2)-(2.3)), the above properties drive the choice of effective iterative algorithms for the solution to the finite dimensional counterpart of (2.3) (see Sections 6, 7).
Sections 3, 4, 5 focus on different boundary value problems. For each of them, we will try to follow strictly the guidelines here traced, and we will make precise from case to case the equivalence statement between (2.1) and (2.2)-(2.3), the structure of the operators S, S_1, S_2 as well as their most relevant properties.
Let us introduce now some notations. We will indicate by $H^s(\Omega)$ and $H^s(\Sigma)$, (if Σ is a suitable curve), the usual Sobolev spaces (see e.g. [LM], [BG]). We set

$$H(div;\Omega) := \{v \in L^2(\Omega) \mid div\ v \in L^2(\Omega)\} \ .$$

When Σ_0 is a curve (not closed) contained in a closed curve Σ, we indicate by $H^{1/2}_{00}(\Sigma_0)$ the set of functions such that their extension by zero on Σ belongs to $H^{1/2}(\Sigma)$.
Finally the symbol X' will denote the dual of a (Hilbert) space X.

3. Elliptic equations.

We consider the problem (2.1) with

$$L\,u := -\sum_{i,j} D_{x_j}(a_{ij}\ D_{x_i}\ u) + a_0\ u \tag{3.1}$$

where $a_{ij}, a_0 \in L^\infty(\Omega)$, and for almost each $x \in \Omega$

$$\sum_{i,j} a_{ij}\ (x)\ \xi_i\ \xi_j \geq \alpha\ |\xi|^2 \quad , \quad \alpha > 0 \tag{3.2}$$

$$a_0(x) \geq 0 \quad (a_0(x) \geq \beta > 0 \ \ if \ \ \Gamma_D = \emptyset) \ . \tag{3.3}$$

The boundary conditions are given by

$$u = g_D \ \ on\ \Gamma_D \quad , \quad \partial u/\partial n_L = g_N \ \ on \ \ \Gamma_N \ , \tag{3.4}$$

where $\partial\Omega = \Gamma_D \cup \Gamma_N$ (either Γ_D or Γ_N may be empty), n is the unit outward normal vector to $\partial\Omega$, and

(3.5) $$\partial u/\partial n_L := \sum_j a_{ij} D_{x_i} u \, n_j$$

is the conormal derivative associated to L.
Assuming $g_D \in H^{1/2}(\Gamma_D)$ and $g_N \in [H_{00}^{1/2}(\Gamma_N)]'$, we denote by $G_D \in H^1(\Omega)$ a suitable extension of g_D to Ω; moreover, we set $G_{k,D} := G_{D|\Omega_k}, k = 1,2$.
Introduce now the bilinear forms ($k = 1,2$)

(3.6) $$a_k(u_k, v_k) := \int_{\Omega_k} (a_{ij} \, D_{x_i} \, u_k \, D_{x_j} \, v_k + a_0 u_k \, v_k) \ , \ u_k, v_k \in H^1(\Omega_k) \ ,$$

(3.7) $$a(u,v) := \sum_k a_k(u_{|\Omega_k}, v_{|\Omega_k}) \ , \ u,v \in H^1(\Omega) \ ,$$

and the spaces

(3.8) $$V_k := \{v_k \in H^1(\Omega_k) \mid v_k = 0 \ on \ \Gamma_{k,D} \cap \Gamma\} \ ,$$

(3.9) $$V := \{v \in H^1(\Omega) \mid v = 0 \ on \ \Gamma_D\} \ ,$$

where $\Gamma_{k,D} := \partial\Omega_k \cap \Gamma_D$ (and analogously we will denote $\Gamma_{k,N} := \partial\Omega_k \cap \Gamma_N$).
Moreover, we denote by $H_*^{1/2}(\Gamma)$ the set of functions such that their extension by zero on Γ_D belongs to $H^{1/2}(\Gamma \cup \Gamma_D)$. In particular, if $\Gamma \cap \Gamma_D = \vee$ one has $H_*^{1/2}(\Gamma) = H^{1/2}(\Gamma)$; if both external points of Γ belongs to Γ_D one has $H_*^{1/2}(\Gamma) = H_{00}^{1/2}(\Gamma)$.
For each $\lambda \in H_*^{1/2}(\Gamma)$ define $R_k \, \lambda \in H^1(\Omega_k)$ ($k = 1,2$) to be the solution of the problem

(3.10) $$\begin{cases} a_k(R_k \, \lambda, \varphi_k) = 0 & \forall \, \varphi_k \in V_k \\ R_k \, \lambda = \lambda & on \ \Gamma \\ R_k \, \lambda = 0 & on \ \Gamma_{k,D} \ . \end{cases}$$

$R_k\lambda$ is the harmonic extension of λ to Ω_k. Moreover denote by $w_k^* \in V_k$ ($k = 1,2$) the solution of the problem (depending solely on the data f, g_N, g_D):

(3.11) $$\begin{cases} w_k^* \in V_k \ : \ \forall \, \varphi_k \in V_k \\ a_k(w_k^*, \varphi_k) = \int_{\Omega_k} (f_{|\Omega_k}) \varphi_k + \langle g_{N|\Gamma_{k,N}}, \varphi_{k|\Gamma_{k,N}} \rangle - a_k(G_{k,D}, \varphi_k) \ , \end{cases}$$

where $\langle\cdot,\cdot\rangle$ denotes the duality between $[H^{1/2}_{00}(\Gamma_{k,N})]'$ and $H^{1/2}_{00}(\Gamma_{k,N})$. (The same symbol will be used in the sequel for any other type of duality). Notice that the existence of $R_k\lambda$ and w^*_k is assured from Lax-Milgram's theorem.
Finally define the Steklov-Poincaré operator S as follows: for each $\lambda \in H^{1/2}_*(\Gamma)$, $S\lambda \in [H^{1/2}_*(\Gamma)]'$ is such that

$$\langle S\,\lambda,\mu\rangle := \sum_k a_k(R_k\,\lambda, R_k\,\mu) \quad \forall\,\mu \in H^{1/2}_*(\Gamma)\ . \tag{3.12}$$

Moreover, $\chi = \chi(f, g_D, g_N) \in [H^{1/2}_*(\Gamma)]'$ is given by

$$\langle\chi,\mu\rangle := \int_\Omega f\,R\,\mu + \langle g_N, (R\,\mu)_{|\Gamma_N}\rangle - a(G_D + w^*, R\,\mu) \quad \forall\,\mu \in H^{1/2}_*(\Gamma)\ , \tag{3.13}$$

where we have defined $R\,\mu \in V$ and $w^* \in V$ as follows:

$$R\,\mu := \begin{cases} R_1\,\mu & in\ \Omega_1 \\ R_2\,\mu & in\ \Omega_2 \end{cases} \quad , \quad w^* := \begin{cases} w_1^* & in\ \Omega_1 \\ w_2^* & in\ \Omega_2 \end{cases} .$$

We can now state the following equivalence result:

Proposition 3.1. *Consider the following problems:*
(i) [Steklov-Poincaré problem] Find $\lambda \in H^{1/2}_*(\Gamma)$ such that

$$S\,\lambda = \chi \tag{3.14}$$

(ii) [Variational form of the elliptic boundary value problem] Find $u \in H^1(\Omega), u = g_D$ on Γ_D such that

$$a(u,\varphi) = \int_\Omega f\,\varphi + \langle g_N, \varphi_{|\Gamma_N}\rangle \qquad for\ all \quad \varphi \in V \tag{3.15}$$

They are equivalent in the sense that, having solved (i), the solution to (ii) can be obtained by solving the two independent problems (3.10). As a matter of fact it is

$$u = R\,\lambda + w^* + G_D\ . \tag{3.16}$$

On the contrary, having solved (ii), the solution to (i) is given by

$$\lambda = (u - G_D)_{|\Gamma}\ . \tag{3.17}$$

Proof. Let λ be a solution to (3.11). We show that (3.16) satisfies (3.15). Given $\varphi \in V$, we notice that $\varphi_k := (\varphi - R\ \varphi_{|\Gamma})_{|\Omega_k} \in V_k$. Hence

$$a(u,\varphi) = \sum_k a_k(R_k\ \lambda + w_k^* + G_{k,D}, \varphi_k + R_k\ \varphi_{|\Gamma}) =$$

$$= \int_\Omega f(\varphi - R\ \varphi_{|\Gamma}) + \langle g_N, (\varphi - R\ \varphi_{|\Gamma})_{|\Gamma_N}\rangle +$$

$$+ \sum_k a_k(R_k\ \lambda + w_k^* + G_{k,D}, R_k\ \varphi_{|\Gamma})\ \ (by\ (3.10),(3.11)) =$$

$$= \int_\Omega f(\varphi - R\ \varphi_{|\Gamma}) + \langle g_N, (\varphi - R\ \varphi_{|\Gamma})_{|\Gamma_N}\rangle +$$

$$+ \int_\Omega f\ R\ \varphi_{|\Gamma} + \langle g_N, (R\ \varphi_{|\Gamma})_{|\Gamma_N}\rangle\ \ (by\ (3.14)) = \int_\Omega f\ \varphi + \langle g_N, \varphi_{|\Gamma_N}\rangle\ .$$

We show now that is u is a solution to (3.15), then λ given by (3.17) satisfies (3.14). Let us define $w := (u - G_D) \in V$. It is easily seen that

$$w_{|\Omega_k} - R_k\ w_{|\Gamma} = w_k^*\ , \tag{3.18}$$

since they solve the same equation (3.11). Hence, for each $\mu \in H_*^{1/2}(\Gamma)$

$$\langle S\ \lambda, \mu\rangle = \sum_k a_k(R_k\ w_{|\Gamma}, R_k\ \mu) = a(w, R\ \mu) - a(w^*, R\ \mu)\ (by\ (3.18)) =$$

$$= a(u, R\ \mu) - a(G_D + w^*, R\ \mu) = \langle \chi, \mu\rangle\ (by\ (3.15))\ . \qquad \square$$

Remark 3.2. Let us remind that, if $u_k \in H^1(\Omega_k)$ and $L_k u_k \in L^2(\Omega_k)$, we can define the conormal derivative of u_k and we have (here n^k denotes the unit outward normal vector to $\partial\Omega_k$ on Γ)

$$\frac{\partial\ u_k}{\partial\ n_L^k} := \sum_{ij} a_{ij}\ D_{x_i} u^k n_j^k \in [H_{00}^{1/2}(\Gamma)]' \quad (k = 1,2)\ .$$

It is easily proven that, for each $\mu \in H_{00}^{1/2}(\Gamma)$ and $\lambda \in H_*^{1/2}(\Gamma)$

$$\langle S\ \lambda, \mu\rangle = \sum_k \langle \frac{\partial\ R_k\ \lambda}{\partial n_L^k}, \mu\rangle\ .$$

Moreover

$$\langle \chi, \mu\rangle = -\sum_k \langle \frac{\partial(w_k^* + G_{k,D})}{\partial n_L^k}, \mu\rangle\ .$$

Since the solution u to (3.15) can be written as

$$u = R(u - G_D)_{|\Gamma} + w^* + G_D \ ,$$

equation (3.14) yields

$$\sum_k \partial(u_{|\Omega_k})/\partial n_L^k = 0$$

in the sense of $[H_{00}^{1/2}(\Gamma)]'$. In other words, equation (3.14) amounts to require that the conormal derivative of the solution across Γ is continuous. □

Notice that S is symmetric (i.e., $\langle S\lambda, \mu\rangle = \langle S\mu, \lambda\rangle$) if so are the coefficients a_{ij}. Moreover it is positive; actually there exists $\alpha_0 > 0$ such that

$$\text{(3.19)} \qquad \langle S\lambda, \lambda\rangle \geq \alpha_0 ||R\lambda||^2_{H^1(\Omega)} \quad \forall \, \lambda \in H_*^{1/2}(\Gamma) \ .$$

Since the norm in the right hand side is equivalent to the norm of λ in $H_*^{1/2}(\Gamma)$ (owing to trace and elliptic regularity theorems), the positivity of S follows at once. The following proposition states another relevant property of S.

Proposition 3.3. *For each $\lambda \in H_*^{1/2}(\Gamma)$, define $S_k \ \lambda \in [H_*^{1/2}(\Gamma)]'$ in the following way:*

$$\text{(3.20)} \qquad \langle S_k \ \lambda, \mu\rangle := a_k(R_k \ \lambda, R_k \ \mu) \quad \forall \, \mu \in H_*^{1/2}(\Gamma) \ , \ k = 1, 2 \ .$$

Since there exist $c_1 > 0, c_2 > 0$ such that

$$\text{(3.21)} \qquad c_1 \ ||R_1 \ \lambda||_{H^1(\Omega_1)} \leq ||R_2 \ \lambda||_{H^1(\Omega_2)} \leq c_2 \ ||R_1 \ \lambda||_{H^1(\Omega_1)} \ ,$$

it follows that

$$\text{(3.22)} \qquad C_1 \ \langle S_1 \ \lambda, \lambda\rangle \leq \langle S_2 \ \lambda, \lambda\rangle \leq C_2 \ \langle S_1 \ \lambda, \lambda\rangle \ .$$

Proof. Inequalities (3.21) follow straightforwardly from trace theorems and elliptic estimates concerning $\lambda, R_1 \ \lambda$ and $R_2 \ \lambda$ (e.g., [LM]). Property (3.22) follows at once from (3.20) and (3.21). □

We notice that in the current situation the property (2.4) is satisfied (as it follows from (3.12) and (3.20)). Moreover owing to (3.19) and (3.22) the property (P) is also verified.

4. Stokes equations for viscous flows.

We consider the linear Stokes problem for a viscous flow whose velocity and pressure fields are denoted respectively by u and p. The equation are (e.g. [Te], [GR]):

$$\begin{cases} -\nu \, \Delta \, u + \nabla \, p = f & in \ \Omega \\ div \ u = g & in \ \Omega \\ u = g_D & on \ \partial\Omega \ , \end{cases} \tag{4.1}$$

where $f \in L^2(\Omega), g \in L^2(\Omega), g_D \in H^{1/2}(\partial\Omega)$ and moreover,

$$\int_\Omega g = \int_{\partial\Omega} g_D \cdot n \ .$$

Under these assumptions it is well known (see, e.g., [GR]) that there exists $G \in H^1(\Omega)$ such that

$$\begin{cases} div \ G = g & in \ \Omega \\ G = g_D & on \ \partial\Omega \ . \end{cases}$$

We will denote in the sequal $G_k := G_{|\Omega_k}$.
Following the procedure introduced for the elliptic equations, let us define the bilinear forms ($k = 1, 2$)

$$a_k(u_k, v_k) := \nu \int_{\Omega_k} \nabla \, u_k \cdot \nabla \, v_k \quad , \quad u_k, v_k \in H^1(\Omega_k) \ , \tag{4.2}$$

$$b_k(v_k, p_k) := - \int_{\Omega_k} p_k \ div \ v_k \quad , \quad v_k \in H^1(\Omega_k), p_k \in L^2(\Omega_k) \ , \tag{4.3}$$

$$a(u, v) := \sum_k a_k(u_{|\Omega_k}, v_{|\Omega_k}) \quad , \quad u, v \in H^1(\Omega) \ , \tag{4.4}$$

$$b(v, p) := \sum_k b_k(v_{|\Omega_k}, p_{|\Omega_k}) \quad , \quad v \in H^1(\Omega), p \in L^2(\Omega) \ . \tag{4.5}$$

For each $\lambda \in \hat{H}^{1/2}_{00}(\Gamma) := \{\lambda \in H^{1/2}_{00}(\Gamma) \mid \int_\Gamma \lambda \cdot n = 0\}$, we define $U_k\lambda \in H^1(\Omega_k), P_k\lambda \in L^2(\Omega_k)/\mathbb{R}$ as the solution of

$$(4.6)\qquad \begin{cases} a_k(U_k\lambda,\varphi_k)+b_k(\varphi_k,P_k\lambda)=0 & \forall\ \varphi_k\in H_0^1(\Omega_k)\\ b_k(U_k\lambda,\psi_k)=0 & \forall\ \psi_k\in L^2(\Omega_k)/\mathbb{R}\\ U_k\lambda=\lambda & on\ \Gamma\\ U_k\lambda=0 & on\ \partial\Omega_k\setminus\Gamma\ , \end{cases}$$

and indicate by $U_k^*\in H_0^1(\Omega_k), P_k^*\in L^2(\Omega_k)/\mathbb{R}$ the solution of

$$(4.7)\qquad \begin{cases} a_k(U_k^*,\varphi_k)+b_k(\varphi_k,P_k^*)=\int_{\Omega_k}(f_{|\Omega_k})\varphi_k-a_k(G_k,\varphi_k) & \forall\varphi_k\in H_0^1(\Omega_k)\\ b_k(U_k^*,\psi_k)=0 & \forall\psi_k\in L^2(\Omega_k)/\mathbb{R} \end{cases}$$

By well known results (see, e.g., [GR]) these problems are uniquely solvable.
For each $\lambda\in\hat{H}_{00}^{1/2}(\Gamma)$ introduce the Steklov-Poincaré operator S: $S\ \lambda\in[\hat{H}_{00}^{1/2}(\Gamma)]'$ satisfies

$$(4.8)\qquad \langle S\lambda,\mu\rangle:=\sum_k a_k(U_k\lambda,U_k\mu)\qquad\forall\ \mu\in\hat{H}_{00}^{1/2}(\Gamma)\ .$$

Moreover $\chi=\chi(f,g,g_D)\in[\hat{H}_{00}^{1/2}(\Gamma)]'$ is given by

$$(4.9)\qquad \langle\chi,\mu\rangle:=\int_\Omega f\cdot U\mu-a(G+U^*,U\mu)\qquad\forall\ \mu\in\hat{H}_{00}^{1/2}(\Gamma)\ ,$$

where

$$(4.10)\qquad U\mu:=\begin{cases}U_1\ \mu & in\ \Omega_1\\ U_2\ \mu & in\ \Omega_2\end{cases}\ ,\qquad U^*:=\begin{cases}U_1^* & in\ \Omega_1\\ U_2^* & in\ \Omega_2\end{cases}\ .$$

The following equivalence result holds:

Proposition 4.1. *Consider the following problems:*
(i) [Steklov-Poincaré problem] Find $\lambda\in\hat{H}_{00}^{1/2}(\Gamma)$ *such that*

$$(4.11)\qquad S\ \lambda=\chi$$

(ii) [Variational form of the Stokes problem] Find $u\in H^1(\Omega), p\in L^2(\Omega)/\mathbb{R}$ *such that* $u=g_D$ *on* $\partial\Omega$ *and*

$$(4.12)\qquad \begin{cases} a(u,\varphi)+b(\varphi,p)=\int_\Omega f\cdot\varphi & \forall\ \varphi\in H_0^1(\Omega)\\ b(u,\psi)=-\int_\Omega g\psi & \forall\ \psi\in L^2(\Omega)/\mathbb{R}\ . \end{cases}$$

These problems are equivalent in the sense that, having solved (i), the solution to (ii) is given by

$$u = U\lambda + U^* + G \quad , \quad p = P\lambda + P^* \tag{4.13}$$

where

$$P\,\lambda := \begin{cases} P_1\,\lambda & in\ \Omega_1 \\ P_2\,\lambda & in\ \Omega_2 \end{cases} \quad , \quad P^* := \begin{cases} P_1^* & in\ \Omega_1 \\ P_2^* & in\ \Omega_2 \end{cases} \quad .$$

On the contrary, having solved (ii), the solution to (i) is given by

$$\lambda = (u - G)_{|\Gamma} \quad . \tag{4.14}$$

Proof. *(i) ⇒ (ii). Given $\varphi \in H_0^1(\Omega)$, we set $\varphi_k := (\varphi - U\ \varphi_{|\Gamma})_{|\Omega_k} \in H_0^1(\Omega_k)$. Then we obtain*

$$a(u,\varphi) + b(\varphi,p) = \sum_k [a_k(U_k\ \lambda + U_k^* + G_k, \varphi_k + U_k\ \varphi_{|\Gamma}) +$$

$$+ b_k(\varphi_k + U_k\ \varphi_{|\Gamma}, P_k\ \lambda + P_k^*)] = \int_\Omega f(\varphi - U\ \varphi_{|\Gamma}) +$$

$$+ \sum_k a_k(U_k\ \lambda + U_k^* + G_k, U_k\ \varphi_{|\Gamma}) \ \ (by\ (4.6),(4.7)) = \int_\Omega f\ \varphi \ \ (by\ (4.11)) \quad .$$

Moreover for each $\psi \in L^2(\Omega)/\mathbb{R}$,

$$b(u,\psi) = \sum_k b_k(U_k\ \lambda + U_k^* + G_k, \psi_{|\Omega_k}) = -\int_\Omega g\ \psi$$

since $div(U_k\ \lambda + U_k^* + G_k) = g_{|\Omega_k}$.
(ii) ⇒ (i). Define $w := (u - G) \in H_0^1(\Omega)$, which satisfies $div\ w = 0$. It is verified at once that

$$w_{|\Omega_k} - U_k\ w_{|\Gamma} = U_k^* \quad , \quad p_{|\Omega_k} - \hat{p}_k - P_k\ w_{|\Gamma} = P_k^*$$

(where $\hat{p}_k := |\Omega_k|^{-1} \int_{\Omega_k} p_{|\Omega_k}$) since they satisfy the same equation (4.7). Hence, for each $\mu \in \hat{H}_{00}^{1/2}(\Gamma)$

$$\langle S\ \lambda, \mu\rangle = \sum_k a_k(U_k\ w_{|\Gamma}, U_k\ \mu) = a(w, U\mu) - a(U^*, U\mu)\ (by\ (4.15)) =$$

$$= a(u, U\mu) - a(G + U^*, U\mu) = \langle\chi,\mu\rangle - b(U\mu, p)\ (by\ (4.12)) = \langle\chi,\mu\rangle$$

since div $U\mu = 0$. □

Remark 4.2. Let us notice that equation (4.11) is equivalent to

$$\sum_k \left(\nu \frac{\partial\, U_k\, \lambda}{\partial\, n^k} - P_k\, \lambda\, n^k\right) = -\sum_k \left[\nu\, \frac{\partial(U_k^* + G_k)}{\partial\, n^k} - P_k^*\, n^k\right]$$

in the sense of $[\hat{H}_{00}^{1/2}(\Gamma)]'$, i.e., a solution (u,p) to (4.12) satisfies

$$\text{(4.15)} \quad \sum_k \sigma_k(u_{|\Omega_k}, p_{|\Omega_k}) = 0 \quad \text{on } \Gamma\ , \quad \text{where } \sigma_k(v,q) := \nu\, \frac{\partial v}{\partial n^k} - q\, n^k\ .$$

In other words, equation (4.11) requires that both the normal and shear stresses of the solution must be continuous across Γ. □

It easily follows from (4.8) that S is symmetric.
In order to check the positivity of S, we notice that

$$\text{(4.16)} \ \forall\, \lambda \in \hat{H}_{00}^{1/2}(\Gamma)\ ,\ ||U_k\lambda||_{H^1(\Omega_k)} \ \textit{is equivalent to}\ ||\lambda||_{H_{00}^{1/2}(\Gamma)} \quad \textit{for } k = 1,2$$

As a matter of fact, the first norm is bounded from below by a constant time the second norm owing to the trace theorem since $U_k\lambda_{|\Gamma} = \lambda$. The converse is also true. Indeed, let $G_k\lambda$ be the function of $H^1(\Omega_k)$ satisfying *div* $G_k\lambda = 0$ in Ω_k, $G_k\lambda = \lambda$ on Γ, $G_k\lambda = 0$ on $\partial\Omega_k \setminus \Gamma$, and moreover

$$\text{(4.17)} \qquad ||G_k\lambda||_{H^1(\Omega_k)} \le C_1\ ||\lambda||_{H_{00}^{1/2}(\Gamma)}$$

(the existence of such a function is shown, e.g., in [GR]).
Taking $\varphi_k = U_k\lambda - G_k\lambda$ in (4.6) one gets easily that $||U_k\lambda||_{H^1(\Omega_k)} \le C\ ||G_k\lambda||_{H^1(\Omega_k)}$, whence the desired result follows using (4.17).
From the definition (4.8) and from (4.16) it follows that there exists $\alpha_0 > 0$ such that

$$\text{(4.18)} \qquad \langle S\, \lambda, \lambda\rangle \ge \alpha_0 ||\lambda||^2_{H_{00}^{1/2}(\Gamma)} \qquad \forall\, \lambda \in \hat{H}_{00}^{1/2}(\Gamma)\ .$$

Furthermore, the following result holds. It is the counterpart for the current problem of the result stated in the Proposition 3.3 for the elliptic problem.

Proposition 4.3. *For each* $\lambda \in \hat{H}_{00}^{1/2}(\Gamma)$, *define* $S_k\, \lambda \in [\hat{H}_{00}^{1/2}(\Gamma)]'$ *as*

$$\text{(4.19)} \qquad \langle S_k\, \lambda, \mu\rangle := a_k(U_k\, \lambda, U_k\, \mu) \ \textit{for all} \ \ \mu \in \hat{H}_{00}^{1/2}(\Gamma)\ .$$

There exist $c_1 > 0, c_2 > 0$ *such that*

$$c_1 \,||U_1\,\lambda||_{1,\Omega_1} \leq ||U_2\,\lambda||_{1,\Omega_2} \leq c_2||U_1\,\lambda||_{1,\Omega_1} \quad , \tag{4.20}$$

and $C_1 > 0, C_2 > 0$ *such that*

$$C_1\,\langle S_1\,\lambda,\lambda\rangle \leq \langle S_2\,\lambda,\lambda\rangle \leq C_2\,\langle S_1\,\lambda,\lambda\rangle \quad . \tag{4.21}$$

Proof. (4.20) follows at once from (4.16). In turn, (4.21) can be deduced from (4.19) and (4.20). □

From (4.8) and (4.19) it follows that the splitting formula (2.4) holds; moreover, the property (P) is satisfied owing to (4.18) and (4.21).

5. Viscous and inviscid generalized Stokes equations.

The results presented in Section 4 can be extended to the viscous generalized Stokes equations

$$\begin{cases} \alpha\,u - \nu\,\Delta\,u + \beta\,\nabla\,p = f & in\ \Omega \\ \alpha\,p + div\ u = g & in\ \Omega \\ u = g_D & on\ \partial\Omega \quad , \end{cases} \tag{5.1}$$

where α and β are positive constants, $f \in L^2(\Omega), g \in L^2(\Omega)$ and $g_D \in H^{1/2}(\partial\Omega)$ (see [BGMP], [QSV]). In fact, we define in this case (for $k = 1, 2$)

$$\begin{aligned} a_k^\nu[(u_k,p_k),(v_k,q_k)] := \int_{\Omega_k} &(\beta\,\alpha\,p_k\,q_k + \beta\,div\,u_k\,q_k + \\ &+\alpha\,u_k\cdot v_k + \nu\,\nabla\,u_k\cdot\nabla\,v_k - \beta\,p_k\,div\,v_k) \quad , \end{aligned} \tag{5.2}$$

where (u_k,p_k) and (v_k,q_k) belongs to $H^1(\Omega_k) \times L^2(\Omega_k)$. Moreover, for each $\lambda \in H_{00}^{1/2}(\Gamma)$ we indicate by $E_k\,\lambda \in H^1(\Omega_k) \times L^2(\Omega_k)$ the solution to

$$\begin{cases} a_k^\nu[(E_k\,\lambda,(v_k,q_k)] = 0 & \forall\ (v_k,q_k) \in H_0^1(\Omega_k) \times L^2(\Omega_k) \\ (E_k\,\lambda)_1 = \lambda & on\ \Gamma \\ (E_k\,\lambda)_1 = 0 & on\ \partial\Omega_k \setminus \Gamma \quad , \end{cases} \tag{5.3}$$

and by $(w_k^*,p_k^*) \in H_0^1(\Omega_k) \times L^2(\Omega_k)$ the solution to

$$(5.4) \qquad \begin{cases} a_k^\nu[(w_k^*,p_k^*),(v_k,q_k)] = \int_{\Omega_k}[(f_{|\Omega_k})\cdot v_k + \beta(g_{|\Omega_k})q_k] - \\ \qquad -a_k^\nu[(G_{k,D},0),(v_k,q_k)] \quad \forall\,(v_k,q_k)\in H_0^1(\Omega_k)\times L^2(\Omega_k) \ , \end{cases}$$

where $G_D \in H^1(\Omega)$, is a suitable extension of g_D on Ω, and $G_{k,D} := (G_D)_{|\Omega_k}$. Both problems (5.3) and (5.4) admit a solution owing to the Lax-Milgram theorem.
In this case the Steklov-Poincaré operator S^ν is given by

$$(5.5) \qquad \langle S^\nu\ \lambda,\mu\rangle := \sum_k a_k^\nu[E_k\ \lambda, E_k\ \mu] \ ,$$

for each $\lambda,\mu \in H_{00}^{1/2}(\Gamma)$. Moreover, defining $\chi^\nu = \chi^\nu(f,g,g_D) \in [H_{00}^{1/2}(\Gamma)]'$ as

$$(5.6) \qquad \langle \chi^\nu,\mu\rangle := \int_\Omega f\cdot E\ \mu - a^\nu[(G_D + w^*,p^*),E\ \mu] \ ,$$

where $(E\ \mu)_{|\Omega_k} := E_k\ \mu$ and $a^\nu := a_1^\nu + a_2^\nu$, problem (5.1) is equivalent to solve

$$(5.7) \qquad S^\nu\ \lambda = \chi^\nu \ .$$

Actually, the solution to (5.1) is precisely given by $(u,p) = E\ \lambda + (w^*,p^*) + (G_D,0)$. As before, (5.7) amounts to prescribe that both the normal and shear stresses of the solution to (5.1) are continuous across Γ.
The Steklov-Poincaré operator satisfies

$$(5.8) \qquad \langle S^\nu\ \lambda,\lambda\rangle \ge \alpha_0\ \|\lambda\|^2_{H_{00}^{1/2}(\Gamma)} \quad , \quad \alpha_0 > 0 \ .$$

Furthermore, though the bilinear forms a_k^ν are not symmetric, one can easily verify that

$$(5.9) \qquad a_k^\nu[E_k\ \lambda, E_k\ \mu] = a_k^\nu[E_k\ \mu, E_k\ \lambda] \qquad (k=1,2) \ ,$$

hence the operator S is symmetric. Even in this case, a result like the one stated in Proposition 4.3 holds. Therefore the property (P) is verified from the operator S^ν.

If we take $\nu = 0$ in (5.1), we are left with the inviscid generalized Stokes equations

$$(5.10) \qquad \begin{cases} \alpha\ u + \beta\ \nabla\ p = f & in\ \Omega \\ \alpha\ p + div\ u = g & in\ \Omega \\ u\cdot n = g_N & on\ \partial\Omega \ , \end{cases}$$

where the boundary condition has been modified too, and we assume $g_N \in [H^{1/2}(\partial\Omega)]'$.

Taking $\nu = 0$ in (5.2), we define a_k^0 for each (u_k, p_k) and (v_k, q_k) in $H(div;\Omega_k) \times L^2(\Omega_k)$.
In this case we will consider a different geometrical situation (that is more pertinent to the applications of the model at hand), described in Fig. 5.1.

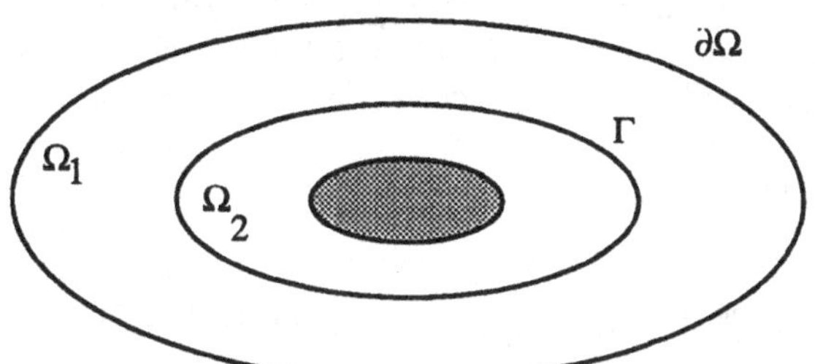

Fig. 5.1. Alternative decomposition of Ω.

As it well-known (see, e.g., [GR]), if $u \in H(div;\Omega)$ then $u \cdot n^k \in [H^{1/2}(\Gamma)]'(k = 1,2)$. Moreover, for each $\lambda \in [H^{1/2}(\Gamma)]'$, we indicate by $F_k\lambda \in H(div;\Omega_k) \times H^1(\Omega_k)$ the solution to

$$\begin{cases} a_k^0[F_k\lambda, (v_k, q_k)] = 0 & \forall\ (v_k, q_k) \in H_0(div;\Omega_k) \times L^2(\Omega_k) \\ (F_k\lambda)_1 \cdot n^k = \lambda & on\ \Gamma \\ (F_k\lambda)_1 \cdot n^k = 0 & on\ \partial\Omega_k \setminus \Gamma\ , \end{cases} \tag{5.11}$$

and by $(z_k^*, r_k^*) \in H_0(div;\Omega_k) \times H^1(\Omega_k)$ the solution to

$$\begin{cases} a_k^0[(z_k^*, r_k^*), (v_k, q_k)] = \int_{\Omega_k} [(f_{|\Omega_k} \cdot v_k + \beta(g_{|\Omega_k}) q_k] - \\ \qquad -a_k^0[(G_{k,N}, 0), (v_k, q_k)] \quad \forall\ (v_k, q_k) \in H_0(div;\Omega_k) \times L^2(\Omega_k) \end{cases} \tag{5.12}$$

where $H_0(div;\Omega_k) := \{v_k \in H(div;\Omega_k) \mid v_k \cdot n^k = 0 \text{ on } \partial\Omega_k\}$, $G_N \in H(div;\Omega)$ is a suitable extension of g_N on Ω, and $G_{k,N} := (G_N)_{|\Omega_k}(k = 1,2)$. For the solvability of (5.11) and (5.12) see [QV].
The Steklov-Poincaré operator $S^0 : [H^{1/2}(\Gamma)]' \to H^{1/2}(\Gamma)$ is given by

$$\langle \mu, S^0\lambda \rangle := \sum_k a_k^0[F_k\lambda, F_k\mu] \tag{5.13}$$

for each $\lambda, \mu \in [H^{1/2}(\Gamma)]'$, and problem (5.10) is equivalent to

$$S^0\lambda = \chi^0\ , \tag{5.14}$$

where $\chi^0 = \chi^0(f, g, g_N) \in H^{1/2}(\Gamma)$ is defined as

$$\langle \mu, \chi^0 \rangle := \int_\Omega f \cdot F\mu - a^0[(G_N + z^*, r^*), F\mu] \ , \tag{5.15}$$

for each $\mu \in [H^{1/2}(\Gamma)]'$.
In this case (5.14) is equivalent to prescribe that the solution p to (5.10) is continuous across Γ. In fact it is easily, verified that $S^0\lambda = \sum_k \beta(F_k\lambda)_2 n^k$ and $\chi^0 = -\sum_k \beta r_k^* n^k$, and the solution to (5.10) is given by $(u,p) = F\lambda + (z^*, r^*) + (G_N, 0)$. Since we have $(F_k\lambda)_2 \in H^1(\Omega_k), r_k^* \in H^1(\Omega_k)$, condition (5.14) assures moreover that $p = [(F\lambda)_2 + r^*] \in H^1(\Omega)$.
In [QV] it is proved that

$$\langle \lambda, S^0\lambda \rangle \geq \alpha_0 ||\lambda||^2_{[H^{1/2}(\Gamma)]'} \quad \forall\, \lambda \in [H^{1/2}(\Gamma)]' \tag{5.16}$$

and that

$$a_k^0[F_k\lambda, F_k\mu] = a_k^0[F_k\mu, F_k\lambda] \ ,$$

hence S^0 is a symmetric and positive operator.
Moreover, having defined S_k^0 as follows

$$\langle \mu, S_k^0\lambda \rangle := a_k^0[F_k\lambda, F_k\mu] \ , \ k = 1, 2 \ , \tag{5.17}$$

as in Proposition 4.3 we get that the norms of $F_1\lambda$ in $H(div;\Omega_1) \times L^2(\Omega_1)$ is equivalent to the norm of $F_2\lambda$ in $H(div;\Omega_2) \times L^2(\Omega_2)$, whence there exist $C_1 > 0, C_2 > 0$ such that

$$C_1\langle \lambda, S_1^0\lambda \rangle \leq \langle \lambda, S_2^0\lambda \rangle \leq C_2\langle \lambda, S_1^0\lambda \rangle \ . \tag{5.18}$$

6. Finite dimensional approximation and iterative methods for the Steklov-Poincaré problem.

Finite dimensional approximations to boundary value problems by Domain Decomposition Methods (briefly, DDM) is a natural place where Steklov-Poincaré operators (in a discrete form) are called into play.
By DDM the computational domain Ω is preliminary partitioned into subdomains Ω_i $i = 1, ..., M$ (that are supposed here to be non overlapped). Then the problem at hand (refer, e.g., to the elliptic problem (3.15)) is approximated by a numerical method (such as finite elements or spectral methods) for which the numerical unknowns are naturally split into two vectors. The former, say u_h, contains the solution values at

all gridpoints internal to the subdomains. The latter, say λ_h, includes the remaining unknowns pertaining to the gridpoints lying on Γ, the union of all subdomain interfaces. The fashion in which the matching conditions for the solutions pertaining to the several different subdomains are enforced is a matter of which kind of numerical method is being used. Actually, matching conditions are generally enforced directly (by specific equations) in the frame of either finite difference or collocation methods. Otherwise they are implicitly incorporated in the global formulation of the numerical problem by those methods moving from a variational approach like, e.g., finite element or spectral element methods.
In all cases, it is possible to obtain for λ_h an autonomous problem of the form

$$S_h \, \lambda_h = \chi_h \tag{6.1}$$

that can be considered as the finite dimensional counterpart of the Steklov-Poincaré problem (2.3). The matrix S_h is known in the literature as the capacitance (or Schur-complement) matrix.
Solving the interface system (6.1) is the crucial task of any domain decomposition approach. As a matter of fact, once the interface unknowns λ_h are available, the overall solution can be achieved by solving a family of reduced numerical boundary-value problems, representing the numerical counterparts of problems like those in (2.2). Moreover, since the latter are independent one from the other, they can be solved simultaneosly within a parallel computing environment.
Since the Steklov-Poincaré operator is generally unbounded, the associated matrix S_h turns out to be ill conditioned. Iterative procedures with preconditioning are therefore in order to solve the interface system (6.1). In this regard, based upon complete characterization of S_h and its properties, an effective iterative technique can be chosen.
It is possible to define S_h as follows (we denote by $(\cdot,\cdot)$ the euclidean scalar product)

$$(S_h \, \lambda, \mu) := \sum_{k=1}^{M} a_{h,k}(T_{h,k} \, \lambda, T_{h,k} \, \mu) \quad , \tag{6.2}$$

in analogy with (3.12), (4.8), (5.5) and (5.13). Here $a_{h,k}(\cdot,\cdot)$ denotes the finite dimensional bilinear form, corresponding to $a_k(\cdot,\cdot)$ while $T_{h,k}$ indicates a discrete extension operator which is a suitable approximation to the operators defined in (3.10), (4.6), (5.3) and (5.11), respectively. In the right hand side we have denoted the finite dimensional functions on Γ whose values at the gridpoints on Γ are given by the components of the vectors λ and μ. We notice that all such functions give raise to a finite dimensional space which from now on will be denoted by Λ_h.
Most often, $a_{h,k}(\cdot,\cdot)$ retains the symmetry property of its infinite-dimensional counterpart $a_k(\cdot,\cdot)$ (though this is not the case if the Chebyshev collocation approximation is used). Moreover, $a_{h,k}(\cdot,\cdot)$ is certainly positive if $a_k(\cdot,\cdot)$ is positive.

The splitting (2.4) suggests a similar one on S_h. Actually using e.g. a black and white ordering of the subdomains (see e.g. Fig. 6.1) and denoting by $\Omega_{(1)} = \bigcup_{\substack{j=1 \\ j\ odd}}^{M} \Omega_j$ and $\Omega_{(2)} = \bigcup_{\substack{k=2 \\ k\ even}}^{M} \Omega_k$,

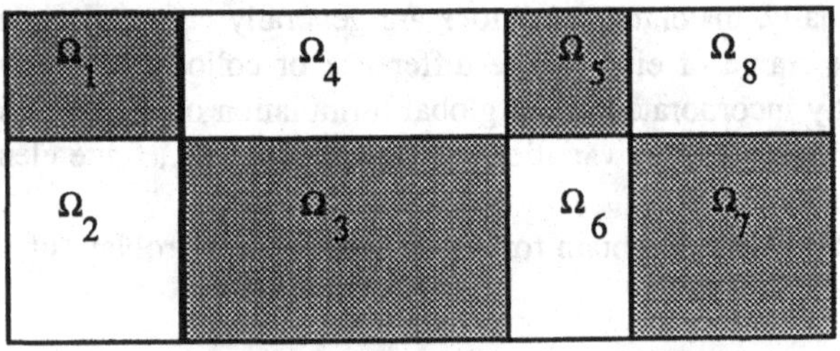

Fig. 6.1 Black and white partition of Ω .

one has

$$(6.3)\quad S_h = S_{h,1} + S_{h,2}\ ,\ (S_{h,i}\ \lambda, \mu) := \begin{cases} \sum_{k\ odd} a_{h,k}(T_{h,k}\ \lambda, T_{h,k}\ \mu) & if\ i = 1 \\ \sum_{k\ even} a_{h,k}(T_{h,k}\ \lambda, T_{h,k}\ \mu) & if\ i = 2 \end{cases}$$

The analogy we have established between finite dimensional and differential case can be carried out furtherly. However, a difficult task in the finite dimensional case is to show that there exist two positive constants c_1, c_2 independent of h so that

$$(6.4)\quad c_1 \sum_{k\ odd} ||T_{h,k}\ \lambda||_k^2 \leq \sum_{k\ even} ||T_{h,k}\ \lambda||_k^2 \leq c_2 \sum_{k\ odd} ||T_{h,k}\ \lambda||_k^2 \quad \forall\, \lambda \in \Lambda_h\ .$$

Here $||\cdot||_k$ stands for a proper norm defined on Ω_k (typically the norm of $H^1(\Omega_k)$). The above is an equivalence result (uniform in h) for finite dimensional extension operators. It has been proven true for all cases considered in Sections 3, 4, 5 in the framework of finite element approximations (see, e.g., [BPS], [BW], [MQ], [QSV]) and partly for spectral collocation approximations (e.g. [QS1]). Indeed, we notice that for a decomposition with internal cross-ponts (i.e., vertices which are shared by more than two subdomains and do not belong to $\partial\Omega$) uniformity is achieved up to a factor $log(h^{-1})$.
We will assume hereafter that (6.4) is satisfied. An easy consequence is that:

$$(6.5)\qquad C_1(S_{h,1}\ \lambda, \lambda) \leq (S_{h,2}\ \lambda, \lambda) \leq C_2(S_{h,1}\ \lambda, \lambda) \quad \forall\ \lambda \in \Lambda_h$$

for suitable constants C_1, C_2 independent of h.
It follows from (6.2), (6.3) and (6.5) that the matrix S_h with its splitting (6.3) satisfies the same property (P) of Section 2 enjoyed by the Steklov-Poincaré operator S.

Property (6.5) suggests a way to find a preconditioner P_h for S_h which is *optimal* in the sense that

$$\exists\, \overline{C} > 0 \text{ independent of } h: \quad k(P_h^{-1}\, S_h) \leq \overline{C} \ , \tag{6.6}$$

where $k(B)$ denotes the condition number of B (see below for its precise definition). As a matter of fact, taking either $P_h = S_{h,1}$ or $P_h = S_{h,2}$ ensures (6.6). This can be easily shown as follows (as pointed out, e.g., in [BPS]): we first notice that

$$[\lambda, \mu]_h := (P_h\ \lambda, \mu) \tag{6.7}$$

is a scalar product since P_h is symmetric and positive definite. Moreover the product $P_h^{-1}\, S_h$ is positive definite and symmetric with respect to (6.7), since

$$[P_h^{-1}\, S_h\ \lambda, \mu]_h = (S_h\ \lambda, \mu) = (\lambda, S_h\ \mu) = (P_h\ \lambda, P_h^{-1}\, S_h\ \mu) = [\lambda, P_h^{-1}\, S_h\ \mu]_h$$

by the symmetry of P_h and S_h. Hence, denoting by

$$||B||_h := \sup_{\lambda \neq 0} \frac{[B\lambda, B\lambda]_h^{1/2}}{[\lambda, \lambda]_h^{1/2}} \ ,$$

and setting $k(B) := ||B||_h\ ||B^{-1}||_h$ the associated condition number, it follows that

$$k(P_h^{-1}\, S_h) = \eta_{max}(P_h^{-1}\, S_h)/\eta_{min}\,(P_h^{-1}\, S_h) \tag{6.8}$$

where we have denoted by η the eigenvalues of $P_h^{-1}\, S_h$.
Owing to (6.5), the property (6.6) follows from (6.8) with $\overline{C}$ depending solely on C_1, C_2. Another convenient choice of P_h is

$$P_h := \alpha\ S_{h,1} + \beta\ S_{h,2} \quad , \quad \alpha, \beta \geq 0, \alpha + \beta = 1 \tag{6.9}$$

which includes the previous cases. The optimality property (6.6) is fulfilled even in this case; this can be proven by applying the Sherman-Morrison-Woodbury formula to compute the inverse of P_h (see [GVL, pag.3]) which allows the reduction to the case $\alpha\ \beta = 0$ dealt with previously.
Suitable iterative procedures on the preconditioned matrix $P_h^{-1}\, S_h$ yields a convergence rate that does not deteriorate as h decreases.

Iterative methods for problem (6.1) of Richardson type (with preconditioning) can be stated as follows: let λ_h^0 be given, and define $r^0 = \chi_h - S_h\ \lambda_h^0, z^0 = P_h^{-1}\ r^0$. For $n \geq 1$ consider the recursion

$$(6.10)\qquad \begin{cases} \lambda_h^{n+1} & = \lambda_h^n + \alpha\, z^n \\ r^{n+1} & = r^n - \alpha\, S_h\, z^n \\ z^{n+1} & = P_h^{-1}\, r^{n+1} \end{cases}$$

The constant $\alpha > 0$ is the acceleration parameter. The variational form of (6.10) is

$$(6.11)\qquad (P_h[\lambda_h^{n+1} - \lambda_h^n], \mu_h) = \alpha(\chi_h - S_h\, \lambda_h^n, \mu_h) \quad \forall\, \mu_h \in \Lambda_h$$

In some cases (e.g. for initial boundary-value problems of hyperbolic type, see [Q1]) one has $||I - P_h^{-1}\, S_h|| < 1$ for a suitable matrix norm, hence the iterations (6.10) converge taking $\alpha = 1$.
The iteration matrix of (6.10) is $G_h = I - \alpha\, P_h^{-1}\, S_h$. A sufficient and necessary condition for convergence is that $I - G_h$ is non singular, moreover $(I - G_h)^{-1}(I + G_h)$ is an N-stable matrix, i.e., its eigenvalues have positive real parts (see [HY]). In the current situation this amounts to require that both P_h and S_h be nonsingular, and moreover, denoting by η_i the eigenvalues of $P_h^{-1}\, S_h$,

$$(6.12)\qquad \alpha < \frac{2\mathrm{Re}(\eta_i)}{|\eta_i|^2} \quad \forall\, i = 1, ..., N$$

If $P_h^{-1}\, S_h$ is a symmetric, positive definite matrix with respect to some scalar product, then it has positive real eigenvalues and condition (6.12) reduces to $\alpha < 2/\eta_{max}$. In this case it is easily seen that the optimal value of α is $\alpha_{opt} = 2/(\eta_{min} + \eta_{max})$, and the corresponding error reduction factor per iteration is $\rho_{opt} = \frac{k-1}{k+1}$, where k is the spectral condition number of the matrix $P_h^{-1}\, S_h$ defined in (6.8).
We already noticed that if both S_h and P_h are positive definite and symmetric, then $P_h^{-1}\, S_h$ is also positive definite and symmetric with respect to the inner product (6.7). In this case we deduce from (6.6) that the convergence rate is independent of h.
A *dynamical choice* of the acceleration parameter α in (6.10) can be accomplished according to several different strategies. The cyclic Chebyshev acceleration (e.g. [HY]) would require to fix the cycle length but needs also an a priori estimate for η_{min} and η_{max}.
Strategies which do not require the knowledge of the extreme eigenvalues are, e.g., the minimum residue algorithm (MR) and the steepest descent algorithm (SD). According to MR, which can be used provided P_h is symmetric and positive definite and S_h is positive definite, in (6.10) we take $\alpha = \alpha^n$, where α^n minimizes $||z^{n+1}||_h^2 := [z^{n+1}, z^{n+1}]_h$. It follows easily that

$$(6.13)\qquad \alpha^n = \frac{(S_h z^n, z^n)}{(P_h^{-1} S_h z^n, S_h z^n)}$$

Notice that since S_h is positive definite, $\alpha^n > 0$ for all n unless $r^n = 0$. The residue at each step reduces as follows: $||z^{n+1}||_h \leq \left(\frac{k-1}{k+1}\right) ||z^n||_h$, with k given by (6.8).
If in addition S_h is symmetric, one might use the SD algorithm which requires that

$$\alpha^n = \frac{(z^n, r^n)}{(S_h z^n, z^n)} \tag{6.14}$$

With this choice, one minimizes the norm $|||e^{n+1}|||^2 := (S_h e^{n+1}, e^{n+1})$ with $e^n := \lambda^n - \lambda$. By the SD algorithm at each iteration the error reduces according to the following law: $|||e^{n+1}||| \leq \left(\frac{k-1}{k+1}\right) |||e^n|||$ (as usual, k is the one defined in (6.8)).
If the matrices S_h and P_h are symmetric and positive definite, the preconditioned conjugate gradient (PCG) algorithm can be used instead of (6.10) to solve (6.1). This method, in the style of Concus, Golub and O'Leary, is defined as follows (see [GVL, p.375]):

$$\lambda_h^0 \text{ given} \; ; \; r^0 = \chi_h - S_h \, \lambda_h^0 \; ; \; z^0 = P_h^{-1} \, r^0 \quad ,$$

$$\alpha_0 = \frac{(z^0, P_h z^0)}{(z^0, S_h z^0)} \; , \; \lambda_h^1 = \lambda_h^0 + \alpha_0 \, z^0 \; , \; \omega_1 = 1 \quad ,$$

and, for $n \geq 1$,

$$\begin{cases} P_h \, z^n = r^n = \chi_h - S_h \, \lambda_h^n \\ \alpha_n = \frac{(z^n, P_h \, z^n)}{(z^n, S_h \, z^n)} \\ \omega_{n+1} = \left[1 - \frac{\alpha_n}{\alpha_{n-1}} \frac{(z^n, P_h \, z^n)}{(z^{n-1}, P_h \, z^{n-1})} \frac{1}{\omega_n}\right]^{-1} \\ \lambda_h^{n+1} = \lambda_h^{n-1} + \omega_{n+1}(\alpha_n \, z^n + \lambda_h^n - \lambda_h^{n-1}) \end{cases} \tag{6.15}$$

One of the nicest features of the PCG algorithm is that it converges in at most d iterations if $P_h^{-1} \, S_h$ has d distinct eigenvalues (e.g., [GVL]). This property is quite useful in practice, since it occurs in some significant domain decomposition situations.

7. Interpretation of the iterative methods for the capacitance problem as iteration-by-subdomains procedure.

It is possible to rewrite either algorithms (6.10) and (6.15) as iterative procedures involving the solution of finite dimensional boundary value problems approximating (2.2) within the subdomains $\Omega_k, k = 1, ..., M$.
Let us show that this equivalence holds for instance for the elliptic problem considered in Section 3. For simplicity of notation we take $g_D = 0$; moreover, let us confine

ourselves to the case of a decomposition by two subdomains only, as the one depticted in Fig. 2.1. We will moreover assume that

$$P_h = S_{h,2} \tag{7.1}$$

Then (6.10) yields:

$$\lambda_h^{n+1} = (1-\alpha_n)\,\lambda_h^n + \alpha_n[S_{h,2}^{-1}\,\chi - S_{h,2}^{-1}\,S_{h,1}\,\lambda_h^n] \tag{7.2}$$

Owing to the definition of χ, (7.2) amounts to accomplish the following sweeps on Ω_1 and Ω_2:

(i) given λ_h^n, provide a finite dimensional solution $u_{h,1}^n$ to the Dirichlet problem:

$$\begin{cases} L_h\, u_{h,1}^n = f & in\ \Omega_1 \\ u_{h,1}^n = \lambda_h^n & on\ \Gamma \end{cases} \tag{7.3}$$

plus the given boundary conditions (3.4) on $\partial\Omega_1 \cap \partial\Omega$;

(ii) solve in Ω_2 the Neumann problem:

$$\begin{cases} L_h\, u_{h,2}^n = f & in\ \Omega_2 \\ \frac{\partial u_{h,2}^n}{\partial n_{L_h}^2} = -\frac{\partial u_{h,1}^n}{\partial n_{L_h}^1} & on\ \Gamma \end{cases} \tag{7.4}$$

which should also be supplemented with boundary conditions (3.4) on $\partial\Omega_2 \cap \partial\Omega$;

(iii) set

$$\lambda_h^{n+1} = \alpha_n\, u_{h,2|\Gamma}^n + (1-\alpha_n)u_{h,1|\Gamma}^n \tag{7.5}$$

and restart from (i).

(7.3)-(7.5) is called a Dirichlet-Neumann sweep. In (7.3) and (7.4) L_h is used (formally) to denote the finite dimensional approximation to the elliptic operator L given in (3.1) and $\frac{\partial}{\partial n_{L_h}^k}$ is the approximate form of the conormal derivative $\frac{\partial}{\partial n_L^k}$ defined in (3.5). When $M(>2)$ subdomains are used (see Fig. 6.1)), the Dirichlet problems are solved within every odd subdomain, while the Neumann problems are solved within the remaining subdomains. In this case all Dirichlet subproblems in the subdomains Ω_i, i odd, are independent one from the other, hence they can be solved simultaneously within a

parallel computing environment. The same substantially holds also for the Neumann problems in the subdomains Ω_k for k even.
The precise equivalence statement between the iterations (6.10) and the Dirichlet-Neumann algorithm (7.3)-(7.5) is provided in [MQ] for finite element approximations, and in [QS1] for spectral collocation approximations. In the same references, other strategies for selecting the acceleration parameters α^n are also indicated.

An analogous statement holds for the Stokes problem of Section 4 as well. In this case, still using the preconditioner (7.1), and recalling that this time λ_h provides the value of the discrete velocity field u_h on Γ, one iteration step (7.2) amounts to accomplish the following sweep:

(i) given λ_h^n, provide a finite dimensional solution $u_{h,1}^n, p_{h,1}^n$ to the problem:

$$(7.6)\qquad \begin{cases} -\nu\, \Delta_h\, u_{h,1}^n + \nabla_h\, p_{h,1}^n = f & in\ \Omega_1 \\ div_h\, u_{h,1}^n = g & in\ \Omega_1 \\ u_{h,1}^n = \lambda_h^n & on\ \Gamma \end{cases}$$

supplemented with the given boundary condition $u_{h,1}^n = g_D$ on $\partial\Omega_1 \cap \partial\Omega$;

(ii) solve in Ω_2 the following Stokes problem to provide the solution $u_{h,2}^n, p_{h,2}^n$:

$$(7.7)\qquad \begin{cases} -\nu\, \Delta_h\, u_{h,2}^n + \nabla_h\, p_{h,2}^n = f & in\ \Omega_2 \\ div_h\, u_{h,2}^n = g & in\ \Omega_2 \\ \sigma_{2,h}(u_{h,2}^n, p_{h,2}^n) = -\sigma_{1,h}(u_{h,1}^n, p_{h,1}^n) & on\ \Gamma \end{cases}$$

plus the prescribed boundary condition $u_{h,2}^n = g_D$ on $\partial\Omega_2 \cap \partial\Omega$ (we recall that stresses on Γ are defined in (4.15); here $\sigma_{1,h}$ and $\sigma_{2,h}$ denote their finite dimensional analogs);

(iii) finally set the new value of the velocity field on Γ by (7.5), and restart from (7.6).

As before, for ease of notation we have denoted by Δ_h, ∇_h and div_h the finite dimensional approximation to the corresponding differential operators.
We refer to [MQ] and [Q2] for precise details and generalizations.
A similar interpretation can be provided for the generalized Stokes problem presented in Section 5 (we refer the reader to [QV] and [QSV]).
Parallel to the algorithm (6.10), the PCG iterations can be shown to be equivalent to solving a sequence of boundary-value-problems like (2.2) within each subdomain. For instance, for the elliptic problem discussed in Section 3, going from λ_h^n to λ_h^{n+1} through (6.15) amounts to solve the Dirichlet problem (7.3) in Ω_1 and the Neumann problem (7.4) in Ω_2. Furthermore, one has to solve two other independent Dirichlet problems (with right hand side $f = 0$), one in Ω_1 and the other in Ω_2 (for the precise statement we refer to [QS2] where finite dimensional approximations based on the spectral collocation method are analyzed).

In the case of decompositions by many subdomains, at each iteration one has to solve a family of independent subproblems, making it possible to take advantage from parallel computations.

We emphasize that by the approach followed in this section, it is clear that iterations on the capacitance system (6.1) can be carried out by means of standard finite dimensional solvers for the given boundary-value-problem within each subdomain. The interface conditions provide different kinds of boundary conditions for each subproblem.

In this way, there is no need of handling explicitly with the capacitance matrix S_h, nor to build up its preconditioner P_h, making it easy to adapt existing single-domain codes to the multi-domain situation.

References

[Ag] V.I.Agoshkov, Poincaré-Steklov's operators and domain decomposition methods in finite dimensional spaces, in [GGMP], pp.73-112.

[BG] F.Brezzi and G.Gilardi, *Finite Element Handbook*, H.Kardestuncer and D.H.Norrie, Eds., Ch. 1,2,3, Mc Graw-Hill, New York, 1987.

[BGMP] M.O.Bristeau, R.Glowinski, B.Mantel and J.Periaux, Numerical Methods for the Navier-Stokes Equations. Applications to the Simulation of Compressible and Incompressible Viscous Flows, preprint 1987 (partly issued from Finite Elements in Physics, R.Gruber Ed., Computer Physics Report 144, North Holland, Amsterdam, 1987).

[BPS] J.Bramble, J.Pasciak and A.Schatz, An Iterative Method for Elliptic Problems on Regions Partitioned into Substructures, Math. Comput. 46 (1986), pp.361-369.

[BW] P.Bjorstad and O.Widlund, Iterative Methods for the Solution of Elliptic Problems on Regions Partitioned into Substructures, SIAM J.Numer.Anal. 23 (1986), pp.1097-1120.

[CGPW] T.F.Chan, R.Glowinski, J.Periaux and O.B.Widlund, Eds., *Domain Decomposition Methods*, SIAM, Philadelphia, 1989.

[GGMP] R.Glowinski, G.H.Golub, G.A.Meurant and J.Periaux, Eds., *Domain Decomposition Methods for Partial Differential Equations*, SIAM, Philadelphia, 1988.

[GR] V.Girault and P.A.Raviart,*Finite Element Methods for the Navier-Stokes Equations: Theory and Algorithms*, Springer, Berlin, 1986.

[GVL] G.H.Golub and C.F.Van Loan, *Matrix Computation*, Johns Hopkins University Press, Baltimore, 1983.

[HY] L.A.Hageman and D.M.Young, *Applied Iterative Methods*, Academic Press, New York, 1981.

[LM] J.L.Lions and E.Magenes, *Nonhomogeneous Boundary Value Problems and Applications, I*, Springer-Verlag, Berlin, 1972.

[MQ] L.D.Marini and A.Quarteroni, A Relaxation Procedure for Domain Decomposition Methods Using Finite Elements, Numer.Math. 55 (1989), pp.575-598.

[Q1] A.Quarteroni, Domain Decomposition Methods for Systems of Conservation Laws:

Spectral Collocation Approximations, Publ. n.662, I.A.N.-C.N.R., Pavia (to appear in SIAM J.Sci.Stat.Comput.).

[Q2] A.Quarteroni, Domain Decomposition Algorithms for the Stokes Equations, in [CGPW], pp.431-442.

[QS1] A.Quarteroni and G.Sacchi Landriani, Domain Decomposition Preconditioners for the Spectral Collocation Method, J.Scient.Comput. 3 (1988), pp.45-75.

[QS2] A.Quarteroni and G.Sacchi Landriani, Parallel Algorithms for the Capacitance Matrix Method in Domain Decompositions, Calcolo 25 (1988), pp.75-102.

[QSV] A.Quarteroni, G.Sacchi Landriani and A.Valli, Coupling of Viscous and Inviscid Stokes Equations via a Domain Decomposition Method for Finite Elements, submitted to Numer. Math.

[QV] A.Quarteroni and A.Valli, Domain Decomposition for a Generalized Stokes Problem, Quaderno n.1/89, Univ.Cattolica Brescia; in Proc. 2nd ECMI Conference, Glasgow 1988, in press.

[Te] R.Temam, *Navier-Stokes Equations. Theory and Numerical Analysis*, North-Holland, Amsterdam, 1977.

ON THE PROBLEM OF NATURAL CONVECTION

S. Rionero
Dept. di Matematica Univ.di Napoli, 80134 Napoli (Italy)
and
B. Straughan
Dept. of Mathematics,Univ. of Glasgow, G128QW Glasgow U.K.

ABSTRACT. The problem of natural convection in a horizontal layer of viscous thermally conducting fluid heated from below (Bènard problem) is relevant in many physical situations and, in particular, in geophysics and astrophysics. Our goal is to give an account of the questions arising in the determination of the critical value that the Rayleigh number must exceed before the convection (instability) can manifest itself. We limit ourselves in considering the classical Bènard problem, the magnetic Bènard problem and the convection in a porous medium with internal heat source and variable gravity effects in the framework of linear and non-linear stability.

INTRODUCTION

Consider a horizontal layer of fluid in which an adverse temperature gradient ($\beta>0$) is maintained by heating the underside. Because of the thermal expansion, the fluid at the bottom will be lighter than the fluid at the top and this is a top-heavy arrangement which is potentially unstable [1]. The fundamental problem of natural convection (Bènard problem) consists in determining:

i) the critical value that β must exceed before the convection (instability) can manifest itself;

ii) the motions which ensue on surpassing the critical temperature gradient.

If one enlarges the problem to include other effects (i.e. rotation, magnetic field, porosity, heat source, variable gravity, etc..) then the fluid is subject to conflicting tendencies and many striking aspects of fluid behavior, disclosed by the mathematical theory in these connexions, have been experimentally demonstrated. Of course the main field of interest are geophysics and astrophysics. The Bènard problem has been studied, in the past as nowadays, by many Authors. Our aim is to give an account of the question i) in the light of linear and non-linear stability, limiting ourselves to the classical simple Bènard problem (Part I), Magnetic Bènard problem (Part II) and convection in a porous medium with internal heat source and variable gravity effects (Part III).

R. Spigler (ed.), Applied and Industrial Mathematics, 205–216.

PART I - Classical simple Bènard problem (B.P.)

1.1. Relevant equations in the Boussinesq scheme

Let us consider a horizontal layer of viscous thermally conducting fluid between two horizontal parallel planes under the action of a vertical gravity field $\underline{g}$, in which an adverse temperature gradient ($\beta>0$) is maintained by heating the underside. The basic flow m_0 whose stability is to be investigated is the motionless state with a linear temperature profile

$$m_0 = \{ \tilde{\underline{v}} = 0 \ , \ \tilde{T} = -\beta z + \tilde{T}_0 \} \qquad 0 \leq z \leq d \tag{1.1}$$

where β (>0) and $\tilde{T}_0$ are constants.

Introducing the symbols

$\underline{u} = (u, v, w)$	perturbation to the velocity field
θ	perturbation to the temperature field
p	perturbation to the pressure field
$P_r = \dfrac{\nu}{K}$	Prandtl number
$R^2 = \dfrac{g\alpha_T\beta d^4}{\nu K}$	Rayleigh number
ν	Kinematic viscosity
α_T	coefficient of thermal expansion
K	coefficient of heat conduction,

the Boussinesq equations governing the perturbations ($\underline{u}$, θ, p) in the velocity, temperature and pressure fields are found to be

$$\begin{cases} \underline{u}_t + \underline{u}\cdot\nabla\underline{u} = -\nabla p + R\theta\,\underline{k} \\ P_r\theta_t + P_r\underline{u}\cdot\nabla\theta = Rw + \Delta\theta \\ \nabla\cdot\underline{u} = 0 \end{cases} \tag{1.2}$$

where $\underline{k}$ is the unity vector in the z-direction (opposite to the gravity direction) and under the initial conditions

$$\underline{u}(\underline{x}, 0) = \underline{u}_0(\underline{x}) \ , \quad \theta(\underline{x}, 0) = \theta_0(\underline{x}) \tag{1.3}$$

with $\nabla\cdot\underline{u}_0 = 0$.

As concerns the boundary conditions, they depend on the nature of the planes bounding the layer according to the fact that a plane is rigid or free. It is found that the boundary conditions are [1]

$$\underline{u} = \theta = \frac{\partial w}{\partial z} = \underline{k} \times \nabla \times \underline{u} = 0 \qquad \text{on a rigid plane,} \tag{1.4}$$

$$w = \theta = \frac{\partial u}{\partial z} = \frac{\partial v}{\partial z} = \frac{\partial^2 w}{\partial z^2} = \frac{\partial}{\partial z}(\underline{k} \times \nabla \times \underline{u}) = 0 \text{ on a free plane.} \tag{1.5}$$

1.2 Linear stability

Disregarding the non-linear terms appearing in (1.2), it follows:

$$\begin{cases} \underline{u}_t = -\nabla p + R\vartheta\,\underline{k} + \Delta\,\underline{u} \\ P_r\vartheta_t = Rw + \Delta\vartheta \\ \nabla\cdot\,\underline{u} = 0 \end{cases} \tag{1.6}$$

Because (1.6.) is autonomous, one can look for solutions of the kind:

$$\vartheta = \hat{\vartheta}(\underline{x})\ e^{-\sigma t}\ ,\quad \underline{u} = \hat{\underline{u}}(\underline{x})e^{-\sigma t}\ ,\quad p = \hat{p}(\underline{x})e^{-\sigma t} \tag{1.7}$$

where σ is a priori a complex parameter. Consequently one is led to solving the spectral problem:

$$\begin{cases} -\sigma\,\hat{u} = -\nabla\hat{p} + R\hat{\vartheta}\,\underline{k} + \Delta\hat{\underline{u}} \\ -\sigma P_r\hat{\vartheta} = R\hat{w} + \Delta\hat{\vartheta} \\ \Delta\cdot\hat{\underline{u}} = 0 \end{cases} \tag{1.8}$$

under appropriate boundary conditions according to (1.4)-(1.5).

Assuming that the perturbation fields are periodic function of x and y of periods $2\pi/a_x$, $2\pi/a_y$ (a_x, a_y >0) and requiring the "average" velocity conditions:

$$\int_\Omega u\, d\Omega = \int_\Omega v\, d\Omega = 0 \tag{1.9}$$

where

$$\Omega = \left[0, \frac{2\pi}{a_x}\right] \times \left[0, \frac{2\pi}{a_y}\right] \times \left[0,1\right]$$

is the periodicity cell, the following results hold [1]:

i) the principle of the exchange of stabilities is valid, i.e. σ is real and the marginal states are characterized by $\sigma = 0$

ii) the parameters characterizing the marginal case are :
(R^2_c and a respectively denote the critical values of the Rayleigh number and the wave numbers).

Nature of the bounding planes	R^2_c	$a=\sqrt{a^2_x + a^2_y}$
Both free	657.511	2.2214
Both rigid	1707.762	3.117
One rigid and one free	1100.65	2.682

TABLE.1 Critical Rayleigh and wave numbers

1.3 Non-linear stability

Choosing as Liapunov function the energy norm

$$V = \frac{1}{2} \int_{\Omega} (u^2 + P_r \theta^2)\, d\Omega \tag{1.10}$$

and evaluating it's derivative along the solutions to (1.2), then it follows [2]

$$\dot{V} = \left[R \frac{2 \int_{\Omega} \theta w \, d\Omega}{\int_{\Omega} [\nabla \underline{u} : \nabla \underline{u} + (\nabla \theta)^2]\, d\Omega} - 1 \right] \int_{\Omega} [\nabla \underline{u} : \nabla \underline{u} + (\nabla\theta)^2]\, d\Omega \tag{1.11}$$

Setting

$$\frac{1}{R^2_E} = 2 \max_{S} \frac{\int_{\Omega} \theta w \, d\Omega}{\int_{\Omega} [\nabla \underline{u} : \nabla \underline{u} + (\nabla\theta)^2]\, d} \tag{1.12}$$

$$S = \left\{ \begin{array}{l} \underline{u}, \theta : \nabla \underline{u} = 0, \ \underline{u}, \theta \text{ periodic in the x and y directions} \\ \text{and verifyng (1.4)-(1.5)} \end{array} \right\} \tag{1.13}$$

by means of Dirichelet inequality it follows that

$$R \le R^2_E \longrightarrow V \le V_0 \exp \gamma (R - R^2_E)\, t \tag{1.14}$$

with γ positive constant and R^2_E critical Rayleigh number of energy stability. But, as easily seen, the Eulero-Lagrange equations associated to the variational problem (1.12) are

$$\left\{ \begin{array}{l} R\theta \underline{k} + \Delta \underline{u} = \Delta p \\ R \underline{u} \cdot \underline{k} + \Delta\theta = 0 \\ \nabla \cdot \underline{u} = 0 \end{array} \right. \tag{1.15}$$

under the relevant boundary conditions, therefore because (1.15) are identical with (1.8) for $\sigma = 0$, the principle of the exchange of stability assures the non-existence of sub critical instability and the validity of the following principle of linearization [2].

Theorem 1.1

There is coincidence between the results of linear stability and the results of non-linear stability in the energy norm :

$$R^2_L = R^2_E$$

PART II - Magnetic Bènard problem (M.B.P)

2.1 Notations, basic equations and hypotheses

Let us consider an infinite horizontal layer of homogeneous viscous electrically conducting fluid permeated by an imposed uniform magnetic field $\underline{H}$ normal to the layer and in which an adverse temperature gradient $\beta>0$ is maintained. As in part I, we introduce a cartesian coordinate system $Oxyz = (O, \underline{i}, \underline{j}, \underline{k})$ with the z-axis pointing vertically upwards and we assume that the fluid is confined between the planes $z = 0$ and $z = d > 0$, with assigned temperatures

$$\tilde{T}\ (z=0) = \tilde{T}_0,\ \tilde{T}(z=d) = -\beta z + \tilde{T}_0.$$

We want to study in the Boussinesq scheme the stability of the rest state

$$m_0 = \left\{ \tilde{\underline{v}} = 0,\ \tilde{\underline{H}} = H\ \underline{k}\ ,\ \tilde{T} = -\beta z + \tilde{T}_0\ ,\ \tilde{p} \right\}.$$

In the sequel we need the further following symbols

$\underline{h} = (h_1, h_2, h\ )$	perturbation in the magnetic field
$P_m = \nu/\eta$	magnetic Prandtl number
$Q^2 = \mu^2 H^2 d^2 / 4\pi\rho\nu\eta$	Chandrasekhar number
η	resistivity
μ	magnetic permeability
ρ	density

The non-dimensional equations for a perturbation $(\underline{u}, \underline{k}, \theta, p)$ to m_0 are [1]:

$$\begin{cases} \underline{u}_t + \underline{u}\cdot\nabla\underline{u} - P_m\ \nabla\underline{h} = -\nabla p + R\theta\ \underline{K} + \Delta\underline{u} + Q\ \underline{h}_z \\ \nabla\cdot\underline{u} = 0 \\ P_m\ (\underline{h}_t + \underline{u}\cdot\nabla\underline{h} - \underline{h}\cdot\nabla\underline{u}) = Q\ \underline{u}_z + \Delta\underline{h} \\ \nabla\cdot\underline{h} = 0 \\ P_r\,(\theta_t + \underline{u}\cdot\nabla\theta) = Rw + \Delta\theta \end{cases} \tag{2.1}$$

in the space-time cylinder $\mathbb{R}^2 \times (0,1) \times [0,\infty)$.

Limiting ourselves to the stress free electrically non-conducting boundary case [1], with system (2.1), we adopt the initial-boundary conditions :

$$\underline{u}(\underline{x},0) = \underline{u}_0(\underline{x})\ ,\ \underline{h}(\underline{x},0) = \underline{h}_0(\underline{x})\ ,\ \theta(\underline{x},0) = \theta_0(\underline{x}) \tag{2.2}$$

$$w(\underline{x},t)=\theta(\underline{x},t)=\underline{h}(\underline{x},t)=u_z(\underline{x},t)=v_z(\underline{x},t)=0, \qquad z=0,1;\ t \geq 0 \tag{2.3}$$

where $\underline{u}_0(\underline{x})$, $\underline{h}_0(\underline{x})$ and $\theta_0(\underline{x})$ are assigned fields with $\nabla\cdot\underline{u}_0 = \nabla\cdot\underline{h} = 0$.

As in part I, we assume that:

i) perturbation fields are periodic functions of x and y;

ii) $\Omega = \left[0, \frac{2\pi}{a_x}\right] \times \left[0, \frac{2\pi}{a_y}\right] \times \left[0,1\right]$ is the periodicity cell.

Moreover , in order to exclude any other rigid solution, we require the "average velocity condition" (1.7).

2.2 Results of linear stability

The results of linear stability for M.B.P. can be found in [1]. In the case $P_m < P_r$ the critical value R_L^2 for the Rayleigh number obtained in the linear analysis are the following ones:

Q^2	R^2_L
0	657.511
5	796.573
20	1114.19
100	2563.71
105.1789	2741.3
10^3	15207
10^4	119832
4×10^4	445507

TABLE 2. Critical Rayleigh number for M.B.P.

Remark 2.1

From the linear analysis it follows that the magnetic field has a strong inhibiting effect on the onset of convection

Remark 2.2

The inhibition of convection by a magnetic field discovered by the linear mathematical theory in the years 1951-52 [3] [4], has been successively (1955) experimentally verified by Nagakawa [5] and Sirlow [6].

2.3 Non-linear stability of M.B.P.

Non-linear stability of M.B.P. has been studied by many authors and in particular by Rionero [7], Galdi-Straughan [8], Galdi [9], Rionero-Mulone [10]. The fundamental problem is how to choose the Liapunov function in order to reach results as close as possible to those of linear stability under computable conditions on the initial data. In fact, choosing as Liapunov function the standard energy (L^2-norm) one obtains only the non-destabilizing effect of the magnetic field (Rionero 1968). But choosing as Liapunov function a "generalized energy" (i.e. a norm stronger than the L^2-norm) it is possible to obtain the stabilizing effect of the magnetic field [Galdi-Straughan (1985), Galdi (1985)] and also the coincidence between the linear and non-linear stability parameters, at least in the case $P_m < P_r$, $Q^2 \leq 105.18$,

$R^2 \leq 2741.3$ [Rionero-Mulone (1988)]. In order to reach the aforesaid coincidence.

We split the Liapunov function V into two parts

$$V = V_o + V_1 \tag{2.4}$$

and we require that:

i) V_o has to dominate the linear problem

ii) V_o has to depend on field variables (essential fields) X_1, X_2, ... which represent physical causes inhibiting or promoting instability and on balances

$$\ell_j = X_i - c_{ij}X_j \qquad (c_{ij} = \text{const.}) \tag{2.5}$$

of opposite causes.

iii) V_1 has to dominate the non-linear terms.

Because the instability occurs at the onset of convection, a physical cause inhibits or promotes instability according to the way it inhibits or promotes convection. Consequently

1) the heat conduction $-\theta_z \underline{k}$ may promote instability ,

2) $\zeta = \underline{k} \times \nabla \times \underline{u}$ i.e. the local angular velocity about the vertical axis, inhibits the onset of convection,

3) $\zeta^{(m)} = \underline{k} \times \nabla \times \underline{h}$ and h - by the Lorentz force - inhibit the onset of convection ,

4) w may promote instability.

In Rionero-Mulone (1988) we assume respectively $\theta_z, \zeta, \zeta^{(m)}, h, w$ as essential fields; $\ell = h - c\,\theta_z$ (c = constant) as balance and

$$V_o = 1/2 \left[\|\nabla w\|^2 + \|\zeta\|^2 + P_m \|\zeta^{(m)}\|^2 + a_1\|\theta_z\|^2 + a_2\|\nabla\ell\|^2 \right] \tag{2.6}$$

($\|\cdot\|$ = L_2-norm ; c, a_1, a_2 = constants to be determined later) as first part of the Liapunov function.

Setting

$$(\varphi_1, \varphi_2) = \int_\Omega \varphi_1\varphi_2 \, d\Omega \ , \quad \Delta_1 = \frac{\partial^2}{\partial x^2} + \frac{\partial^2}{\partial y^2} \ . \tag{2.7}$$

$$I_o = -(\Delta_1\theta, w) + 1/R \left[a_2 \left(\frac{Q}{P_m} - \frac{cR}{P_r} \right) - Q \right] (\Delta\ell_z, w) + \\ + \left(\frac{c\,Q}{R} - \frac{a_1}{P_r} \right) (w_z, \Delta\theta_z) - \frac{a_2\,c}{R} \left[1/P_m - 1/P_r \right] (\Delta\theta_z, \Delta\ell) \tag{2.8}$$

$$D_o = \|\nabla\zeta\|^2 + \|\nabla w\|^2 + \|\nabla\zeta^{(m)}\|^2 + (a_1/P_r)\|\nabla\theta_z\|^2 + (a_2/P_m)\|\nabla\ell\|^2, \tag{2.9}$$

the derivative $\dot{V}_o$ along the solution to (2.1)+(2.2)+(2.3) is found to be (cf.Rionero-Mulone 1988)

$$\dot{V}_o = R\,I_o - D_o = R \left(\frac{I_o}{D_o} - \frac{1}{R} \right) D_o \tag{2.10}$$

Introducing the variational problem

$$\frac{1}{R_E} = \max_{\mathcal{S}} \frac{I_o}{D_o} \tag{2.11}$$

where $\mathscr{S} = \Big\{ w, \xi, \xi^{(m)}, \theta, \ell$ which are periodic in x and y

and satisfy boundary conditions of M.B.P. $\Big\}$

is the class of kinematical perturbations . By use of some Poincare'-type inequalities then it follows (Rionero-Mulone 1988)

Theorem 2.1
The condition

$$R^2 < R_E^2 \tag{2.12}$$

assures the linear asymptotic exponential stability according to

$$V_o(t) \le V_o(0) \exp\left[-\lambda_o \frac{2\pi^2}{P_r P_m}\left(1 - \frac{R}{R_E}\right)\right] t \tag{2.13}$$

where

$$\lambda_o = \min\left\{ P_r P_m, P_r, P_m \right\} \tag{2.14}$$

In table 3 are collected the critical Rayleigh numbers R_E^2 of linear stability obtained solving (1.10) in the case $P_m < P_r$

Q^2	R_E^2
0	657.511
5	796.573
20	1154.19
100	2563.71
105.1789	2741.3
10^3	10350
10^4	35025
4×10^4	71160

TABLE 3. Critical Rayleigh numbers R_E^2 against the Chandrasekhar numbers in the case $P_m < P_r$.

Remark 2.3
Let us note that comparing Table 2 and 3 then it follows

$$R_L^2 = R_E^2 \quad \text{for } P_m < P_r \quad Q^2 \le 105.1789$$

In order to obtain conditions assuring non-linear stability one has to choose the second part V_1 of the Liapunov function in such a way that also the non-linear terms can be dominated. Choosing (b=positive constant)

$$V_1 = b/2 \left[\|\nabla u\|^2 + P_m \|\nabla h\|^2 + P_r \left(\|\nabla \theta_x\|^2 + \|\nabla \theta_y\|^2 \right) \right]$$

in Rionero-Mulone (1988) has been proved that:

<u>Theorem 2.7</u>

<u>In the case</u> ($P_m < P_r \quad Q^2 \le 105.1789$) <u>the condition which assures linear stability</u> $R^2 < R_L^2$, <u>assures also non-linear stability in the V-norm under computable restrictions on the initial data.</u>

PART III. Convèction in a porous medium with internal heat source and variable gravity effects.

3.1 Notations and basic equations in the Boussinesq scheme.

Let us consider the problem of Bènard convection in a fluid saturated porous material with an internal heat source, allowing the gravity g to depend on the vertical coordinate z. For convective motion of an incompressible fluid in a porous solid, the relevant equations may be derived as in Joseph [2], and are

$$\begin{cases} \nabla p = -\rho_0 \, [1-\alpha(T-T_0)] \, g(z) \, \underline{k} - \dfrac{\mu}{K_*} \underline{v} \\ \nabla \cdot \underline{v} = 0 \\ T_t + \underline{v} \cdot \nabla T = K \, \Delta T + Q(z) \end{cases} \tag{3.1}$$

where the following further symbols have been introduced

$\underline{v}$	seepage velocity
T	temperature field
T_0 = const	reference temperature
ρ_0 = const	density at the temperature T_0
μ	dynamic viscosity
K_*	permeability of the porous medium
Q	internal heat source

In the sequel we shall consider the equations (1.3) in the layer $\{z \in (0,d)\} \times R^2$. The conduction solution which satisfies the boundary conditions

$$T = T_l \ , \ z = 0; \qquad T = T_v, \ z = d \qquad (T_l > T_v) \tag{3.2}$$

is

$$\bar{v} \equiv 0 \ , \ \bar{T} = - \kappa^{-1} \int_0^z \int_0^\xi Q(\eta) \, d\eta \, d\xi - cz + T_l \qquad (3.3)$$

where the constant c is given by

$$c = d^{-1} (T_l - T_u) - \frac{1}{\kappa d} \int_0^d \int_0^\xi Q(\eta) \, d\eta \, d\xi \qquad (3.4)$$

and where the hydrostatic pressure $\bar{p}$ is determined from (3.1).

The appropriate equations for the perturbations $\{\underline{u}, \ \theta, \ p\}$ in non-dimensional form are (further details are given in [11])

$$\begin{cases} \nabla p = R \, H(z) \theta \underline{k} - \underline{u} \\ \nabla \cdot \underline{u} = 0 \\ P_r \ (\theta_t + \underline{u} \cdot \nabla \theta) = R \, N(z) w + \Delta \theta \end{cases} \qquad (3.5)$$

where

$$\begin{cases} \delta \, q(z) = \dfrac{F(z)}{c} \ ; & F(z) = \kappa^{-1} \int_0^z Q(\xi) \, d\xi \\ N(z) = 1 + \delta \, q(z) \ ; & \varepsilon \, h(z) = \dfrac{g(z) - \bar{g}}{\bar{g}} \\ H(z) = 1 + \varepsilon \, h(z) \ ; & \delta, \ \varepsilon, \ \bar{g} = \text{constants} \end{cases} \qquad (3.6)$$

Equations (3.3) hold on the region $\{ (x, y) \in R^2\} \times \{z \in (0,1)\}$ and the boundary conditions to be satisfied are:

$$\theta = 0, \quad z = 0,1; \qquad w = 0, \quad z = 0,1 \qquad (3.7)$$

We also assume the functions $\underline{u}$, θ, p are spatially periodic in the sense that they satisfy

$$\Delta_1 \, \theta = -a^2 \theta$$

for a wave number a.

We develop an energy analysis for (3.5)-(3.7). The energy identities are derived as:

$$\frac{1}{2} P_r \frac{d}{dt} \| \theta \|^2 = R \langle N w \theta \rangle - D(\theta)$$

$$R \langle H w \theta \rangle = \| \underline{u} \|^2$$

where $\langle \ . \ \rangle$ and $\|.\|$ denote integration over, and the L^2-norm on Ω, respectively; Ω here is the cell of solution periodicity.

For $\lambda (>0)$ to be chosen we form the identity:

$$\frac{dE}{dt} = R \langle M(z) w \theta \rangle - D(\theta) - \lambda \, \|\underline{u}\|^2$$

where

$$E = \frac{1}{2} P_r \, \| \theta \|^2$$

$$M(z) = 1 + \delta q + \lambda(1 + \varepsilon h)$$

If we define R_E by

$$R_E^{-1} = \max_H \frac{I}{D} \tag{3.8}$$

with $I = \langle Mw\theta \rangle$, $D = D(\theta) + \lambda \|\underline{u}\|^2$, where, in (3.8) H is the space of admissible functions, then it is straightforward to show $E(t) \to 0$ at least exponentially provided $R < R_E$, cf Rionero [12]. The existence of a maximizing solution to problems like (3.8) is discussed in Rionero [12]: an explicit proof may be derived by a slight modification of the analysis in Galdi et al. [13].

The maximization problem (3.8) is solved numerically; in fact we determine

$$\max_\lambda \min_{a^2} R_E(a^2;\lambda) \tag{3.9}$$

Some values of the critical Rayleigh and wave numbers for non-linear energy stability, R_E^2, a_E, and for linear instability, R_L^2, a_L, for various ε, δ, are now given when $h = -z$, $q = z$. Further values, including those for other choices of h and q are given in Rionero-Straughan [11].

R_L^2	R_E^2	a_L^2	a_E^2	ε	δ	$\bar{\lambda}$
39.478	39.478	9.870	9.870	0.0	0.0	1.000
77.020	74.971	10.209	10.178	1.0	0.0	1.500
132.020	122.374	12.314	11.928	1.5	0.0	2.565
63.114	61.563	10.026	10.110	1.0	0.5	2.318
111.930	101.513	11.887	11.562	1.5	0.5	3.201
35.384	35.092	9.835	9.870	0.5	1.0	2.000
53.375	51.715	9.918	10.032	1.0	1.0	2.797

TABLE 4. Critical Rayleigh and wave numbers for $h = -z$, $q = z$.
($\bar{\lambda}$ denotes the best value of λ in (3.9))

Remark 3.1

Let us underline that the critical Rayleigh numbers R_E^2 of non-linear stability are very close to the which ones R_L^2 of linear stability.

References

[1] Chandrasekhar S.: Hydrodynamic and Hydromagnetic stability, Dover Publications, Inc. New York, 1981

[2] Joseph D.D., Stability of fluid motions I-II, Springer-Verlag, 1976

[3] Thomson W.B., Thermal convection in a magnetic field, Phil. Mag. Ser. 7, 42, 1417-32, 1951

[4] Chandrasekhar S.:On the inhibition of convection by a magnetic field, Phil. Mag., Ser. 43, 501-32, 1952

[5] Nakagawa Y.: An experiment on the inhibition of thermal convection by a magnetic field, Nature, 175, 417-19, 1955

[6] Jirlow K.: Experimental investigation of the inhibition of convection by a magnetic field, Tellus, 8, 252-3, 1956

[7] Rionero S.: Sulla stabilita' magnetoidrodinamica in media con vari tipi di condizioni al contorno, Ricerche di Matematica, 17, 1968

[8] Galdi G.P.-Straughan B.: Arch. Rat. Mech. Anal. 89, 211, 1985

[9] Galdi G.P.: Arch. Rat. Mech. Anal. 87, 167, 1985

[10] Rionero S.-Mulone G.: Arch. Rat. Mech. Anal. 103, 4, 1988

[11] Rionero S.-Straughan B.: Convection in a porous medium with internal heat source and varible gravity effects. Int. J. Eng. Sci (to appear)

[12] Rionero S.: Metodi variazionali per la stabilita asintotica in media in magnetoidrodinamica. Ann. Mat. Pura Appl., 78, 339-364, 1968

[13] Galdi G.P., Payne L. E., Proctor M.R.E; Straughan B.: Convection in thawing subsen permafrost. Proc. Roy. Soc. London A, 414, 83-102, 1987

Selected Contributed Papers

Non-Newtonian Phenomena in Shear Flow

John A. Nohel
Center for the Mathematical Sciences
University of Wisconsin-Madison
610 Walnut Street
Madison, Wisconsin 53705 U.S.A.

ABSTRACT. We present mathematical results that are needed to analyse novel phenomena occurring in dynamic shearing flows of highly elastic and viscous non-Newtonian fluids. The key property of solutions to the time-dependent, quasilinear partial differential equations that are used to model such flows is a non-monotonic relation between the steady shear stress and strain rate. The phenomena discussed may lead to material instabilities that could disrupt polymer processing.

Introduction. Viscoelastic materials with fading memory, e.g. polymers, suspensions and emulsions, exhibit behavior that is intermediate between the nonlinear hyperbolic response of purely elastic materials and the strongly diffusive, parabolic response of viscous fluids; they incorporate a subtle dissipative mechanism induced by effects of the fading memory (see extensive discussion and survey of mathematical results in [13], [2]).

The purpose of this lecture, primarily based on recent joint work with D. Malkus and B. Plohr (see [7], [8], [9]), is to present mathematical results that are needed to describe novel phenomena in dynamic shearing flows of non-Newtonian fluids of importance to advanced materials engineering and process design. Examples of such materials are high strength polymers and lubricant additives used in spinning of synthetic fibers and injection molding. The materials are often highly elastic and very viscous; consequently, their dynamic response involes multiple time-scales. Understanding the qualitative and quantitative behavior of the equations of motion coupled with various constitutive assumptions has proved to be of significant physical, mathematical, and computational interest. The phenomena discussed below appear to be relevant to material instabilities that can disrupt polymer processing. Details of the theory developed can be found in the papers cited above and in related research with R. Pego and A. Tzavaras in [10].

One striking phenomenon, called "spurt", was observed by Vinogradov *et al.* [14] in the flow of highly elastic and viscous, monodispersive polyisoprenes through capillaries. They found that under quasistatic loading, the volumetric flow rate increased dramatically at a critical stress that was independent of molecular weight. Until recently, spurt had been overlooked or dismissed by rheologists because no plausible mechanism was capable to

R. Spigler (ed.), Applied and Industrial Mathematics, 219–229.

explain it in the context of steady flows. Spurt was associated with failure of the polymer to adhere to the wall and was lumped together with instabities such as "slip", "apparent slip", "melt fracture" (see [1], [12]). The inability to to reproduce spurt by applying known computational techniques to the steady governing equations was generally attributed to a change of type from elliptic to hyperbolic (at a high Weissenberg or Deborah number) resulting in Hadamard instability.

Satisfactory modeling and explanation of spurt and related phenomena require studying the full dynamics of the equations of motion and constitutive equations. The key feature of constitutive models that exhibit spurt is a non-monotonic relation between the steady shear stress and strain rate. This allows jumps in the steady strain rate to form when the driving pressure gradient exceeds a critical value; such jumps correspond to the sudden increase in volumetric flow rate observed in the experiments of Vinogradov *et al.* [14]. Earlier, Hunter and Slemrod [3] studied the qualitative behavior of these jumps in an elementary, one-dimensional viscoelastic model of rate type and they predicted shape memory and hysteresis effects related to spurt. A salient feature of their model is linear instability and loss of evolutionarity in a region of state space. in a region of state space. By contrast, the time-dependent system of quasilinear PDE's we use to model the dynamic behavior of shear flows through a slit die remains linearly stable and evolutionary, as one expects of realistic models. The governing system exhibits spurt, as well as related phenomena of shape memory and hysteresis under cyclic loading and unloading observed in numerical experiments. Our analysis also predicts other effects: latency, normal stress oscillations and molecular weight dependence of hysteresis; these remain be tested in rheological experiment.

The problem described below to model these flows under isothermal and incompressible conditions derives from a three-dimensional constitutive equation (see [4], [11]) that embodies the response characteristics of the fluid. Essential properties of constitutive relations are exhibited in simple planar Poiseuille shear flow. We study shear flow of a non-Newtonian fluid between parallel plates, located at $x = \pm h/2$, with the flow aligned along the y-axis, symmetric about the center line, and driven by a constant pressure gradient $\overline{f}$. We restrict attention to the simplest model of a single relaxation-time differential model that possesses steady state solutions exhibiting a non-monotone relation between the total steady shear stress and strain rate, and thereby reproduces spurt and related phenomena discussed below. The total shear stress T is decomposed into a polymer contribution and a Newtonian viscosity contribution. When restricted to one space dimension the initial-boundary value problem, in non-dimensional units with distance scaled by h, governing the flow can be written in the dimensionless form:

$$\begin{aligned} \alpha v_t - \sigma_x &= \varepsilon v_{xx} + \overline{f} \,, \\ \sigma_t - (Z+1)v_x &= -\sigma \,, \\ Z_t + \sigma v_x &= -Z \end{aligned} \tag{JSO}$$

on the interval $[-1/2, 0]$, with boundary conditions

$$v(-1/2,t) = 0 \quad \text{and} \quad v_x(0,t) = 0 \tag{BC}$$

and initial conditions

$$v(x,0) = v_0(x)\,, \quad \sigma(x,0) = \sigma_0(x)\,, \quad \text{and} \quad Z(x,0) = Z_0(x)\,, on \; -1/2 \le x \le 0; \tag{IC}$$

symmetry of the flow and compatibility with the boundary conditions requires that $v_0(-1/2) = 0$, $v_0'(0) = 0$ and $\sigma_0(0) = 0$. The first equation in (JSO) describes the balance of linear momentum. The evolution of σ, the polymer contribution to the shear stress, and of Z, a quantity proportional to the normal stress difference, are governed by the second and third equations in system (JSO). As a result of scaling motivated by numerical simulation and introduced in [8], there are only three essential parameters: α is a ratio of Reynolds number to Deborah number, ε is a ratio of viscosities, and $\overline{f}$ is the constant pressure gradient. Observe that in this formulation the incompressibility condition is automatically satisfied; also, the usual convective terms in the balance of linear momentum vanish.

1. Summary of Results. When $\varepsilon = 0$, and $Z + 1 \geq 0$, system (JSO) is hyperbolic, with characteristics speeds $\pm[(Z+1)/\alpha]^{1/2}$ and 0. Moreover, for smooth intial data in the hyperbolic region and compatible with the boundary conditions, techniques in [13] can be used to establish global well-posedness (in terms of classical solutions) if the data are small, and finite-time blow-up of classical solutions if the data are large. If $\varepsilon > 0$, system (JSO) for any smooth or piece-wise smooth data; indeed, general theory developed in [10] (see Sec. 3 and particularly Appendix A) yields global existence of classical solutions for smooth initial data of arbitrary size, and also existence of almost classical, strong solutions with discontinuities in the initial velocity gradient and in stress components; the latter result allows one to prescribe discontinuous initial data of the same type as the discontinuous steady states studied in this paper. The asymptotic behavior of solutions is difficult to determine due to a lack of available a priori estimates. In unpublished work with Pego and Tzavaras, we are showing that every solution of (JSO) tends to equilibrium as $t \to \infty$ if α is sufficiently small, as suggested by numerical simulation in [5], [6], [7], [8].

The steady-state solutions of system (JSO) play an important role in our discussion. Such a solution, denoted by $\overline{v}$, $\overline{\sigma}$, and $\overline{Z}$, can be described as follows. The stress components $\overline{\sigma}$ and $\overline{Z}$ are related to the strain rate $\overline{v}_x$ through the relations

$$\overline{\sigma} = \frac{\overline{v}_x}{1+\overline{v}_x^2}\,, \quad \overline{Z} + 1 = \frac{1}{1+\overline{v}_x^2}\,. \tag{1.1}$$

Therefore, the steady total shear stress $\overline{T} := \overline{\sigma} + \varepsilon\overline{v}_x$ is given by $\overline{T} = w(\overline{v}_x)$, where

$$w(s) := \frac{s}{1+s^2} + \varepsilon s\,. \tag{1.2}$$

The properties of w, the steady-state relation between shear stress and shear strain rate, are crucial to the behavior of the flow. By symmetry, it suffices to consider $s \geq 0$. For all $\varepsilon > 0$, the function w has inflection points at $s = 0$ and $s = \sqrt{3}$. When $\varepsilon > 1/8$, the function w is strictly increasing, but when $\varepsilon < 1/8$, the function w is not monotone. Lack of monotonicity is the fundamental cause of the non-Newtonian behavior; hereafter we assume that $\varepsilon < 1/8$.

The graph of w is shown in Fig. 1. Specifically, w has a maximum at $s = s_M$ and a minimum at $s = s_m$, where it takes the values $\overline{T}_M := w(s_M)$ and $\overline{T}_m := w(s_m)$ respectively. As $\varepsilon \to 1/8$, the two critical points coalesce at $s = \sqrt{3}$. Evidently, there is a unique steady solution for each $\overline{T} < \overline{T}_m$ and for each $\overline{T} > \overline{T}_M$; however, there are three possible steady

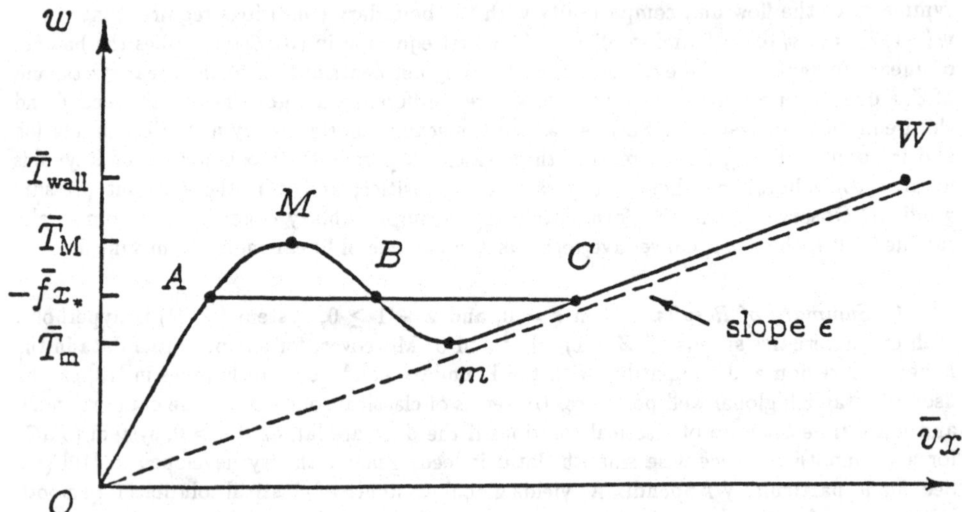

Fig. 1: Total steady shear stress $\overline{T}$ *vs.* shear strain rate $\overline{v}_x$ for steady flow. The case of three steady solutions is illustrated.

states for each $\overline{T}_m < \overline{T} < \overline{T}_M$. If $\overline{T} = \overline{T}_m$ or $\overline{T} = \overline{T}_M$ (respectively called "bottom" and "top" jumping), there are two possible steady states.

The momentum equation, together with the boundary condition at the centerline, implies that the steady total shear stress satisfies $\overline{T} = -\overline{f}x$ for every $x \in [-\frac{1}{2}, 0]$. Therefore, the steady velocity gradient can be determined as a function of x by solving

$$w(\overline{v}_x) = -\overline{f}x \,. \tag{1.3}$$

Equivalently, a steady state solution $\overline{v}_x$ satisfies the cubic equation $P(\overline{v}_x) = 0$, where

$$P(s) := \varepsilon s^3 - \overline{T} s^2 + (1+\varepsilon)s - \overline{T} \,. \tag{1.4}$$

The steady velocity profile is obtained by integrating $\overline{v}_x$ and using the boundary condition at the wall. However, because the function w is not monotone, there might be up to three distinct values of $\overline{v}_x$ that satisfy Eq. (1.3) for any particular x on the interval $[-1/2, 0]$. Consequently, $\overline{v}_x$ can suffer jump discontinuities, resulting in kinks in the velocity profile in marked contrast to the smooth velocity profile in classical Newtonian Poiseuille flow. Indeed, a steady solution must contain such a jump if the total stress $\overline{T}_{wall} = \overline{f}/2$ at the wall exceeds the total stress $\overline{T}_M$ at the local maximum M in Fig. 1. The larger area under the steady velocity profile for a spurting solution yields a correspondingly larger volumetric flow rate in the Vinogradov experiment. We also remark that the flow problem discussed here can also be modelled by a system based on a differential constitutive law with two widely spaced relaxation times but no Newtonian viscosity contribution; it is shown in Sec. 2. of [13] that with an appropriate choice of relevant parameters, the resulting governing system exhibits the same steady states and the same qualitative properties as (JSO).

Effective numerical methods were developed in [5], [6], [7], [8] for simulating transient, one-dimensional shear flows in a slit die at high Weissenberg (Deborah) number and very low Reynolds number (appropriate for highly elastic and very viscous fluids). When restricted to flows of monodisperse polyisoprenes, results of numerical simulation using these methods agreed qualitatively and quantitatively with results of the Vinogradov loading experiment exhibiting spurt [5], [6], [7]. Furthermore, numerical calculations also exhibited the following non-Newtonian phenomena in cyclic loading and unloading: latency preceding spurt, and shape memory, hysteresis, normal stress oscillations, as well as molecular weight dependence of hysteresis when loading is followed by unloading; these latter phenomena require additional testing and verification in rheological experiment in order to identify the underlying physical process responsible for the non-Newtonian characteristics.

A great deal of information about the structure of solutions of system (JSO) can be garnered by studying a one parameter family of quadratic system of ordinary differential equations that approximates it in a certain parameter range. Motivation for this approximation comes from the following observation: in experiments of Vinogradov *et al.* [14], α is of the order 10^{-13} while ε is of the order 10^3; thus the term αv_t in the momentum equation of system (JSO) is negligible even when v_t is moderately large. This leads us to approximate the system (JSO) by setting $\alpha = 0$. Then the momentum equation in system (JSO) can be integrated and one sees that the total shear stress $T := \sigma + \varepsilon v_x$ coincides with the steady total stress $\overline{T}(x) = -\overline{f}x$. Thus $T = \overline{T}(x)$ is a function of x only, but σ and v_x are functions of both x and t. Consequently, the remaining equations of system (JSO) yield the one-parameter family of autonomous, quadratic, planar systems of ordinary differential equations:

$$\begin{aligned} \dot{\sigma} &= (Z+1)\left(\frac{\overline{T}-\sigma}{\varepsilon}\right) - \sigma\,, \\ \dot{Z} &= -\sigma\left(\frac{\overline{T}-\sigma}{\varepsilon}\right) - Z\,, \end{aligned} \tag{Q}$$

where x is the parameter. Here, the "dot" denotes the derivative d/dt. We emphasize that for each $\overline{f}$, a different dynamical system is obtained at each x on the interval $[-1/2, 0]$ because $\overline{T} = -\overline{f}x$. By symmetry, we focus attention on the case $\overline{T} > 0$; also recall that $\varepsilon < 1/8$. The dynamical system (Q) can be analyzed completely by phase-plane analysis as outlined below; the reader is referred to Sec. 3 in [9] for details. Here we state the main results.

The critical points of system (Q) satisfy the algebraic system

$$\begin{aligned} &(Z+1+\varepsilon)\left(\frac{\sigma}{\overline{T}} - 1\right) + \varepsilon = 0\,, \\ &\frac{\overline{T}^2}{\varepsilon}\frac{\sigma}{\overline{T}}\left(\frac{\sigma}{\overline{T}} - 1\right) - Z = 0\,. \end{aligned} \tag{1.5}$$

These equations define, respectively, a hyperbola and a parabola in the σ-Z phase plane. The critical points are intersections of these curves and they lie in the strip $0 < \sigma < \overline{T}$. Eliminating Z in equations (1.5) shows that the σ-coordinates of the critical points satisfy the cubic equation $Q(\sigma/\overline{T}) = 0$, where

$$Q(\xi) := \left[\frac{\overline{T}^2}{\varepsilon}\xi(\xi-1)+1+\varepsilon\right](\xi-1)+\varepsilon\,. \tag{1.6}$$

A straightforward calculation using Eqs. (1.4), (1.6) shows that

$$P(\overline{v}_x) = P\left(\frac{\overline{T}-\sigma}{\varepsilon}\right) = -\frac{\overline{T}}{\varepsilon}Q(\sigma/\overline{T})\,. \tag{1.7}$$

Thus each critical point of the system (Q) defines a steady-state solution of system (JSO): such a solution corresponds to a point on the steady total-stress curve (see Fig. 1) at which the total stress is $\overline{T}(x)$.

The algebraic signs of the trace, the determinant and the discriminant of the matrix of the linearization of system (Q) at each critical point, and standard perturbation theory of ODE's yields the following *local structure* in neighborhoods of critical points:

- For each $\overline{f}$, each x, and each $0 < \varepsilon < 1/8$, there are three possibilities:

(1) There is a single critical point A when $\overline{T} < \overline{T}_m$; A is an attracting node (classical attractor).

(2) There is a single critical point C if $\overline{T} > \overline{T}_M$; C is an attracting spiral point (spurt attractor).

(3) There are three critical points A, B, and C when $\overline{T}_m < \overline{T} < \overline{T}_M$; B is a saddle point and A, C as above. For $\overline{T}$ close to $\overline{T}_m$, C is an attracting node.

The next task is to determine the *global structure* of the flow. For this purpose, the key result, proved by applying Bendixson's theorem to a rescaled form of system (Q)), is:

- System (Q) has neither periodic orbits nor separatrix cycles.

  Another important ingredient of the analysis is the fact that:

- System (Q) is endowed with the identity

$$\frac{d}{dt}\{\sigma^2 + (Z+1)^2\} = -2\left[\sigma^2 + (Z+\tfrac{1}{2})^2 - \tfrac{1}{4}\right]\,. \tag{L}$$

  The function $V(\sigma, Z) := \sigma^2 + (Z+1)^2$ serves as a Lyapunov function for (Q). It is crucial that identity (L) is independent of $\overline{T}$ and ε. Let Γ denote the circle on which the right side of identity (L) vanishes; note that the critical points of (Q) are restricted to lie on Γ for every value of $\overline{T}$. Let C_r denote the circle of radius r centered at $\sigma = 0$ and $Z = -1$, i.e. $C_r := \{(\sigma, Z) : V(\sigma, Z) = r, r > 0\}$; each circle C_r is a level set of V. It is easy to see that:

- Each closed disk bounded by $C_r, r \geq 1$, is positively invariant for (Q).

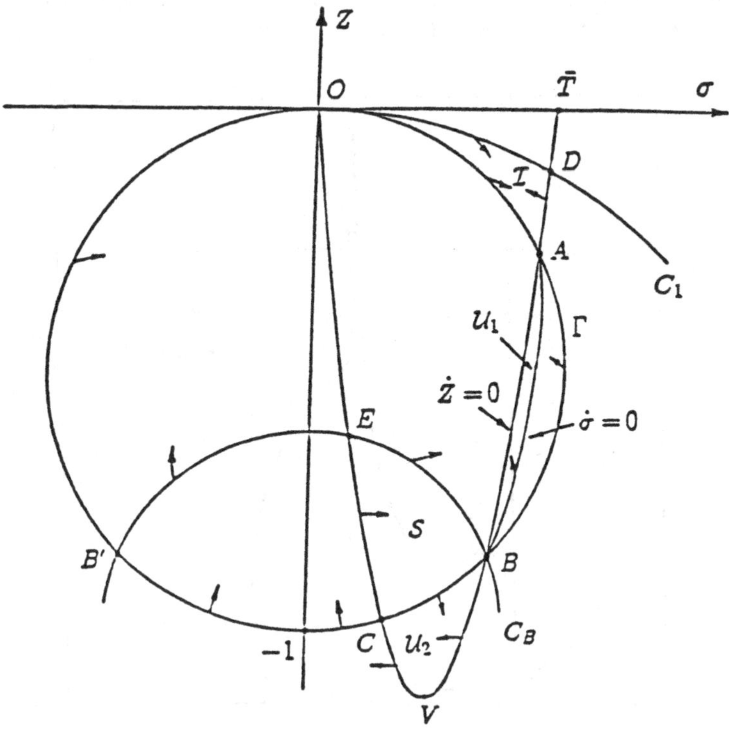

Fig. 2: Invariant Regions (3 critical points).

Invariant sets, such as those shown in Fig. 2 are used to determine the global structure of the orbits and the stable and unstable manifolds of the saddle point at B.

Elementary considerations detailed in [9] yield:

- **Structure of the Flow for** $\overline{T}_m < \overline{T} < \overline{T}_M$: There are three critical points. The basin of attraction of A, *i.e.*, the set of points that flow toward the classical attractor A as $t \to \infty$, comprises those points on the same side of the stable manifold of B as is A; points on the other side are in the basin of attraction of the spurt attractor C. Moreover, the arc of the circle Γ through the origin in Fig. 2, between B and its reflection B' is contained in the basin of attraction of A. In particular, the stable manifold for B cannot cross Γ on the arc $B'0B$. The resulting structure of the flow is shown in Fig. 3.

A similar (simpler) analysis yields:

- **Structure of the Flow for** $\overline{T} > \overline{T}_M$: All the orbits flow towards the spurt attractor (spiral point) at C as $t \to \infty$ (the figure is omitted).

- **Structure of the Flow for $\overline{T} < \overline{T}_m$:** All the orbits flow towards the classical attractor (attracting node) at A as $t \to \infty$ (the figure is omitted).

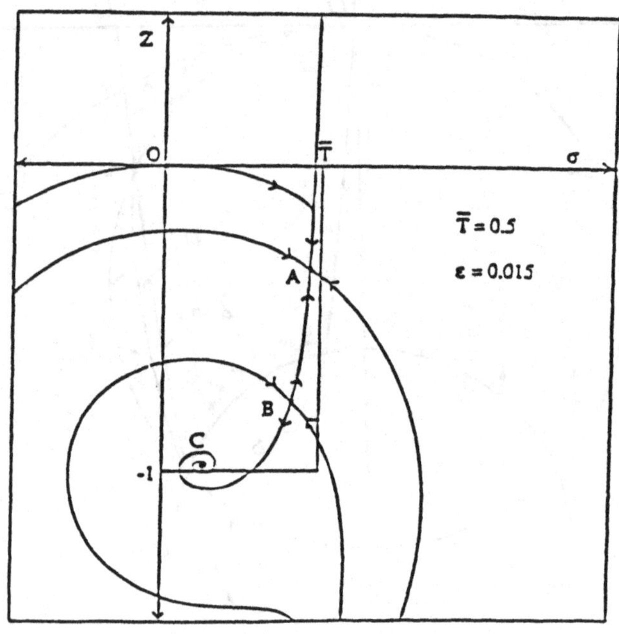

Fig. 3: Flow for Three Critical Points

We remark that from the point of view of phase plane analysis, the top and bottom jumping steady solutions corresponding to $\overline{T} = \overline{T}_M$ and $\overline{T} = \overline{T}_m$ respectively are saddle-node bifurcations.

The results of the phase plane analysis can be used to explain various loading and unloading experiments. Start with steady flow at initial forcing $\overline{f}_0$; each x at $\overline{f}_0$ is either a classical or a spurt attractor. Load or unload forcing to $\overline{f} = \overline{f}_0 + \Delta\overline{f}$. Phase-plane analysis implies:

- A classical point A_0 for the initial forcing $\overline{f}_0$ lies in the domain of attraction of the classical attractor A for $\overline{f}$, provided that A exists (*i.e.*, $|\overline{f}x| < \overline{T}_M$).
- A spurt point C_0 for the initial forcing $\overline{f}_0$ lies in the domain of attraction of the spurt attractor C for $\overline{f}$ unless (a) C does not exist (*i.e.*, $|\overline{f}x| < \overline{T}_m$); or (b) C lies on the classical side of the stable manifold of the saddle. *Remark:* Case (b) never occurs on loading, and also not if $|\Delta\overline{f}|$ is sufficiently small.

Starting with $\overline{f}_0 = 0$, load to $\overline{f} > 0$. The initial state for each x lies at the origin $\sigma = 0$, $Z = 0$. Therefore, each $x \in [-1/2, 0]$ such that $\overline{f}|x| < \overline{T}_M$ is a classical attractor; each x for which $\overline{f}|x| > \overline{T}_M$ is a spurt attractor. We conclude:

- If the forcing is subcritical (*i.e.*, $\overline{f} < \overline{f}_{\text{crit}} := 2\overline{T}_M$), the asymptotic steady flow is entirely classical.

- If the forcing is supercritical ($\overline{f} > \overline{f}_{\text{crit}}$), there is a single kink in the velocity profile, located at $x_* = -\overline{T}_M/\overline{f}$; those $x \in [-1/2, x_*)$, near the wall, are spurt points, whereas $x \in (x_*, 0]$, near the centerline, are classical ("top jumping" – the stress $\overline{T}_* = \overline{T}_M$ at the kink as large as possible; as close as possible to wall).

These results are used in [9] to explain quasistatic, cyclic loading and unloading leading to spurt (preceded by latency), shape memory, and hysteresis that were observed in numerical experiments in [6], [7], [8] as shown in Fig. 4. A somewhat more complicated complicated argument in [9] also explains possible occurrence of steady flows with multiple kinks and also flow reversal.

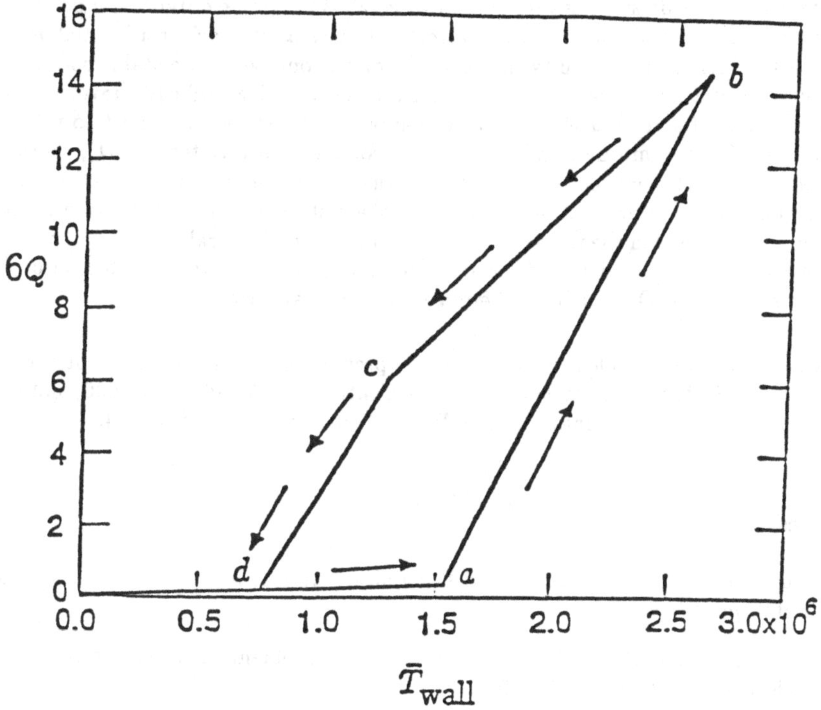

Fig. 4: Hysteresis under cyclic load.

We believe that the results described above are not limited to the specific model discussed above. As already mentioned, they hold for the Johnson-Segalman constitutive law with two widely separated relaxation times. In addition, numerical simulation in [5] using the Giesekus differential constitutive law with a single relaxation time and a small Newtonian viscosity also suggests the occurence of spurt solutions.

The analysis of the complete dynamics of the approximating quadratic system (Q) raises several challenging mathematical issues. Numerical simulation in [6], [7], [8] suggests that at least for $\alpha > 0$ sufficiently small, the dynamics of the full system (JSO) of governing

PDE's are the same as those of (Q). In this direction, a collaboration with R. Pego and A. Tzavaras [10] identifies nonlinearly stable, discontinuous steady states for a model initial-boundary value problem in one space dimension for incompressible, isothermal shear flow of a non-Newtonian fluid between parallel plates, driven by a constant pressure gradient. The non-Newtonian contribution to the shear stress is assumed to satisfy a single, much simpler differential constitutive law than is the case for system (JSO). The model incorporates several features of the more complex problem described above (e.g., for a particular choice, the two systems have the same steady states). Again, the key property is a non-monotone relation between the total steady shear stress and steady shear strain rate that results in steady states having, in general, discontinuities in the strain rate that also lead to steady velocity profiles with kinks, We explain why every solution, regardless of the size of the data and of α tends to a steady state as $t \to \infty$, and we establish a simple criterion for determining which steady states are nonlinearly asymptotically stable if a set of arbitrarily small measure that includes points of discontinuity in the strain rate is omitted. The stable solutions are local minimizers of a nonconvex functional, and the key technical difficulty involves finding invariant intervals of a related ODE that are uniform in length with respect to position x in the channel. For the more complex problem associated with system (JSO), research in progress employing different techniques indicates that discontinuous steady states that take their values on increasing parts of the steady total shear stress vs strain rate curve are stable in the same sense, provided the ratio α of Reynolds number to Deborah number is sufficiently small. Another approach under study involves resolving the singular perturbation problem that relates systems (JSO) and (Q), where α is the singular parameter.

Acknowledgement. This research and the preparation of this manuscript were supported by the U. S. Army Research Office under Contract DAAL 03-87-K-0036, and by the National Science Foundation under Grants DMS-8712058 and DMS-8907264.

References

1. M. Denn, "Issues in Viscoelastic Fluid Dynamics," *Annual Reviews of Fluid Mechanics* **22** (1990), pp. 22–34.

2. W. Hrusa, J. Nohel, and M. Renardy, "Initial Value Problems in Viscoelasticity," *Appl. Mech. Rev.* **41** (1988), pp. 371–378.

3. J. Hunter and M. Slemrod, "Viscoelastic Fluid Flow Exhibiting Hysteretic Phase Changes," *Phys. Fluids* **26** (1983), pp. 2345–2351.

4. M. Johnson and D. Segalman, "A Model for Viscoelastic Fluid Behavior which Allows Non-Affine Deformation," *J. Non-Newtonian Fluid Mech.* **2** (1977), pp. 255–270.

5. R. Kolkka and G. Ierley, "Spurt Phenomena for the Giesekus Viscoleastic Liquid Model," *J. Non-Newtonian Fluid Mech.*, 1989. To appear.

6. R. Kolkka, D. Malkus, M. Hansen, G. Ierley, and R. Worthing, "Spurt Phenomena of the Johnson-Segalman Fluid and Related Models," *J. Non-Newtonian Fluid Mech.* **29** (1988), pp. 303–325.

7. D. Malkus, J. Nohel, and B. Plohr, "Time-Dependent Shear Flow Of A Non-Newtonian Fluid," pp. 91–109 in *Conference on Current Problems in Hyberbolic Problems: Riemann Problems and Computations*, ed. B. Lindquist, Contemporary Mathematics. Amer. Math. Soc., Providence, R. I., 1989.

8. D. Malkus, J. Nohel, and B. Plohr, "Dynamics of Shear Flow of a Non-Newtonian Fluid," *J. Comput. Phys.*, 1989. To appear; also CMS Tech. Summary Rep. #89-14.

9. D. Malkus, J. Nohel, and B. Plohr, "Analysis of New Phenomena in Shear Flow of Non-Newtonian Fluids," *SIAM J. Appl. Math.*, 1989. Submitted; also CMS Technical Summary Report #90-6.

10. J. Nohel, R. Pego, and A. Tzavaras, "Stability of Discontinuous Shearing Motions of Non-Newtonian Fluids," *Proc. Royal Soc. Edinburgh*, 1989. To appear..

11. J. Oldroyd, "Non-Newtonian Effects in Steady Motion of Some Idealized Elastico-Viscous Liquids," *Proc. Roy. Soc. London* **A 245** (1958), pp. 278–297.

12. J. Pearson, *Mechanics of Polymer Processing*, Elsevier Applied Science, London, 1985.

13. M. Renardy, W. Hrusa, and J. Nohel, *Mathematical Problems in Viscoelasticity*, Pitman Monographs and Surveys in Pure and Applied Mathematics V. 35, Longman Scientific & Technical, Essex, 1987.

14. G. Vinogradov, A. Malkin, Yu. Yanovskii, E. Borisenkova, B. Yarlykov, and G. Berezhnaya, "Viscoelastic Properties and Flow of Narrow Distribution Polybutadienes and Polyisoprenes," *J. Polymer Sci., Part A-2* **10** (1972), pp. 1061–1084.

NUMERICAL ANALYSIS FOR COMPRESSIBLE VISCOUS ISOTHERMAL STATIONARY FLOWS

C. Bernardi (+), M.O. Bristeau (*), O. Pironneau (+*), M.G. Vallet (*)

INRIA BP 105 Le Chesnay FRANCE 78153 (*)
Université Paris 6 (+)

Abstract

In this paper we address the problem of approximation of viscous compressible flows by the finite element method: should one use the same approximation for the density and the velocity as in Euler flows or should one use two different spaces as for the Stokes problem? When the pressure is a function of the density (isothermal flow as an academic example) we show theoretically and numerically convergence of the approximations if the density and the velocity are approximated as in the Stokes problem with two different grids or if artificial viscosity is used in the equation of conservation of mass.

1. INTRODUCTION

For the incompressible Navier-Stokes equation it is well known that the velocity u and the pressure p cannot be approximated in the same finite element space if an optimal error estimate is desired (cf. Girault-Raviart [10] for example). This fact was discovered probably independently by Harlow et al (the Marker in cell staggered grid finite difference method [12]) and by Fortin [7], Hood-Taylor [13] for the finite element methods, Bernardi-Maday [3] for spectral methods ; Brezzi [5] gave a clear explanation of the problem and showed that even for a finite element approximation of the Stokes operator one needs an Inf-Sup condition that forces u and p to be approximated differently. Penalization as in Bercovier [2], Girault-Raviart [10] and regularization as in Hughes et al. [14], Fortin et al.[8]) are ways around this problem but they are also special tools that must be injected in the discretization of the system.

The basic question we would like to address in this presentation is the following: Should u, ρ, θ or S (velocity density temperature or reduced entropy) be approximated differently for the numerical simulation of compressible fluids ? Is it better to work with u or ρu ?

As shown by Bristeau et al. [6], Rogé [17], Rostand [27],Verfurth[20], the pressure is computed more acurately if it is approximated on a grid diffrerent from the one used for u, as in the incompressible case. This phenomenon can be explained intuitively by noting that when u is small, the density becomes almost constant and so $\nabla.u$ is small and the fluid is almost incompressible; Geymonat-Leyland [9], Soulaimani et al [19] have also reached the same conclusion in their study of the linearized system. Our conclusion is similar: *if ρ and u are approximated as p and u in the*

R. Spigler (ed.), Applied and Industrial Mathematics, 231–243.

Stokes problem then convergence can be shown under certain regularity conditions and the numerical results are much better, furthermore upwinding is not compulsory.

2. ISO-THERMAL VISCOUS FLOWS.

As in the incompressible case, we study the stationary system first. The transient case is likely to be well posed when p, u, ρ, θ are approximated on the same grid but then the problems arise when the convergence to the stationary case is desired. In mathematical terms if the system $A(w) = 0$ has a continuum of solution, the solution of

$$\frac{dw}{dt} - A(w) = 0, \quad w(0) = w^o \tag{1}$$

may not converge to the desired stationary solution.

The Navier-Stokes equations are :

$$\partial_t(\rho u) + \nabla.(\rho u \otimes u) - \eta\Delta u - \lambda\nabla(\nabla.u) + \nabla p = f \tag{2}$$

$$\partial_t \rho + \nabla.\rho u = 0 \tag{3}$$

$$\rho\partial_t\theta + \rho u\nabla\theta + (\gamma - 1)\rho\theta\nabla.u - C_v\kappa\Delta\theta = \frac{\eta}{2}|\nabla u + \nabla u^T|^2 + (\lambda - \eta)|\nabla.u|^2 \tag{4}$$

$$\frac{p}{\rho} = R\theta \tag{5}$$

with $\gamma = 1.4$ for air, R is the perfect gas constant and η, λ, κ are the viscosities and thermal diffusion.

In some cases it is possible to neglect the variations of temperature; then (2)(3) become self-contained:

$$\partial_t(\rho u) + \nabla.(\rho u \otimes u) - \eta\Delta u - \lambda\nabla(\nabla.u) + c\nabla\rho = f \tag{6}$$

$$\partial_t \rho + \nabla.\rho u = 0, \tag{7}$$

where $c = R\theta$ is *constant*. Even though this is a much simpler problem than the full Navier-Stokes system, it is complicated enough to have shocks when η, λ are zero and boundary layers, shocks/boundary-layer interactions otherwise. A similar problem was studied in [4] for the shallow water equation and the following results can be shown.

3. APPROXIMATION AND RESULTS.

In order to approximate the full problem (1)(5) we will first study problem (6)(7). On this second problem we will also study first the stationary low Reynolds number approximation, then the stationary high Reynolds number case then the full problem.

3.1 The low Reynolds number case

The stationary subproblem obtained by neglecting $\nabla.(\rho u \otimes u)$ (small Reynolds number)

$$-\eta\Delta u - \lambda\nabla(\nabla.u) + \nabla\rho = f, \quad \nabla.(\rho u) = 0 \quad \textit{in} \quad \Omega \tag{8}$$

is well posed (it has at least one regular solution) with the following boundary conditions:

u is given at the boundary Γ of Ω and ρ given on Γ^-,

where Γ^- is the part of Γ where u.n < 0, provided that $\int_\Gamma u.nd\gamma \geq 0$ (there cannot be more fluid coming in than coming out).

Problem (8) can be approximated by the finite element method with compatible approximations for u and ρ (u quadratic, ρ linear for instance) and convergence can be established. Artificial viscosity $\epsilon\log\rho-\epsilon\Delta\log\rho$ (or upwinding in practice) must be used in the equation of conservation of mass to establish the result. This means that one can show that there is a solution (u_h,ρ_h) of

$$\eta\int_\Omega \nabla u_h\nabla v_h+\lambda\int_\Omega \nabla.u_h\nabla.v_h+\int_\Omega v_h\nabla\rho_h=\int_\Omega fv_h,\quad \forall v_h\in V_{oh};\quad u_h-u_{\Gamma h}\in V_{oh} \tag{9}$$

$$-\int_\Omega \rho_h u_h\nabla q_h+\epsilon\int_\Omega q_h\log\rho_h+\epsilon\int_\Omega \nabla q_h\nabla\log\rho_h+\int_{\Gamma-}\rho_\Gamma u_\Gamma.nq_h+\int_{\Gamma-\Gamma-}\rho_h u_\Gamma.nq_h=0\quad\forall q_h\in Q_h \tag{10}$$

$$V_{oh}=v_h \text{ continuous piecewise quadratic on the triangulation, } v_{h|\Gamma}=0 \tag{11}$$

$$Q_h=\{q_h \text{ continuous piecewise linear on the triangulation }\} \tag{12}$$

Each solution is isolated (unique in some neighborhood) and $(u_h,\rho_h)\to(u,\rho)$ when $\epsilon,h\to0$.

3.2 The stationary case

If the Reynolds number is large the same is probably true but there is a technical difficulty when $u.n\neq0$ so we state the result only when $u.n=0$; the driven cavity problem is one instance of such flows.

Problem (6)(7) has a stationary solution when u is given on the boundary with u.n=0 (no data on ρ are needed). Let $\Pi_h(e^{\sigma_h})$ denote the linear interpolate of e^{σ_h} ; the problem

$$\eta\int_\Omega \nabla u_h\nabla v_h+\lambda\int_\Omega \nabla.u_h\nabla.v_h+\int_\Omega v_h\nabla\rho_h+\int_\Omega v_h\nabla.(\rho_h u_h\otimes u_h)=\int_\Omega fv_h,\quad \forall v_h\in V_{oh};\quad u_h-u_{\Gamma h}\in V_{oh} \tag{13}$$

$$-\int_\Omega \rho_h u_h\nabla\Pi_h(e^{\sigma_h})+\epsilon\int_\Omega q_h\sigma_h+\epsilon\int_\Omega \nabla q_h\nabla\sigma_h+\int_{\Gamma-}\rho_\Gamma u_\Gamma.nq_h+\int_{\Gamma-\Gamma-}\Pi_h(e^{\sigma_h})u_\Gamma.nq_h=0\quad\forall q_h\in Q_h \tag{14}$$

has at least one solution $(u_h,\rho_h=e^{\sigma_h})$, each solution is isolated and it converges to a stationary solution (u,ρ) of (6)(7) when $\epsilon,h\to0$.

3.3 The full problem.

The need for artificial viscosity in the mass equation seems to be more theoretical than practical as we shall demonstrate in the next section. Thus the following formulation has been studied (see [4] for more details)

$$\sigma_{,t}+u\nabla\sigma+\nabla.u=0 \tag{15}$$

$$u_{,t}+u\nabla u+(\gamma-1)(\theta\nabla\sigma+\nabla\theta)=\eta e^{-\sigma}(\Delta u+\frac{1}{3}\nabla(\nabla.u)) \tag{16}$$

$$\theta_{,t}+u\nabla\theta+(\gamma-1)\theta\nabla.u=e^{-\sigma}(\kappa\Delta\theta+F(\nabla u)) \tag{17}$$

$F(\nabla u)=\eta|\nabla u+\nabla u^T|^2/2-2\eta|\nabla.u|^2/3$ (and θ and κ have been normalized).

where σ is $\log\rho$. An implicit time discretisation is made where the temperature is treated explicitly :

$$\frac{1}{k}\sigma^{n+1}+u\nabla\sigma^{n+1}+\nabla.u^{n+1}=\frac{1}{k}\sigma^n, \tag{18}$$

$$\frac{1}{k}u^{n+1}+\nabla.(\rho u^{n+1}\otimes u^{n+1})-\eta e^{-\sigma^n}(\Delta u^{n+1}+\frac{1}{3}\nabla(\nabla.u^{n+1}))+(\gamma-1)\theta^n\nabla\sigma^{n+1}=\frac{1}{k}u^n-(\gamma-1)\nabla\theta^n, \tag{19}$$

$$(1+\frac{1}{k}(\gamma-1)\nabla.u^n)\theta^{n+1}+u\nabla\theta^{n+1}-e^{-\sigma^n}\kappa\Delta\theta^{n+1}=\frac{1}{k}\theta^n+e^{-\sigma^n}F(\nabla u^n) \tag{20}$$

The discretization in space is done as above:

$$\frac{1}{k}\int_\Omega \sigma_h^{n+1}w_h-\int_\Omega \nabla.(u_h^n w_h)\sigma_h^{n+1}+\int_{\Gamma-\Gamma-}\sigma_h^{n+1}u_\Gamma.nw_h+\int_\Omega \nabla.u_h^{n+1}w_h \tag{21}$$

$$=\frac{1}{k}\int_\Omega \sigma_h^n w_h-\int_{\Gamma-}\sigma_\Gamma u_\Gamma.nw_h,\quad \forall w_h\in Q_h$$

$$\frac{1}{k}\int_\Omega e^{\sigma_h^n}u_h^{n+1}v_h+\int_\Omega e^{\sigma_h^n}v_h\nabla.(\rho_h u_h^{n+1}\otimes u_h^{n+1})+\eta\int_\Omega \nabla u_h^{n+1}\nabla v_h+\frac{\eta}{3}\int_\Omega \nabla.u_h^{n+1}\nabla.v_h \tag{22}$$

$$+(\gamma-1)\int_\Omega e^{\sigma_h^n}\theta_h^n\nabla\sigma_h^{n+1}v_h=\frac{1}{k}\int_\Omega e^{\sigma_h^n}u_h^n-(\gamma-1)\int_\Omega e^{\sigma_h^n}\nabla\theta_h^n v_h,\forall v_h\in V_{oh};\quad u_h-u_{\Gamma h}\in V_{oh}$$

$$\int_\Omega(1+\frac{1}{k}(\gamma-1)\nabla.u_h^n)e^{\sigma_h^n}\theta_h^{n+1}w_h+\int_\Omega e^{\sigma_h^n}u_h\nabla\theta_h^{n+1}w_h+\kappa\int_\Omega \nabla\theta_h^{n+1}\nabla w_h \tag{23}$$

$$=\frac{1}{k}\int_\Omega e^{\sigma_h^n}\theta_h^n w_h+\int_\Omega F(\nabla u_h^n)w_h\quad \forall w_h\in Q_h$$

Like all other non-linear systems in this paper this system is solved with the GMRES algorithm (see Appendix).

4. DISCUSSION

4.1 The inf-sup condition.

The proof of result 1 relies heavily on the fact that the momentum equation implies that ρ is regular if u is regular, i.e. in mathematical words :

$$|\rho-\int_\Omega \rho|_o\le c_1\sup_{v,v|\Gamma=0}\frac{\int\nabla\rho v}{|\nabla v|_o}=c\sup_{v,v|\Gamma=0}\eta\frac{\int_\Omega\nabla u\nabla v+\lambda\int_\Omega\nabla.u\nabla.v-\int_\Omega fv}{|\nabla v|_o} \tag{24}$$

$$\le c_2(|\nabla u|_o+|f|_o)$$

A similar inequality must be true for the discrete system (9)-(10) but while the first inequality in (24) is always true in the continuous case, it is true in the discrete case only for compatible couples Q_h-V_{oh} :

$$|\rho_h-\int_\Omega \rho_h|\le c\sup_{v_h\in V_{oh},v|\Gamma=0}\frac{\int_\Omega\nabla\rho_h v_h}{|\nabla v_h|_o},\quad \forall\rho_h\in Q_h. \tag{25}$$

This is the Inf-Sup or BBL condition.

From the numerical point of view one can argue that the number of unknowns should be equal to the number of equations ; a centered finite difference discretization of (6) could not work there unless ρ is sampled at the center of the cells and u at the mid-sides for instance;

From the physical point of view, we know that near the stagnation point, for example, the velocities are small and the flow is nearly incompressible and so whatever is true of incompressible fluids should hold in such regions.

There are two ways to obtain a bound on ρ from the bound on u; either by taking compatible spaces Q_h-V_{oh} or by regularizing the equation of continuity (see [7], [14]) by adding a second order term in ρ as it is done in (10)). When a time marching algorithm is used, upwinding can be applied to obtain stability ; there is a connection between upwinding and artificial viscosity so this may be why schemes which use the same grid for u_h, ρ_h, θ_h still produce reasonable results at stagnation points or near walls. To prove convergence of (9)-(10) toward (6), we have used both artificial viscosity and compatible spaces $Q_h - V_{oh}$; but artificial viscosity is there only to insure local uniqueness of ρ_h and it can be made to tend to zero before h.

If inequality (25) is indeed the key to convergence then for the general system (θ non constant) it would seem that θ_h defined piecewise linear (same grid as ρ_h) would suffice . Finally, the same argument indicates that u is a better variable than ρu unless some changes are made to the formulation of the viscous terms and work with $\Delta(\rho u)$ instead.

4.2 Artificial viscosity.

It is well known that the P1-bubble/P1 element is equivalent to Hughes' regularization procedure in some cases. For example, on the Stokes problem

$$-\Delta u + \nabla p = f \quad \nabla . u = 0 \quad u|_\Gamma = 0 \tag{26}$$

in variational form

$$(\nabla u, \nabla v) + (\nabla p, v) = (f, v) \quad \forall v \in H_0^1(\Omega) \tag{27}$$

$$(u, \nabla q) = 0 \quad \forall q \in L^2(\Omega) \tag{28}$$

and approximated as follows by the finite element method

$$(\nabla u, \nabla v) + (\nabla p, v) = (f, v) \quad \forall v \in V_{0h} \tag{29}$$

$$(u, \nabla q) = 0 \quad \forall q \in Q_h \tag{30}$$

We can make the following analysis. Denote by u_i and u_b the degree of freedom of u at the vertices and at the center of the ρ triangles respectively. Let v^i, v^b be the corresponding basis functions (hat functions). Then (29) is also

$$\sum u_i(\nabla v^i, \nabla v^b) + \sum u_b(\nabla v^b, \nabla v^b) + (\nabla p, v^b) = (f, v^b) \tag{31}$$

and since $(\nabla v^i, \nabla v^b)$ =0, $(\nabla v^{b_1}, \nabla v^{b_2}) = 0$ we can compute u_b :

$$u_b = \frac{-(\nabla p, v^b) + (f, v^b)}{(\nabla v^b, \nabla v^b)} \tag{32}$$

So (30) becomes

$$(u, \nabla q) = 0 \quad \Leftrightarrow \quad \sum u_i(v^i, \nabla q) + \sum u_b(v^b, \nabla q) = 0 \tag{33}$$

$$\sum u_i(v^i, \nabla q) + \sum - \frac{(\nabla p, v^b) + (f, v^b)}{(\nabla v^b, \nabla v^b)}(v^b, \nabla q) = 0. \tag{34}$$

or

$$(u, \nabla q) + \sum -c_b(\nabla p, \nabla q) = -\frac{(f, v^b)}{(\nabla v^b, \nabla v^b)}(v^b, \nabla q) \tag{35}$$

with

$$c_b = \frac{(v^b, 1)^2}{(\nabla v^b, \nabla v^b)|T_b|}, \tag{36}$$

$|T_b|$ being the area of the triangle which support v^b (the ρ triangle). In turn this is a FEM discretization on the ρ triangles of

$$\nabla.u - \nabla.c(h)\nabla p = \nabla.[c(h)f] \tag{37}$$

Thus the P1-bubble/P1 method contains artificial viscosity in the mass equation and it is probably sufficient to prove the convergence of the method.

5. NUMERICAL SIMULATIONS

To illustrate this study, the following tests were made.

1. Flow around a NACA0012 solution of (8) .

At Reynolds number 200 and Mach number 0.85 either in an unbounded fluid or in a canal. Three approximations were tested:

1. P1/P1: piecewise linear on triangles T for u_h and ρ_h .

2. P1 iso P2/P1: piecewise linear on triangles T for ρ_h and piecewise linear on the triangulation obtained by dividing each T into 4 triangles by the mid-sides

3. P1-bubble P1: i.e. P1/P1 but the each ρ triangle is divided in 3 by its center to generate 3 u-triangles.

No upwinding is applied and the GMRES algorithm is used. The oscillations present with the first discretization disappear within the second and third indicating that these instabilities are not due to the convecting character of the equation of conservation of mass.

2. Flow around a NACA0012 solution of (2)(3)(4)

At Reynolds number 5000 and Mach number 0.85 the results show again that the oscillations with the P1/P1 approximation can be removed when a P1 iso P2/P1 method is used. Here also the GMRES quasi Newton method is used and no upwinding is applied. As shown on figures refining the mesh does not remove the oscillations. These test cases also show that upwinding is not necessary to compute transonic viscous flows.

6. CONCLUSION

Navier-Stokes solvers are more than Euler solvers with added viscosity because such modifications which do not take into account the dramatic difference between Euler and Navier-Stokes may not give good results in regions where the viscosity play an important part like stagnation points. We have shown that the cure to oscillations in these regions is to use staggered grids or different spaces of approximations for velocity and density ; sometimes upwinding can do the trick but situations are likely to occur where it would not be sufficient. Finally, we have suggested a simple model system (eq. (6)) which needs the same tools and for which error analysis can be made.

REFERENCES

[1] D. Arnold, F. Brezzi, M. Fortin: A stable finite element for the Stokes equations. *Calcolo* 21(4) 337-344 ,(1984).

[2] M. Bercovier : Perturbation of mixed variational problems. *RAIRO série rouge* 12(3) 211-236 ,(1978).

[3] C. Bernardi, Y. Maday: A collocation method over staggered grids for the Stokes problem. *Int. J. Numer Meth. in Fluids* . 8 , 537-557 (1988).

[4] C. Bernardi, O. Pironneau: On the shallow water equations at small Reynolds number. Internal report, Université Paris 6, 1989.

[5] F. Brezzi: On the existence, uniqueness and approximation of saddle point problems arising from Lagrangian multiplier. *RAIRO serie Analyse Numerique* R2,129-151, (1974).

[6] M.O. Bristeau , R. Glowinski, B. Mantel, J. Periaux, G. Rogé: Adaptative finite element methods for three dimensional compressible viscous flow simulation in aerospace engineering. Proc. 11th Int. Conf. on Numerical methods in fluid mechanics. Williamsburg, Virginia (1988)(*Springer* , to appear)

[7] M. Fortin: Calcul numérique des écoulements par la méthode des éléments finis. *Thèse* Université Paris 6. (1972).

[8] M. Fortin, A. Soulaimani: Finite element approximation of compressible viscous flows. Proc. computational methods in flow analysis, vol 2, H. Niki and M. Kawahara eds. *Okayama University Science press* . (1988).

[9] G. Geymonat, P. Leyland: Analysis of the linearized compressible Stokes problem (to appear).

[10] V. Girault - P.A. Raviart: *Finite Element methods for Navier Stokes equations* . Springer Series SCM Vol **5** ,(1986).

[11] R. Glowinski: *Numerical methods for nonlinear variationnal problems.* Springer Lectures Notes in Computationnal Physics, (1985).

[12] F.H. Harlow, J.E. Welsh: The Marker and Cell method.*Phys. Fluids* **8** , 2182-2189, (1965).

[13] P. Hood - G. Taylor: Navier-Stokes equations using mixed interpolation. in *Finite element in flow problem* Oden ed. UAH Press,(1974).

[14] T.J.R. Hughes, L.P. Franca, M. Mallet: A new finite element formulation for computational fluid dynamics. *Comp. Meth. in Appl. Mech. and Eng* . **63** 97-112 (1987).

[15] O. Pironneau: *Finite element methods for fluids* . 1989 Wiley.

[16] O. Pironneau, J. Rappaz: Finite element approximation of compressible adiabatic viscous flows. *IMPACT* 1989.

[17] G. Rogé: Approximation mixte et accélération de la convergence pour les équations de Navier-Stokes compressible en éléments finis. *Thèse* , Université Paris 6 (1989).

[18] Ph. Rostand, B. Stoufflet: TVD schemes to compute compressible viscous flows on unstructured meshes. Proc. 2nd Int Conf. on hyperbolic problems, Aachen (FRG) (to appear in *Vieweg*) (1988).

[19] A. Soulaimani, M. Fortin, Y. Ouellet, G. Dhatt, F. Bertrand: Simple continuous pressure element for 2D and 3D incompressible flows. *Comp. Meth. in Appl. Mech. and Eng* . **62** 47-69 (1987).

APPENDIX

The GMRES algorithm.

GMRES is a quasi-Newtonian method: for

$$F(u) = 0, \quad u \in R^N$$

The iterative procedure used is of the type :

$$u_{n+1} = u_n - (J_n)^{-1} F(u_n)$$

where $J_n \approx F'(u_n)$. To avoid the calculation of F' the following approximation may be used:

$$D_\delta F(u; v) \equiv \frac{F(u + \delta v) - F(u)}{\delta} \cong F'(u)v. \tag{1}$$

As in the conjugate gradient method to find the solution v of $Jv = -F$ one considers the Krilov spaces :

$$K_n = Sp\{r^0, Jr^0 ..., J^{n-1}r^0\}$$

where $r^0 = -F - Jv^0$, v^0 is an approximation of the solution to find. In GMRES the approximated solution v^n used is the solution of

$$\min_{v \in K_n} ||r^0 - J(v - v^0)||.$$

Algorithm (GMRES) :

0 . *Initialisation:* Choose the dimension k of the Krylov space; choose u_0. Choose a tolerance ϵ and an increment δ , choose a preconditioning matrix $S \in R^{N \times N}$. Put $n = 0$.

1. a. Compute with (1) $r_n^1 = -S^{-1}(F_n + J_n u_n)$, $w_n^1 = r_n^1 / \, ||r_n^1||$, where $F_n = F(u_n)$ and $J_n v = D_\delta F(u_n; v)$.

b. For $j = 2..., k$ compute r_n^j and w_n^j from

$$r_n^j = S^{-1}[D_\delta F(u_n; w_n^{j-1}) - \sum_{i=1}^{j-1} h_{i,j-1}^n w_n^i]$$

$$w_n^j = \frac{r_n^j}{||r_n^j||}.$$

where $h_{i,j}^n = w_n^{iT} S^{-1} \, D_\delta F(u_n; w_n^j)$

2. Find u_{n+1} the solution of

$$\min_{v = u_n + \sum_0^k a_j w_j^n} ||S^{-1}F(v)||^2$$

$$\cong \min_{a_1, a_2, ... a_k} ||S^{-1}[F(u_n) + \sum_{j=1}^{k} a_j D_\delta F(u_n; w_n^j)]||^2$$

3. If $||F(u_{n+1})|| < \epsilon$ stop else change n into n+1.

The implementation of Y. Saad also includes a back-tracking procedure

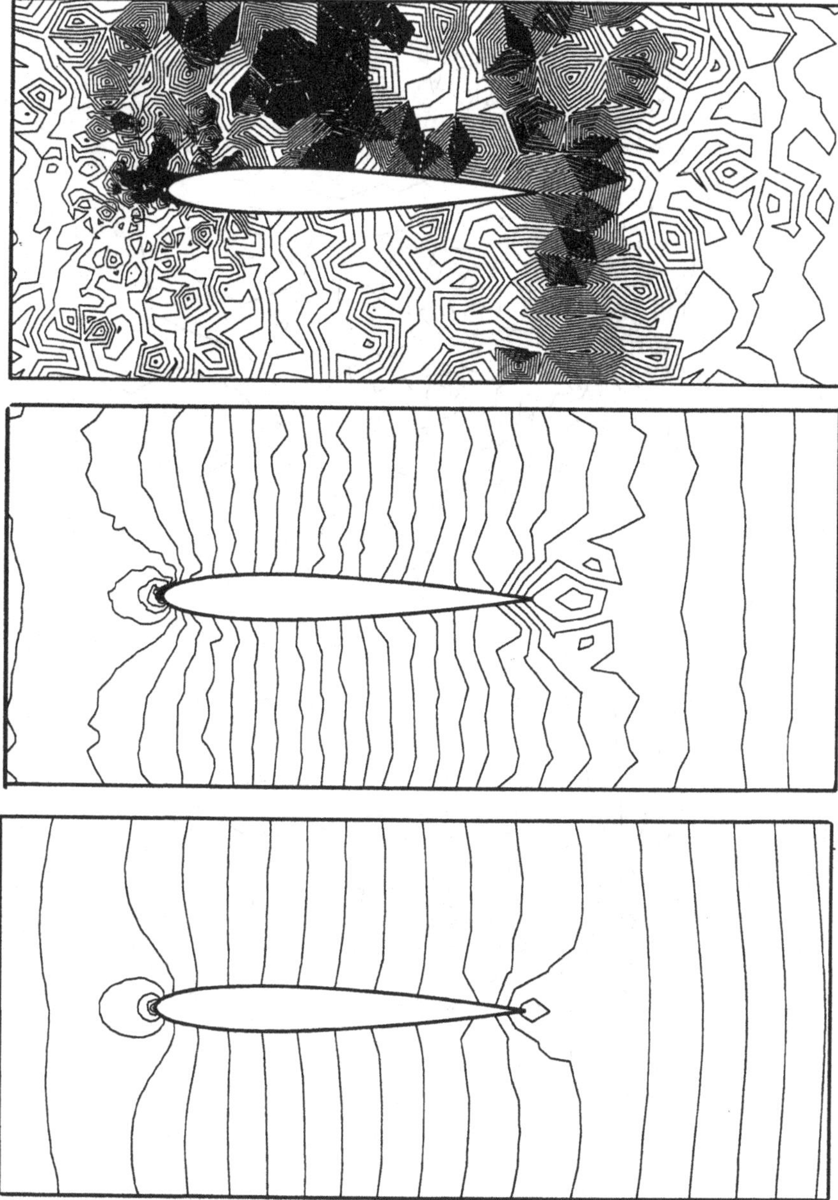

FIGURE 1: Problem (8) :Iso-density lines when the P1/P1 element is used (top), P1-bubble/P1 (middle) P1-iso P2/P1 (bottom). Mach=0.85, Re=200.

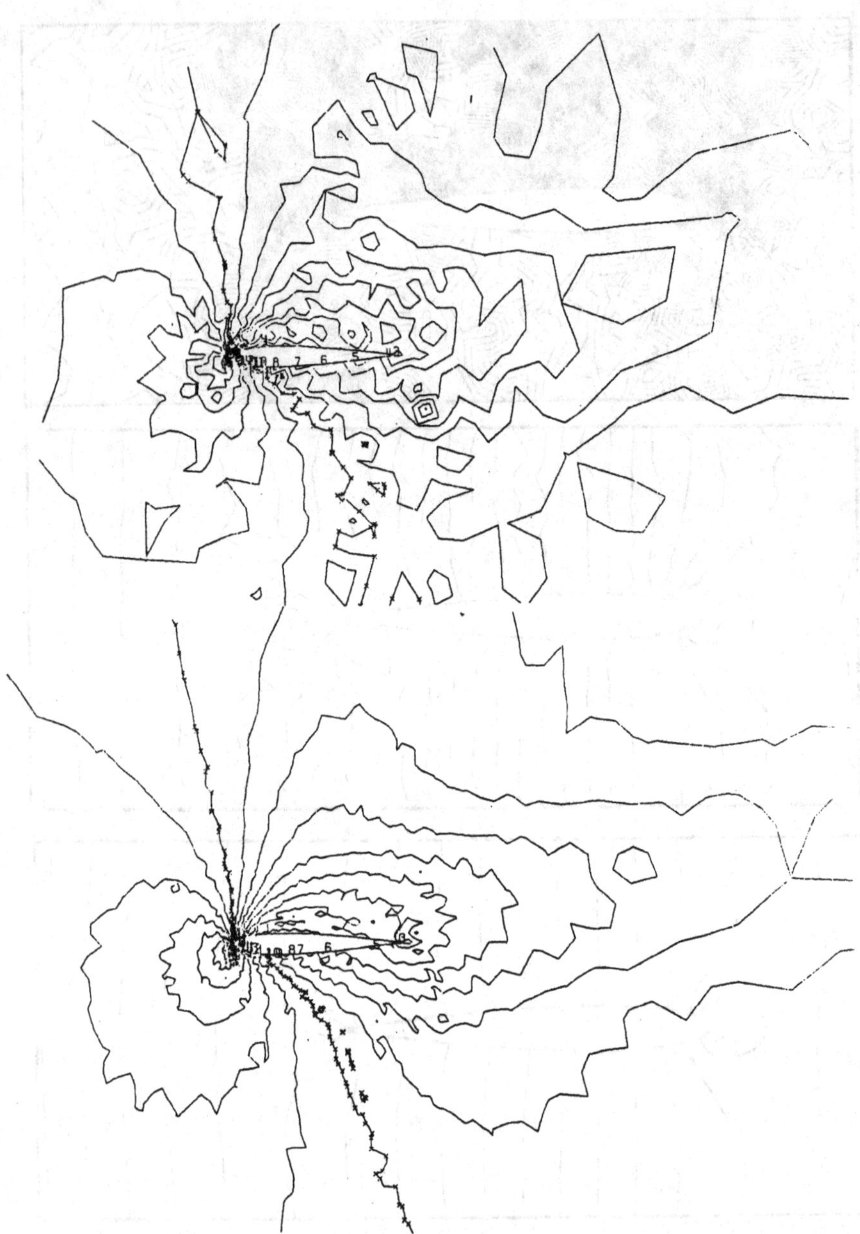

FIGURE 2: Problem (6-7) :Iso-Mach lines when the P1/P1 element is used on a coarse mesh (top) and a fine mesh (bottom). Mach=0.8, Re=73.Incidence angle=10°

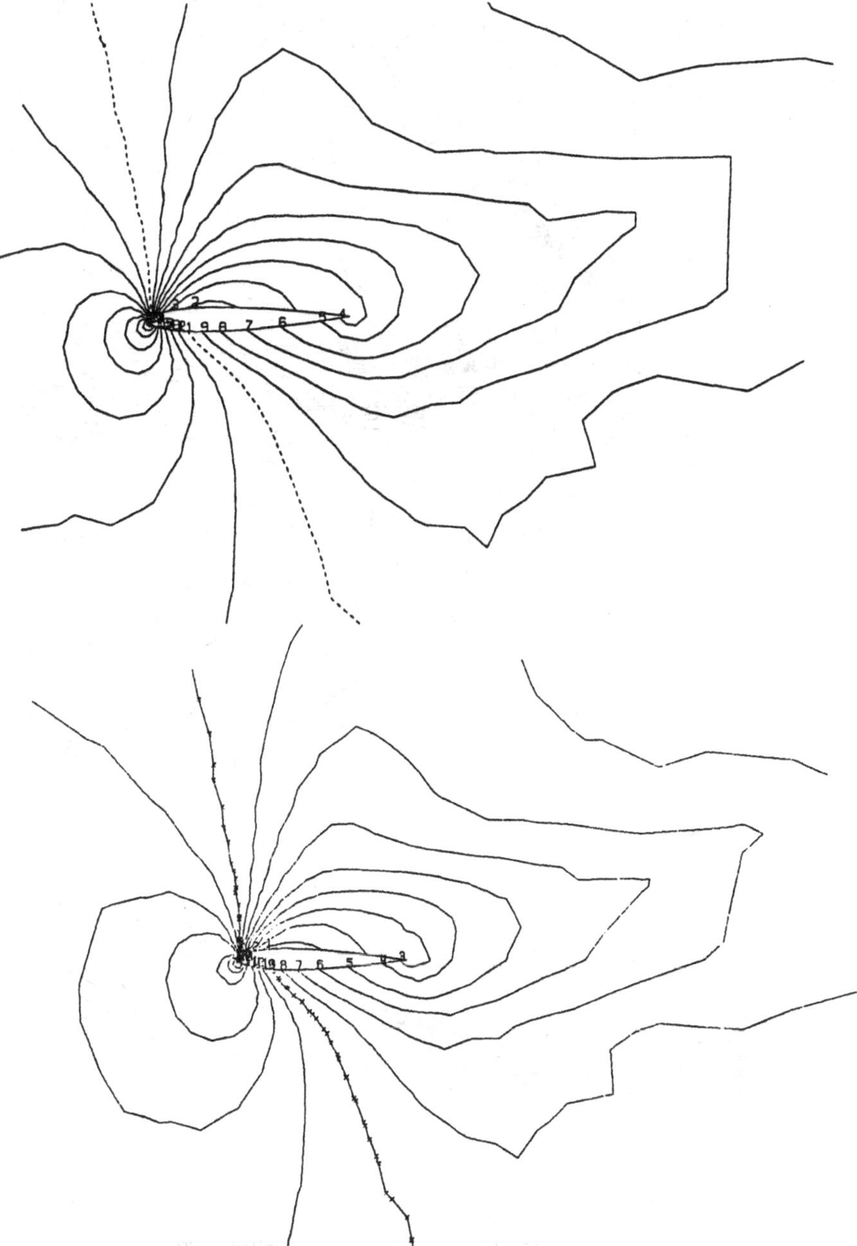

FIGURE 3: Problem (6-7) :Iso-Mach lines when the P1-bubble/P1 (bottom) element is used and the P1-iso P2/P1 element (top). Mach=0.8, Re=73.Incidence angle=10°

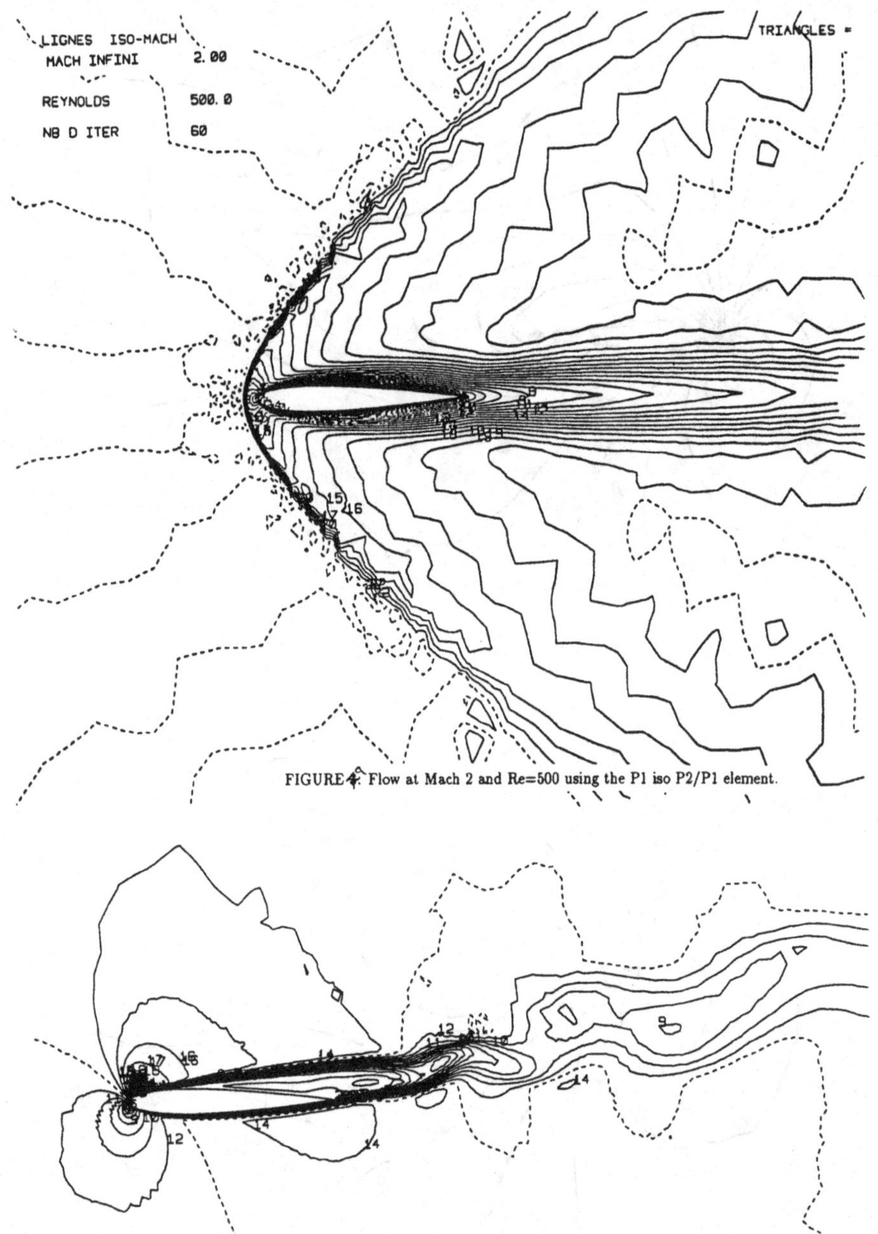

FIGURE 4a: Flow at Mach 2 and Re=500 using the P1 iso P2/P1 element.

FIGURE 4b: Transient flow (Mach 0.6, Re=2000 incidence 10°)

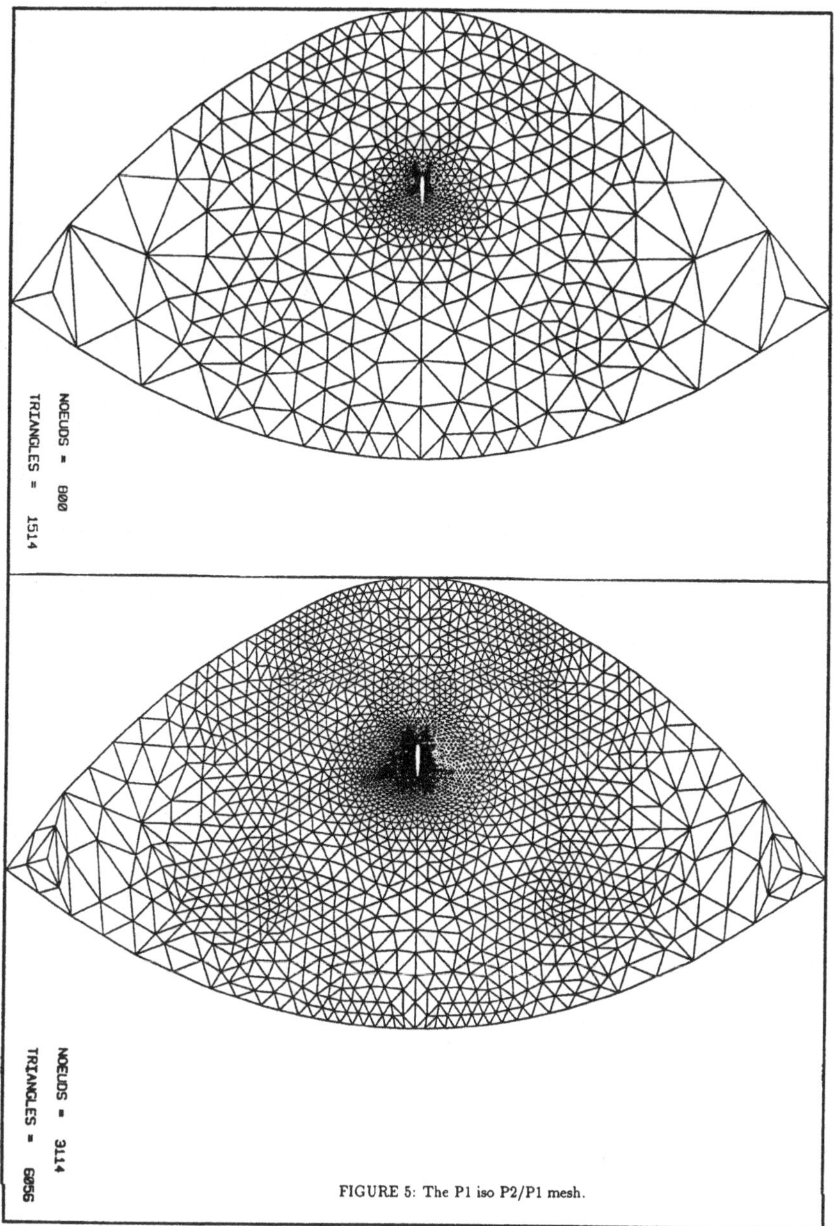

FIGURE 5: The P1 iso P2/P1 mesh.

DROPS OF NEMATIC LIQUID CRYSTAL FLOATING ON A FLUID

D. ROCCATO
E. G. VIRGA
Dipartimento di Matematica
Universita' di Pavia
27100 PAVIA, ITALY

ABSTRACT. We study the equilibrium shapes of a drop of nematic liquid crystal floating on a fluid. We find an instability.

The problem of finding the equilibrium shapes of a drop of nematic liquid crystal floating on an isotropic fluid is fascinating, though hard to solve. We shall lay assumptions that simplify the variational problem and permit us to solve the equilibrium equations.

The drops we consider are in contact with two isotropic fluids: namely, a liquid and a gas above it. According to the theory of Oseen [1] and Frank [2], the **bulk free energy** of the drop is the functional

$$F_{\mathrm{bulk}}[B,\mathbf{n}] := \int_B \sigma(\mathbf{n},\nabla\mathbf{n}), \tag{1}$$

where B is the region of space occupied by the drop and the unit vector $\mathbf{n}$ is the **orientation** of the optical axis. The form of the scalar function σ will not be relevant to our study; we just note here that it depends on the material and the temperature (which is taken as constant throughout this work).

When a liquid crystal is in contact with an isotropic fluid, the **free energy of the interface** is given by the functional

$$F_{\mathrm{surface}}[S,\mathbf{n}] := \int_S w(\mathbf{n}\cdot\nu), \tag{2}$$

where S is the surface of contact and ν is the outer unit normal. We employ for w the formula of Rapini and Papoular [3], which has been confirmed by several

R. Spigler (ed.), Applied and Industrial Mathematics, 245–253.

experiments [4,5,6]. We express this formula in terms of a dimensionless parameter ω, ultimately related to the temperature:

$$w(\mathbf{n}\cdot\nu) = w(\cos\theta) := w_0(1+\omega\cos^2\theta); \tag{3}$$

$w_0\omega$ is known as the **anchoring energy**, θ is often referred to as the **tilt angle.**

For many liquid crystals the surface energy prevails over the bulk energy (see *e.g.* [7] and the references cited therein); *thus henceforth we neglect* F_{bulk} *in the free energy.* Furthemore, for many nematics there is much evidence that the **homeotropic anchoring condition** (which requires $\theta = 0$) holds at the emerging boundary of the drop whenever the gas is air. *If the volume of the submerged part of the drop is much smaller than the total volume, we expect the emerging boundary to be spherical and the orientation of the optical axis to be the radial field* $\mathbf{e}_r$. We call B_0 the ball whose boundary contains the emerging boundary of the drop and whose center is the singular point of $\mathbf{e}_r$.

The problem of minimizing the free energy of the drop reduces to: *Find the surface* S^- *that minimizes*

$$F_{\text{surface}}[S^-,\mathbf{e}_r] = \int_{S^-} w_0(1+\omega\cos^2\theta), \quad \cos\theta = \mathbf{e}_r\cdot\nu, \tag{4}$$

subject to

$$\text{vol}(B) = \beta \qquad \beta > 0, \tag{5}$$

$$\text{vol}(B^-) = \beta^- \qquad \beta^- > 0, \tag{6}$$

where vol denotes the volume measure and B^- is the submerged part of B. The constraint (5) amounts to regard liquid crystals as incompressible fluids, and (6) comes from the law of buoyancy.

We take S^- *to be symmetric about* $\mathbf{e}_z$*, the vertical axis passing through the center of* B_0*; we assume that the intersection of* S^- *and any plane containing* $\mathbf{e}_z$ *may be represented by a function* $\rho(\varphi)$ (see Figure 1).
We presume to may adjust the values of φ_0 and $\rho_0 := \rho(\varphi_0)$ to fulfill the constraints (5) and (6). Here we regard both φ_0 and ρ_0 as prescribed.

For given $\varphi_1 \in [0,\varphi_0[$ we define the class of functions

$$C(\varphi_1) := \{\theta : [0,\varphi_0] \to [-\frac{\pi}{2},\frac{\pi}{2}] \mid \theta|_{[\varphi_1,\varphi_0]} = 0,\ \theta|_{[0,\varphi_1[} \in C^1\}.$$

Each $\theta \in C(\varphi_1)$ describes a surface S^- that adheres to ∂B_0 for $\varphi \in [\varphi_1,\varphi_0]$ and may have an edge along the circle where $\varphi = \varphi_1$, though no extra energy is attached to such an edge (see Figure 1). $C(0)$ is a singleton whose element is $\theta = 0$; the corresponding shape of the drop is the ball B_0 itself. If $\varphi_1 > 0$ all non-negative functions of $C(\varphi_1)$, except $\theta = 0$, represent **bumps** outside B_0; similarly, all non-positive functions of $C(\varphi_1)$, except $\theta = 0$, represent **craters** inside B_0.

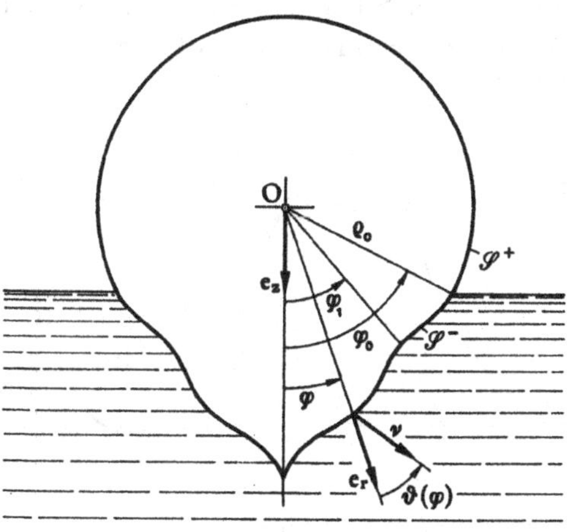

Figure 1

In the classes $C(\varphi_1)$ the free energy of the submerged boundary of the drop becomes a functional of θ which depends on φ_1:

$$F_{\text{surface}}[S^-, \mathbf{e}_r] =: F[\varphi_1, \theta] \quad , \quad \theta \in C(\varphi_1).$$

The **equilibrium shapes** of the drop make F stationary; the **stable shapes** minimize F.

The functional F is stationary with respect to the variations of both φ_1 and θ if

$$h(t(\tau_1); \omega) := \frac{1 + \omega + t^2(\tau_1)}{\sqrt{1 + t^2(\tau_1)}} - 1 - \omega \ = 0, \tag{7}$$

$$t'(\tau) = -\frac{1+t^2}{1+\tau^2} \frac{2\tau(1 + \omega + (2\omega+1)t^2) + t(1 - \omega + t^2)}{\tau(1 - \omega + (2\omega + 1)t^2)} =: f(t, \tau; \omega) \qquad \tau \in \,]0, \tau_1[\,, \tag{8}$$

where the change of variables

$$\tau := tg\varphi \quad , \quad t := \text{tg}\theta$$

has been employed, and $\tau_1 := \text{tg}\varphi_1$. Figure 2 shows the graph of $h(\cdot; \omega)$;

$$\hat{t} := \sqrt{\omega^2 - 1} \qquad , \quad t_0 := \sqrt{\omega - 1}.$$

Thus $t(\tau_1)$ is uniquely determined when $-1 < \omega \leq 1$, and it may take three different values when $\omega > 1$.

For every $\omega > -1$ we seek $\tau_1 \geq 0$ and a solution of (8) such that $t(\tau_1)$ satisfies (7). Of course, $\tau_1 = 0$ and $t(\tau_1) = 0$ represent a trivial solution of (7) and (8): it is the only element of $C(0)$.

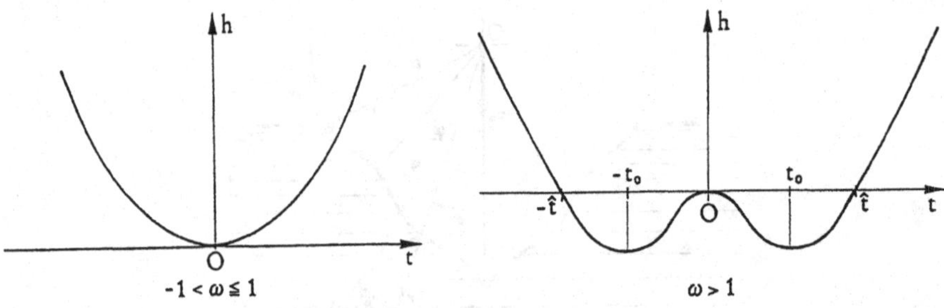

Figure 2

Whether the solution of (8) is monotone or not, it depends on the sign of f; Figures 3,4, and 5 represent the situation when ω is chosen in the intervals $-1 < \omega < -\frac{1}{2}$, $-\frac{1}{2} \leq \omega \leq 1$, and $\omega > 1$, respectively.

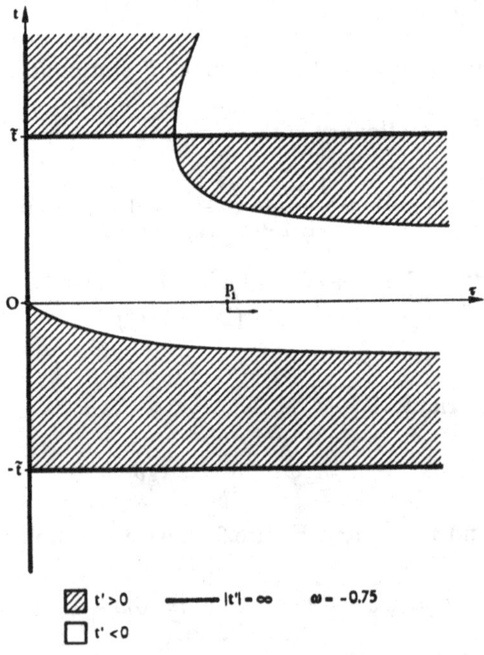

Figure 3

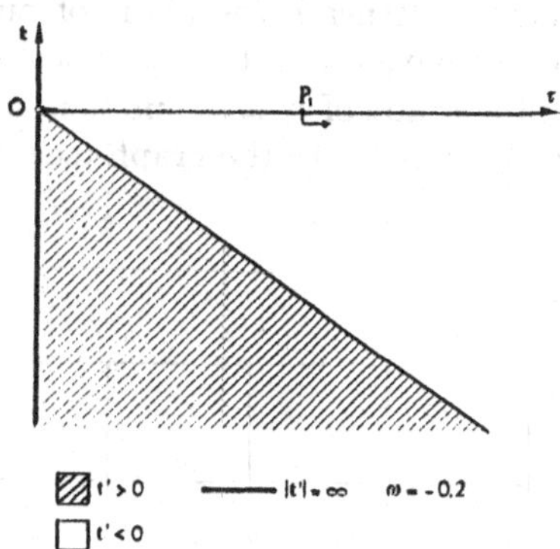

Figure 4

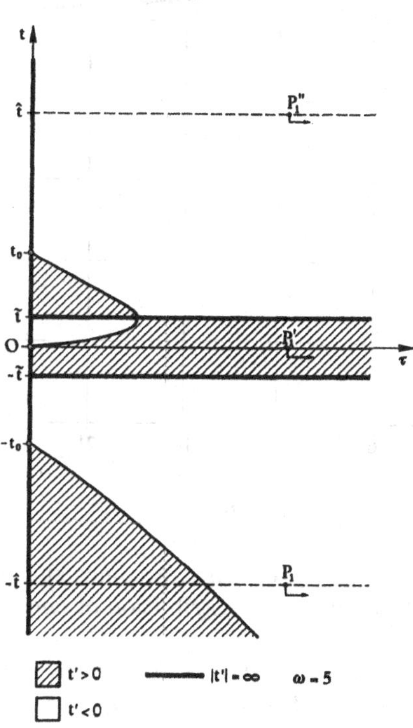

Figure 5

A tedious analysis based upon these graphs led us to conclude that *there is only one non-trivial solution of* (7) *and* (8): *it lives when* $\omega > 1$ *and it is such that* $t(0) = -t_0$ *and* $t(\hat{\tau}_1) = -\hat{t}$ (*cf.* [8] for more details). The corresponding shape of the drop $\hat{B}$ bears a crater underneath the plane of buoyancy: the size of the crater is determined by $\hat{\tau}_1$. We have computed numerically this solution for many values of ω. Figure 6 shows the graph of $\hat{\tau}_1$ as a function of ω. We have proved that the line $\hat{\tau}_1 = \frac{1}{4}$ is indeed an asymptote to the graph (see [8]).

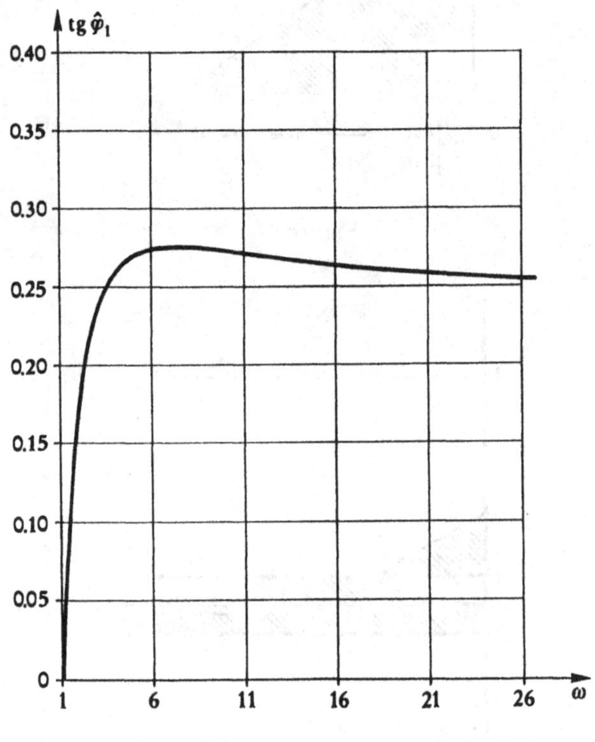

Figure 6

The shape $\hat{B}$ possesses an edge, which becomes sharper as $\hat{t}$ increases. This edge might disappear, if an adequate distribution of energy along it is taken into account.

When $-1 < \omega \leq 1$ the only equilibrium shape of the drop is the ball B_0. When $\omega > 1$ there are two equilibrium shapes, namely B_0 and $\hat{B}$.

It remains to decide which of the two minimizes the energy. Let F_0 and $\hat{F}$ be the values of F for B_0 and $\hat{B}$, respectively. It turns out that

$$F_0 - \hat{F} = 2\pi\rho_0^2 w_0 \sqrt{\omega}\hat{\varphi}_1^2(g(\omega) - 1) + o(\hat{\varphi}_1^2),$$

where $\hat{\varphi}_1 := \text{arctg}\hat{\tau}_1$ and g is the function plotted in Figure 7.

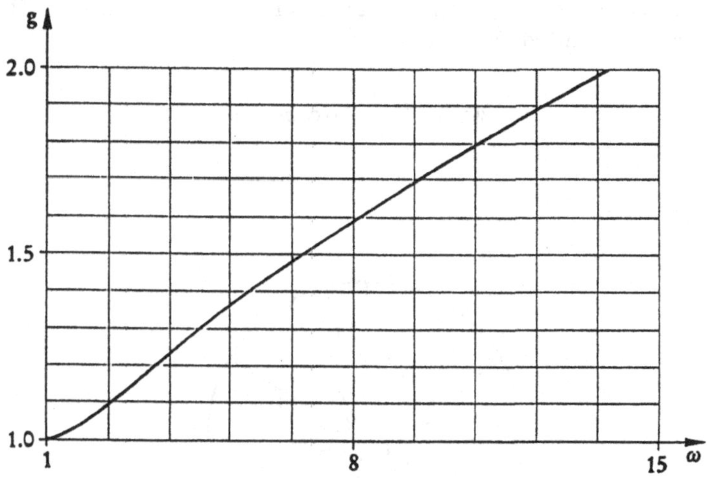

Figure 7

Thus $\hat{B}$ is the stable shape of the drop when $\omega > 1$. A **bifurcation** with exchange of stability occurs at $\omega = 1$ in the class of shapes we have employed for the drop.

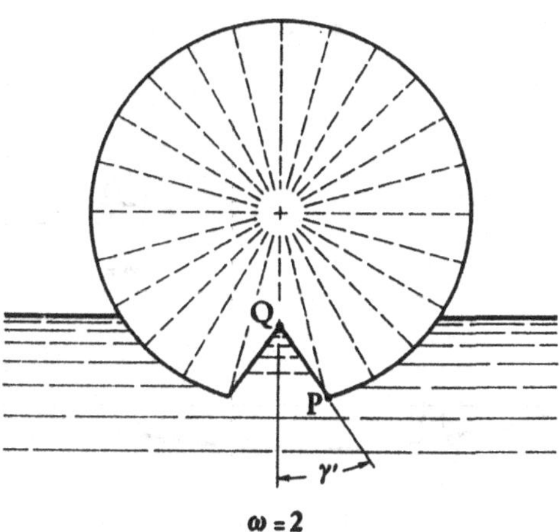

Figure 8

Figures 8 and 9 show the shape $\hat{B}$ that has been determined numerically for $\omega = 2$ and $\omega = 10$. At first glance the crater of $\hat{B}$ looks like a cone.At a closer inspection it differs slightly from a cone. Indeed, the set

$B_0 \backslash \hat{B}$ may either be convex (as for $\omega = 2$) or concave (as for $\omega = 10$). This set possesses always a pointed **tip**, whose amplitude is

$$2(\gamma + \gamma') = 2\arcsin \frac{1}{\sqrt{\omega}}$$

(see Figures 8,9, and 10).

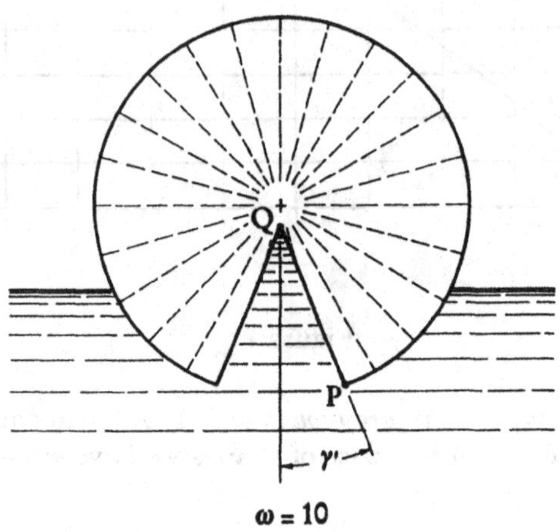

Figure 9

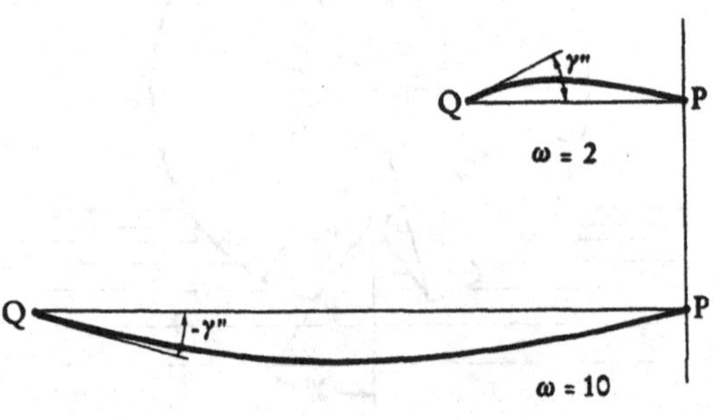

Figure 10

A shape like $\hat{B}$ has apparently been observed by Piéranski (*cf.* plate III on p. XVI of [9]). We are led to predict that rounded shapes resembling B_0 should also be observed when the temperature and the nature of the fluids in contact make $\omega < 1$.

References.

[1] Oseen C.W., The theory of liquid crystals, Trans. Faraday Soc., **29** (1933), 883-889.

[2] Frank F.C., On the theory of liquid crystals, Discuss. Faraday Soc., **28** (1958), 19-28.

[3] Rapini A. and M. Papoular, J.Phys. (Paris), **30** C4 (1969), 54-56.

[4] Naemura S., Measurement of anisotropic interfacial interactions between a nematic liquid crystal and various substrates, Appl. Phys. Lett., **33** (1978), 1-3.

[5] Naemura S., Anisotropic interactions between MBBA and surface-treated substrates, J.Phys. (Paris), **40** C3 (1979), 514-518.

[6] Riviére D., J.Lévy and C. Guyon , Determination of anchoring energies from surface tilt angle measurements in a nematic liquid crystal, J.Phys. (Paris) Lett., **40** (1979), L215-L218.

[7] Virga E.G., Drops of nematic liquid crystals, Arch. Rational Mech. Anal., **107** (1989), 371-390.

[8] Roccato D. and E.G. Virga, Drops of nematic liquid crystals floating on a liquid, *to appear in* Riv. Mat. Univ. Parma (1990).

[9] Kléman M., Points, lines and walls, J. Wiley & Sons, Chichester *etc.*, 1983.

THE KORTEWEG-DE VRIES EQUATION WITH SMALL DISPERSION: HIGHER ORDER LAX–LEVERMORE THEORY

STEPHANOS VENAKIDES
Department of Mathematics
Duke University
Durham, NC 27706 USA

ABSTRACT. We utilize the inverse scattering transformation to study the rapid oscillations which arise in the solution of the initial value problem:

$$u_t + 6uu_x + \epsilon^2 u_{xxx} = 0,$$

$$u(x,0) = v(x),$$

when $\epsilon \to 0$. We refine the Lax-Levermore theory, which gives the weak limit of the solution as $\epsilon \to 0$, by introducing a quantum condition. This allows us to calculate the waveform of the local oscillations up to phase-shifts when ϵ is small.

0. Introduction

The initial value problem for the Korteweg-de Vries Equation (KdV):

$$(1) \qquad u_t + 6uu_x + \epsilon^2 u_{xxx} = 0$$

with small dispersion (small $\epsilon > 0$) is a model for the formation of dispersive shocks. As the solution of the unperturbed problem

$$(2) \qquad u_t + 6uu_x = 0$$

develops a sharp front while approaching a shock, local fast oscillations are generated at the front and propagate in the medium [4]. Shocking of the modulation equations [3] which govern these oscillations may occur at a later time, in which case more complicated oscillations with a higher degree of freedom arise [5]. All this is in sharp contrast to the well-known regularization of the shocks of (2) through a diffusive term in Burger's equation: $u_t + 6uu_x - \epsilon u_{xx} = 0$.

R. Spigler (ed.), Applied and Industrial Mathematics, 255–262.

In this study we perform an asymptotic analysis of the exact solution of the initial value problem (1) as $\epsilon \to 0$ and we derive the structure of the local oscillations. We use positive, decreasing initial data normalized to tend to zero as $x \to +\infty$ and to one as $x \to -\infty$. For such data (2) necessarily shocks.

The solution of (1) is obtained by the inverse scattering transformation [1,2]. As $\epsilon \to 0$, the scattering data are calculated by the W.K.B. method [11]. As in the case studied by Lax and Levermore [4], the reflection coefficient is zero to all W.K.B. orders and is neglected.

We obtain from Dyson's formula:

$$u(x,t,\epsilon) \sim 2 \frac{\partial^2}{\partial x^2} \epsilon^2 \ln \tau(x,t,\epsilon) \tag{3}$$

where τ is a Fredholm determinant.

We write τ in compact form by making three definitions. Firstly, we write sums as integrals by utilizing the distribution:

$$\psi(\eta;\mu_1,\dots,\mu_r) = \pi\epsilon \sum_{k=1}^{r} \delta(\eta - \mu_k) \,. \tag{4}$$

We think of ψ as a generalized function of η which is parametrized by the sequence of atoms $\mu_1,\dots,\mu_r$ in the interval $[0,1]$. Secondly, we let L denote the negative definite [4] integral operator:

$$L\psi(\eta) = \frac{1}{\pi} \int_0^\infty \ln \left|\frac{\eta-\eta'}{\eta+\eta'}\right| \psi(\eta')d\eta' \tag{5}$$

where by definition $\ln\left|\frac{\eta-\eta'}{\eta+\eta'}\right|\Big|_{\eta'=\eta} = \ln \frac{1}{4\eta\pi}$.

Thirdly, we define the function:

$$a(\eta,x,t) = -\eta x + 4\eta^3 t + \eta x_s + \int_{x_s}^{\infty} [\eta - \sqrt{\eta^2 - v(x)}]dx. \tag{6}$$

which depends on the initial data $v(x)$, x_s being the unique solution of $v(x_s) = \eta^2$. The function τ in (3) is given by:

$$\tau(x,t,\epsilon) = 1 + \sum_{r=1}^{\infty} \int_{\Omega_r} \exp\{\frac{1}{\pi\epsilon^2}[(a,\psi) + (L\psi,\psi)]\}d\psi, \tag{7}$$

where the measure space Ω_r is given by:

$$\Omega_r = \{\psi \mid \psi = \pi\epsilon \sum_{k=1}^{r} \delta(\eta - \mu_k), \qquad 0 \le \mu_r \le \dots \le \mu_1 \le 1, \tag{8}$$
$$\text{with Lebesgue measure: } \ d\psi = d\mu_1 \dots d\mu_r\}$$

and

$$(a,\psi) = \int_0^\infty a(\eta)\psi(\eta)d\eta, \tag{9}$$

$$(L\psi,\psi) = \frac{1}{\pi}\int_0^\infty\int_0^\infty \ln\left|\frac{\eta-\eta'}{\eta+\eta'}\right| \psi(\eta)\psi(\eta')d\eta\, d\eta'. \tag{10}$$

Clearly, the rth term of the series in (7) is a multiple integral of order r with integration variables $\mu_1,\dots,\mu_r$. According to the Laplace method of evaluating integrals, the main contribution to $\tau(x,t,\epsilon)$ comes from the neighborhood of ψ's which maximize the exponent $2(a,\psi) + (L\psi,\psi)$. We write:

$$\tau(x,t,\epsilon) \sim \sum_{\psi_m} K_m(\epsilon,\psi_m) \exp\{\frac{1}{\pi\epsilon^2}[2(a,\psi_m) + L(\psi_m,\psi_m)]\}, \tag{11}$$

where summation occurs over all distributions ψ_m of form (4) which <u>locally maximize the exponent</u>. The quantities K_m have physical units of the measure of integration. In the Laplace method they equal a constant divided by the square root of the modulus of the determinant of the Hessian matrix of the exponent evaluated at the corresponding maximizer.

We make the following <u>continuum approximation</u>:

$$\tau(x,t,\epsilon) \sim \sum_{\psi} K(\epsilon,\psi)\exp\{\frac{1}{\pi\epsilon^2}[2(a,\psi) + (L\psi,\psi)]\} \tag{12}$$

where summation occurs over all <u>integrable nonnegative density functions</u>, still denoted by ψ which are local maxima of the functional $2(a,\psi) + (L\psi,\psi)$ and are constrained by the following <u>quantum condition</u>: If the interval I is a topological component of the support of the function ψ then:

$$\frac{1}{\pi\epsilon}\int_I \psi(\eta)d\eta = \text{integer}. \tag{13}$$

By the quantum condition, the mass of ψ on a connected component is an integral multiple of $\pi\epsilon$ as required by (4). We hope that keeping the notation ψ for the density function approximating (4) will not create confusion.

We note that the quantum condition affects maximization at a higher order: an arbitrary $\psi(\eta)$ will satisfy it if corrected by a term of order ϵ. The effects of the quantum condition, however, are far reaching. The solution of the maximization problem is no longer unique in general. We find that it is parametrized by an N–dimensional infinite lattice where $N = N(x,t)$ is the number of gaps separating the connected components of the support of the unique leading order maximizer obtained by neglecting the quantum condition. When $N > 0$ the contributions add up to an infinite sum over the lattice which turns out to be exactly the θ-function representing an N-phase KdV quasiperiodic solution [3] in fast scales.

Our method, produces the local waveforms with one reservation: The phase functions of the oscillations are computed to leading (i.e. $O(1/\epsilon)$) order. It appears that the next term in the phase function is logarithmic in ϵ.

We remark that the key to the solution of the maximization problem is the following observation. If we extend the domain of the variable η to the whole of the real line by:

$$\psi(\eta) = 0 \qquad \text{when} \quad \eta > 1 \tag{14}$$
$$\psi(\eta) = -\psi(-\eta) \quad \text{when} \quad \eta < 0,$$

we have an interesting consequence:

$$\frac{d}{d\eta} L\psi(\eta) = \frac{d}{d\eta}\frac{1}{\pi}\int_0^\infty \ln\left|\frac{\eta-\mu}{\eta+\mu}\right| \psi(\mu)d\mu =$$

$$= \frac{1}{\pi}\int_0^\infty \frac{1}{\eta-\mu}\psi(\mu)d\mu - \frac{1}{\pi}\int_0^\infty \frac{1}{\eta+\mu}\psi(\mu)d\mu =$$

$$= \frac{1}{\pi}\int_{-\infty}^\infty \frac{\psi(\mu)}{\eta-\mu}\, d\mu = H\,\psi(\eta)\,. \tag{15}$$

In other words $\frac{d}{d\eta}L\psi(\eta)$ is the Hilbert transform $H\psi(\eta)$ of $\psi(\eta)$. This allows the use of a Riemann–Hilbert approach to solve the maximization problems in sections 1 and 2.

1. Leading order maximization of the exponent

We obtain the leading order term to the solution of the maximization problem posed in the previous section by maximizing the exponent of (7):

(16) $$\Phi(\psi) = 2(a,\psi) + (L\psi,\psi)$$

$$= 2\int_0^\infty a(\eta)\psi(\eta)d\eta + \frac{1}{\pi}\int_0^\infty\int_0^\infty \ln\left|\frac{\eta-\eta'}{\eta+\eta'}\right| \psi(\eta)\psi(\eta')d\eta d\eta'$$

subject to the conditions:

(17) $$\psi \geq 0,\ \ \text{supp}\ \psi \subseteq [0,1],\ \int_0^\infty \psi(\eta)d\eta < \infty .$$

We neglect the quantum condition. The unique maximizer $\psi^*(\eta)$ depends on (x,t) and is computed by the method of Lax and Levermore [4]. Due to our introduction of the quantum condition, the topological components of the support of the maximizer ψ^* are important. They are $N+1$ closed intervals (which we call bands):

(18) $$[\beta_2,\beta_1],\ [\beta_4,\beta_3],\ldots,[0,\beta_{2N+1}] \subseteq [0,1].$$

The integer N and the β_i's depend on x and t. N can take values $0,1,2,\ldots$. We have ordered the β_i's so that

(19) $$0 \leq \beta_{2N+1} < \beta_{2N} < \cdots < \beta_1 \leq 1 .$$

2. Higher order maximization

In (16) we set:

(20) $$\psi = \psi^* + \epsilon\bar{\psi}$$

and maximize it with respect to $\bar{\psi}$.

After some calculation, which makes use of the fact that ψ^* satisfies the variational equation of the leading order maximization problem, we show that (16) may be replaced by:

(21) $$\Phi(\psi) = (a,\psi^*) + \epsilon^2(L\bar{\psi},\bar{\psi})$$

and maximization with respect to $\bar{\psi}$ must be carried subject to the conditions:

$$\text{(22)}\qquad \begin{aligned} &\text{(a)}\ \bar{\psi}(\eta) = 0 \quad \text{when} \quad \eta \notin \text{supp}\ \psi^* \\ &\text{(b)}\ \frac{1}{\pi\epsilon}\int_{\beta_{2i}}^{\beta_{2i-1}} [\psi^*(\eta) + \epsilon\bar{\psi}(\eta)]d\eta \in \mathbb{Z}_+, \quad i = 1,\ldots,N\,. \\ &\text{(c)}\ \frac{1}{\pi\epsilon}\int_0^{\beta_{2N}} [\psi^*(\eta) + \epsilon\bar{\psi}(\eta)]d\eta \in \mathbb{Z}_+ = \text{set of nonnegative integers.} \end{aligned}$$

The last two conditions are the quantum condition (13).

We omit the calculation [11] and state that essentially the solution set is spanned by an independent set which we call $(\bar{\psi}_j(\eta))_{j=1}^N$. This has the <u>normalization property</u> that the integral of $\bar{\psi}_j$ over the ith band $[\beta_{2i},\beta_{2i-1}]$ equals zero when $i \neq j$ and equals 1 when $i = j$ where $i = 1,\ldots,N$. The quantities $(L\bar{\psi}_j,\bar{\psi}_i) = c_{ji}$, $i = 1,\ldots,N$ are explicitly computed.

Substituting these results in (12) and (3) we obtain after some calculation [11]:

$$\text{(23)}\qquad \begin{aligned} &u(x+\epsilon x_1,t+\epsilon t_1,\epsilon) \sim -\frac{4}{\pi}(\eta,\psi_x^*) \\ &+ 2\frac{\partial^2}{\partial x_1^2}\ln \sum_{\vec{m}\in\mathbb{Z}^N} \exp\{\pi\langle C(\vec{m} - \frac{\vec{\alpha}}{\epsilon} - \vec{\rho}), \vec{m} - \frac{\vec{\alpha}}{\epsilon} - \vec{\rho}\rangle\}, \end{aligned}$$

$\vec{\alpha} = (\alpha_1,\ldots,\alpha_N)$ where α_i equals the integral of ψ^*/π over the ith band. C is the matrix (c_{ij}), $\vec{\alpha}$ and C depend on x and t. $\vec{\rho}$ depends on (x,t) and on the coefficients K of (12). We use Taylor's formula plus the definition of α_i:

$$\text{(24)}\qquad \begin{aligned} &\alpha_i(x+\epsilon x_1,t+\epsilon t_1) \sim \alpha_i(x,t) + \frac{\epsilon}{\pi}(\psi_x^*,1_i)x_1 + \frac{\epsilon}{\pi}(\psi_t^*,1_i)t_1, \\ &\vec{\alpha}(x+\epsilon x_1,t+\epsilon t_1) \sim \vec{\alpha}(x,t) - \frac{\epsilon}{\pi}\vec{k}x_1 - \frac{\epsilon}{\pi}\vec{\omega}t_1 \quad \text{where} \end{aligned}$$

$$\text{(25)}\qquad k_i = -(\psi_x^*,1_i), \quad \omega_i = -(\psi_t^*,1_i), \quad i = 1,\ldots,N\,.$$

The solution takes now its final form:

(26) $$u(x+\epsilon x_1,t+\epsilon t_1,\epsilon) \sim -\frac{4}{\pi}(\eta,\psi_x^*) +$$

$$+ 2\frac{\partial^2}{\partial x_1^2}\ln \sum_{\vec{m}\in\mathbb{Z}^N} \exp <\pi\, C(\vec{m}+\vec{z}),\vec{m}+\vec{z}>$$

where

(27) $$\vec{z} = \frac{1}{\pi}\vec{k}\,x_1 + \frac{1}{\pi}\vec{\omega}\,t_1 - \frac{\vec{\alpha}(x,t)}{\epsilon} - \vec{\beta}(x,t,\epsilon)$$

and the quantities ψ^*, N, C, $\vec{k}$, $\vec{\omega}$, $\vec{\alpha}$ depend only on the slow variables x,t. When $N = 0$ the second term on the right hand side of (27) vanishes and the solution is non-oscillatory.

3. Remarks

1. We recognize (23) as the theta function expression of the N-phase quasiperiodic solution of the KdV equation in the local variables x_1,t_1: In the variable $\lambda = -\eta^2$, the differentials $(-\frac{1}{2}\,\bar{\psi}_j(\eta)d\eta)_{j=1}^N$ are exactly the holomorphic differentials $(\varphi_j)_{j=1}^N$ corresponding to the hyperelliptic curve

(7.1) $$\mathcal{R}(\lambda) = \left\{\prod_{i=1}^{2N+1}(\lambda-\lambda_i)\right\}^{1/2}, \quad \lambda_i = -\beta_i^2,$$

with normalization $\oint_{\alpha_k}\varphi_j = \delta_{jk}$ where α_k is a counterclockwise contour around the kth band $[\lambda_{2k-1},\lambda_{2k}]$, $k = 1,\ldots,N$. The matrix C is the imaginary part of the corresponding period matrix along the contours $(\gamma_i)_{i=1}^N$ which go from λ_{2i} to λ_{2N+1} on the upper Riemann sheet and return from λ_{2N+1} to λ_{2i} on the lower Riemann sheet. C is negative definite and the series (23) converges. Finally $\vec{k}$ and $\vec{\omega}$ correspond to the relevant differentials of the second kind [3].

2. The term $-\frac{4}{\pi}(\eta,\psi_x^*)$ in (23) gives the local average of the solution and is in exact agreement with the result of Lax and Levermore [4]. The second term of (23) gives a fast scale oscillation of mean zero.

Acknowledgement: I thank the National Science Foundation for support this research under grant number DMS 87–02526.

References

[1] Buslaev, V. S., and Fomin, V. N., (1962) 'An inverse scattering problem for the one–dimensional Schrödinger equation on the entire axis', (in Russian), Vestnik Leningrad Uniiv. 17, 56–64.

[2] Cohen, A., and Kappeler, T., (1985) 'Scattering and inverse scattering for steplike potentials in the Schrödinger equation', Indiana U. Math. Jour. 34, 127–180.

[3] Flaschka, H., Forest, M.G., and McLaughlin, D. W., (1980) 'Multiphase averaging and the inverse spectral solutions of the Korteweg–de Vries equation', Comm. Pure Appl. Math. 33, 739–784.

[4] Lax, P.D., and Levermore, C.D., (1983) 'The small dispersion limit of the Korteweg-de Vries equation I, II, III', Comm. Pure Appl. Math. 36, 253–290, 571–593, 809–829.

[5] Levermore, C. D., (1988) 'The hyperbolic nature of the zero dispersion KdV limit', Comm. P.D.E., 495–514.

[6] Venakides, S., (1985) 'The zero dispersion limit of the Korteweg–de Vries equation with nontrivial reflection coefficient', Comm. Pure Appl. Math. 38, 125–155.

[7] Venakides, S., (1985) ' The generation of modulated wavetrains in the solution of the Korteweg–de Vries equation', Comm. Pure Appl. Math. 38, 883–909.

[8] Venakides, S., (1987) 'The zero dispersion limit of the Korteweg–de Vries equation with periodic initial data', AMS Trans. 301, 189–225.

[9] Venakides, S., (1989) 'The continuum limit of theta functions', Comm. PUre Appl. Math., 42, 711–728.

[10] Bender C.M., and Orszag, Steven A., (1978) Advanced Mathematical Methods for Scientists and Engineers, McGraw–Hill Publishers.

[11] Venakides, S., 'The Korteweg–de Vries equation with small dispersion: Higher order Lax–Levermore theory, Comm. Pure Appl. Math., in press.

SOLITONS IN OPTICAL FIBRES

Peter L. Christiansen*
Dipartimento di Matematica e Applicazioni
Università degli Studi
Via Mezzocannone 8, 80125 Napoli
Italy

Optical solitons can propagate on single mode optical fibres without distortion when the self-phase modulation attributable to the intensity-dependent refraction index of some materials is used to compensate for group-velocity dispersion (GVD) due to the frequency dependence of the refractive index. This mechanism was proposed by Hasagawa and Tappart in 1973 [1] and they showed that the slowly varying envelope function for the electromagnetic field satisfies the cubic nonlinear Schrödinger equation (NLS). Shortly before this equation had been solved by Zakharov and Shabat [2] using a generalization of the inverse scattering theory (IST). NLS is of fundamental importance as a prototype equation for strongly dispersive, weakly nonlinear waves.

Since then it has been verified in a number of experiments that so-called <u>bright</u> optical solitons can propagate without distortion in regions of the spectrum for which the fibre has a negative GVD (wavelength $\lambda > 1.3$ μm in standard single mode fibres) [3]. Recently, transmission of 55 psec optical pulses through 4000 km of fibre was achieved by the use of a combination of nonlinear soliton propagation to avoid pulse spreading and Raman amplification to overcome inevitable losses in the fibre material [4]. Such losses can be modelled by an extra term in the NLS equation leading to the extinction of the soliton character of the pulses [5].

For positive GVD ($\lambda < 1.3$ μm), in the visible region, bright pulses cannot propagate as solitons, and the interaction of the nonlinear index with the group-velocity dispersion leads to spectral and temporal broadening of the propagating pulses. This effect is also predicted by solution of the corresponding NLS equation [6]. However, in this case the NLS admits a so-called <u>dark</u> optical soliton which is a localized intensity dip in a constant-intensity background. This solution is an antisymmetric function of time with an abrupt phase shift of π and zero intensity at

* Permanent address:
Laboratory of Applied Mathematical Physics
The Technical University of Denmark
DK-2800 Lyngby
Denmark

R. Spigler (ed.), Applied and Industrial Mathematics, 263–265.

its centre. Other dark solitons with a reduced contrast and a lesser, more gradual phase modulation also exist. Recently Weiner et al. [7] generated 100-200 fsec dark antisymmetric pulses on background pulses of red light, 1-4 psec in duration. These were passed through a single-mode optical fibre 1.4 m long, five times the characteristic length for soliton propagation of these pulses. When the pulses were tailored to the right soliton form they were transmitted through the system undistorted. Broadening of the dark pulse at lower powers and narrowing at higher powers were observed in agreement with the numerical simulations based on the NLS model. Also the so-called "wave breaking" effects of solitons [8-10] predicted by this equation have been confirmed experimentally recently [11]. As a result differing frequencies in the pulse and in the carrier wave cause intensity oscillations to appear at the leading edge of the pulse.

For the computational solution of NLS, Fourier spectral methods, leading to algorithms which are easily vectorized, may be applied. This technique requires periodic boundary conditions. In [12-13] the disadvantage of the periodicity, thus imposed for computational purposes, is removed by introducing two absorbing layers close to the ends of the integration interval in the cases of bright and dark solitons.

The soliton effect has been applied for generation of light pulses of temporal width down to the 10 psec region in a two-cavity laser consisting of a synchronously pumped, modelocked color-center laser coupled to an optical single-mode fibre acting as a control cavity. This soliton laser was invented by Mollenauer and coworkers [14-15] and has been modelled computationally in [16-18].

Optical communication systems are currently based on linear techniques. It is an intriguing question whether the future technology will use optical solitons, which are now reality in the laboratory, to improve information transfer rates.

ACKNOWLEDGEMENTS

The author expresses thanks for the warm hospitality of the Dipartimento di Matematica e Applicazioni, Università degli Studi di Napoli and acknowledges support from Julie Damms Studiefond, The Danish Technical Research Council, Denmark, and Consiglio Nazionale delle Ricerche, Rome, Italy.

REFERENCES

[1] Hasagawa, A. and Tappart, F. (1973) 'Transmission of stationary nonlinear optical pulses in dispersive dielectric fibers. I. Anomalous dispersion', Appl. Phys. Lett. 23, 142-144.

[2] Zakharov, V.E. and Shabat, A.B. (1972) 'Exact theory of two-dimensional self-focusing and one-dimensional self-modulation of waves in nonlinear media', Sov. Phys. - JETP 34, 62-69.

[3] Mollenauer, L.F., Stolen, R.H., and Gordon, J.P. (1980) 'Experimental observation of picosecond pulse narrowing and solitons in optical fibers', Phys. Rev. Lett. 45, 1095-1098.

[4] Mollenauer, L.F. and Smith, K. (1988) 'Demonstration of soliton transmission over more than 4000 km in fiber with loss periodically compensated by Raman gain', Opt. Lett. 13, 675-677.
[5] Mørk, J., Christiansen, P.L., Lassen, H.E. and Tromborg, B. (1987) Evolution of solitons from chirped and noisy pulses in single-mode fibres', IEE Proc. 134, 127-134.
[6] Zhakarov, V.E. and Shabat, A.B. (1974) 'Interaction between solitons in a stable medium', Sov. Phys. JETP 37, 823-828.
[7] Weiner, A.M., Heritage, J.P., Hawkins, R.J., Thurston, R.N., Kirschner, E.M., Leaird, D.E., and Tomlinson, W.J. (1988) 'Experimental Observation of the Fundamental Dark Soliton in Optical Fibers', Phys. Rev. Lett. 61, 2445-2448.
[8] Natasuka, H., Grischkowsky, D., and Balant, A.C. (1981) 'Nonlinear Picosecond-Pulse Propagation through Optical Fibers with Positive Group Velocity Dispersion', Phys. Rev. Lett. 47, 910-913.
[9] Tomlinson, W.J., Stolen, R.H., and Johnson, A.M. (1985) 'Optical wave breaking of pulses in nonlinear optical fibers', Opt. Lett. 10, 457-459.
[10] Lassen, H.E., Mengel, F., Tromborg B., Albertsen, N.C., and Christiansen, P.L. (1985) 'Evolution of chirped pulses in single-mode fibers', Opt. Lett. 10, 34-36.
[11] Rothenberg, J.E. and Grischkowsky, D (1989) 'Observation of the Formation of an Optical Intensity Shock and Wave Breaking in the Nonlinear Propagation of Pulses in Optical Fibers', Phys. Rev. Lett. 62, 531-534.
[12] If, F., Berg, P., Christiansen, P.L., and Skovgaard, O. (1987) 'Split-Step Spectral Method for Nonlinear Schrödinger Equation with Absorbing Boundaries', J. Comp. Phys. 72, 501-503.
[13] Geisler, T., Christiansen, P.L., Mørk, J., and Ramanujam, P.S. (1990) 'Split-Step Spectral Method for Nonlinear Schrödinger Equation with Constant Background Intensities', J. Comp. Phys. (accepted).
[14] Mollenauer, L.F. and Stolen, R.H. (1984) 'The soliton laser', Opt. Lett. 9, 13-15.
[15] Mollenauer, L.F. (1985) 'Solitons in optical fibres and the soliton laser', Phil. Trans. R. Soc. Lond. Ser. A 315, 333-345.
[16] Blow, K.J. and Wood, D. (1986) 'Stability and Compression of Pulses in the Soliton Laser', IEEE J. Quantum Electron. QE-22, 1109-1116.
[17] If, F., Christiansen, P.L., Elgin, J.N., Gibbon, J.D., and Skovgaard, O. (1986) 'A Theoretical and Computational Study of the Soliton Laser', Opt. Comm. 57, 350-354.
[18] Berg, P., If, F., Christiansen, P.L., and Skovgaard, O. (1987) 'Soliton laser: A computational two-cavity model', Phys. Rev. A, 35, 4167-4174.

WAVES IN FINELY LAYERED MEDIA

ROBERT BURRIDGE

Schlumberger Doll Research
Old Quarry Road
Ridgefield, CT 06877

A pulse propagates obliquely through a one-dimensional medium consisting of a large number N of homogeneous elastic layers. As it propagates, the directly transmitted principal arrival is reduced by transmission loss at each interface, but close to this arrival is a broad pulse, made up of multiply scattered energy, which ultimately appears to diffuse about a moving center. The broad pulse evolves according to an integrodifferential equation analogous to the Kolmogorov-Feller forward equation in probability theory, the kernel of which depends upon the auto- and cross-covariances of the material properties of the medium.

The basic theory, which is essentially the method of averaging, is introduced and then various extensions are described. Finally a comparison is made with effective medium theory and with Keller's method for wave propagation in a random medium.

The analysis is illustrated by numerical examples.

1. INTRODUCTION

1.1 The purpose of this paper

It is the aim of this paper to show how a pulse evolves as it travels through a highly discontinuous medium consisting of a large number of uniform, isotropic, elastic layers whose density ρ, and P and S wave speeds α and β differ but slightly from the constant values $\bar{\rho}$, $\bar{\alpha}$, and $\bar{\beta}$. Such a system is a model of seismic pulse propagation through a geological formation.

To fix our ideas let us consider what happens when an impulsive plane wave traverses a very finely layered elastic medium. The continuous curves in Figure 1 show a plane S-wave pulse, incident at 25.1°, as it traverses a stack of 10,000 plane homogeneous layers. On entering the stack the pulse is impulsive and has the form of a Dirac delta function of time. We see the pulse shape as a function of time after it has traversed 2,000, 4,000, 6,000, 8,000, and finally 10,000 layers. Notice particularly the progressive broadening of the pulse and its increasing delay. After traversing 2,000 layers there are still considerable high-frequency 'fluctuations', which decrease as the pulse travels. We shall be concerned with understanding the evolution of the smooth pulse. The magnitude of the fluctuations is a measure of how well the smooth approximation fits the fluctuating wave.

R. Spigler (ed.), Applied and Industrial Mathematics, 267–279.

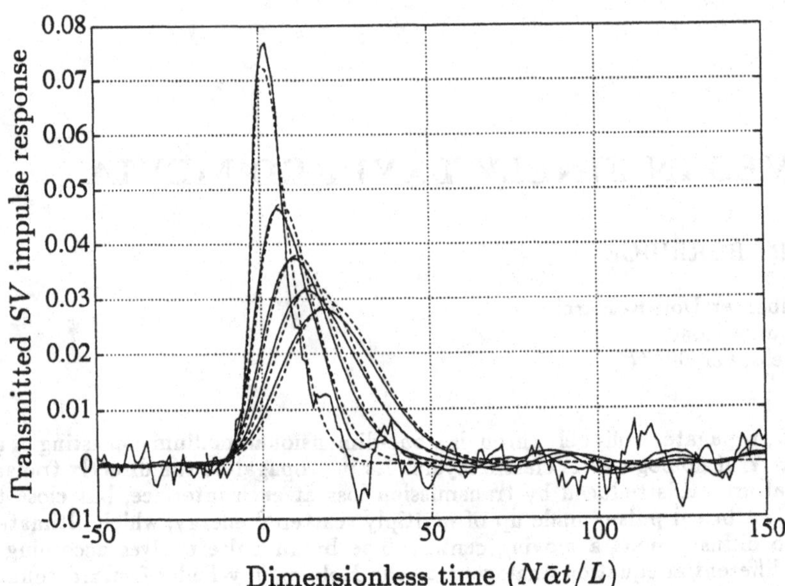

FIGURE 1. Here we show an S-wave pulse, incident at 25.1°, after traveling through 2,000, 4,000, 6,000, 8,000, and 10,000 layers. The simulations are solid curves, the limiting solution is shown dashed.

Our theory predicts that the evolution of the smooth component of the pulse, as shown in the broken curves of Figure 1, depends upon the auto- and cross-covariances of the density and wave speeds of the medium, as described in the following sections.

1.2 Previous related work

In this work we follow a line of research initiated by O'Doherty & Anstey in the now classical paper [1]. They noticed that a pulse is broadened and slightly delayed as it travels through a finely layered medium. In an appendix to their paper they also gave a quantitative explanation of the phenomenon in terms of the statistics of the reflection coefficients for a Goupillaud layered medium, i.e. one in which the transit time across every layer is the same. In the present work, however, we allow our layers to have arbitrary (but small) thicknesses, we consider obliquely incident waves, tunneling waves, and other extensions.

Following O'Doherty & Anstey many authors have elaborated on this phenomenon with numerical simulations and analysis; see Burridge, White, & Papanicolaou [2] and its references for further information; for a fuller account of the present work see [3,4,7,8,9,13]. In particular this work is an extension of [4].

1.3 Outline of this paper

In Section 2 we describe the double scattering process which forms a basis for our approximation. Spatial averaging is introduced in Section 3, which leads to the main result, the integrodifferential equation (15) of Section 4. In Section 5 we compare solutions of the integrodifferential equation with direct numerical simulations. Up to this point we have not admitted critical-angle phenomena – no evanescent waves are present in any layer. In Section 6 we relax this restriction to allow for the effect of tunneling P waves on the propagation of S beyond the critical angle, and we describe some other extensions and applications of the theory. Finally in Section 7 we relate the present work to effective medium theory, and also to Keller's theory of waves in random media.

2. DOUBLE SCATTERING

We shall consider the one-dimensional problem in which the disturbance is made up of plane, or inhomogeneous plane, P and S waves in each layer. Let z and t be the space coordinate, representing depth into the ground, and the time. We shall assume that all waves derive from a single plane wave of P or S type incident on the stack of layers from a uniform half space $z < 0$ above the stack, and so all the plane wave components in every layer have the same transverse slowness (ray parameter) p. We shall investigate the plane transmitted wave which travels in the positive z direction in the same mode as the incident wave.

We now establish some notation following Aki and Richards [5]:

$$\begin{array}{lll} 1 & \text{stands for down-going } P & (\grave{P}), \\ 2 & \text{stands for down-going } S & (\grave{S}), \\ 3 & \text{stands for up-going } P & (\acute{P}), \\ 4 & \text{stands for up-going } S & (\acute{S}). \end{array} \tag{1}$$

We denote the z-component of slowness in mode i by γ_i, and its background value by $\bar{\gamma}_i$:

$$\begin{aligned} \gamma_1 = -\gamma_3 = q_\alpha \equiv \sqrt{\tfrac{1}{\alpha^2(z)} - p^2}, \quad \gamma_2 = -\gamma_4 = q_\beta \equiv \sqrt{\tfrac{1}{\beta^2(z)} - p^2}, \\ \bar{\gamma}_1 = -\bar{\gamma}_3 = q_{\bar{\alpha}} \equiv \sqrt{\tfrac{1}{\bar{\alpha}^2(z)} - p^2}, \quad \bar{\gamma}_2 = -\bar{\gamma}_4 = q_{\bar{\beta}} \equiv \sqrt{\tfrac{1}{\bar{\beta}^2(z)} - p^2}. \end{aligned} \tag{2}$$

We denote the cumulative vertical travel time functions by $T_i(p, z)$:

$$T_i(p, z) = \int_0^z \gamma_i(p, z')\, dz'. \tag{3}$$

We shall let ρ_1, ρ_2, ρ_3 stand for the material properties ρ, α, β, and let S^k stand for the scattering matrix at interface z_k, where entry S^k_{ij} represents the reflection or transmission coefficient from incident mode i into scattered mode j and the waves are normalized by the square root of energy flux in the z-direction. When the contrasts in material properties across the interfaces are small and $j \neq i$, the S^k_{ij} are small and depend linearly upon the contrasts $\Delta\rho^k_\alpha$ in the ρ_α at interface k. When $j = i$ we have

$$S^k_{ii} \cong 1 - \tfrac{1}{2} \sum_{\{j|j\neq i\}} \left(S^k_{ij}\right)^2, \tag{4}$$

which is near unity. See Aki & Richards [5], section 5.2.

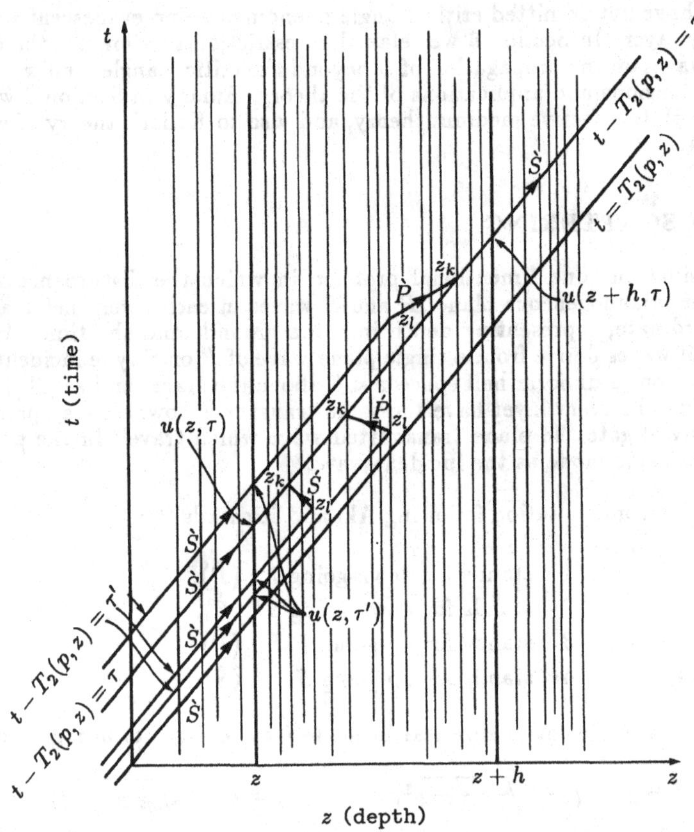

FIGURE 2. This is a space-time diagram showing the kind of double scattering which scatter from $u(z, \tau')$ to $u(z+h, \tau)$. The transmitted pulse is made up of contributions from all such events.

As it travels the pulse is modified by transmission loss and by double scattering events of the following type: an incident wave in mode i, say $u(t - T_i(p, z))$, impinges on interface z_ℓ where it produces a scattered wave in mode $j \neq i$,

$$S^\ell_{ij} u\Big(t - [T_j(p, z) - T_j(p, z_\ell)] - T_i(p, z_\ell)\Big), \tag{5}$$

which produces at interface z_k a scattered wave in the original mode i:

$$\begin{aligned} S^k_{ji} S^\ell_{ij} u\Big(t - [T_i(p, z) - T_i(p, z_k)] - [T_j(p, z_k) - T_j(p, z_\ell)] - T_i(p, z_\ell)\Big) \\ \simeq - S^k_{ij} S^\ell_{ij} u\Big(t - T_i(p, z) - (\bar{\gamma}_i - \bar{\gamma}_j)(z_\ell - z_k)\Big). \end{aligned} \tag{6}$$

Here, the arguments are approximately equal, and $S^k_{ji} \simeq -S^k_{ij}$ follows from reciprocity and the fact that reversing the directions of the waves at an interface changes the sign of the $\Delta\rho^k_\alpha$ and hence changes the sign of the corresponding scattering coefficients in this approximation of small contrasts at the interfaces.

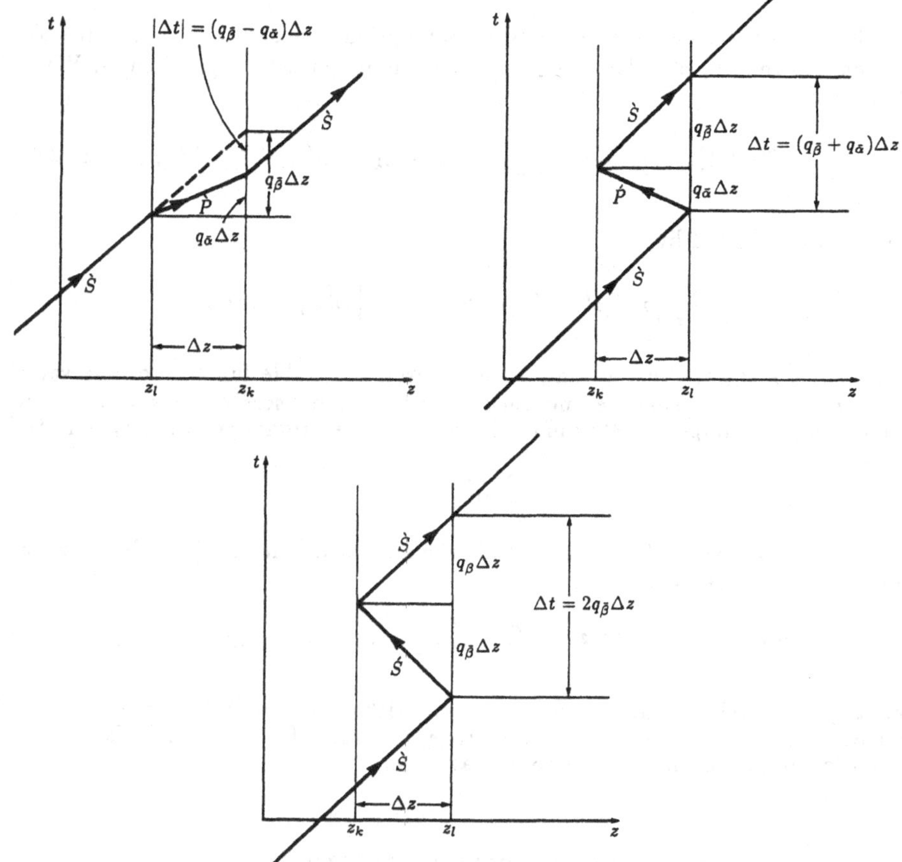

FIGURE 3. A magnified view of the three kinds of double scattering for incident S.

Now let h be small, but large enough that the interval $(z, z+h)$ contains many interfaces, and consider the incremental change to a pulse in mode i which enters the interval $(z,\ z+h)$ at z as $u(z,\ t-T_i(p,z))$ and leaves at $z+h$ as $u(z+h,\ t-T_i(p,z+h)))$. Writing τ for $t - T_i(p,z)$, the increment is approximately

$$\begin{aligned} u(z+h,\ \tau) - u(z,\ \tau) = &-\tfrac{1}{2} \sum_{\{j|j\neq i\}} \sum_{\{\ell|z\leq z_\ell<z+h\}} \left(S_{ij}^{\ell}\right)^2 u(z,\ \tau) \\ &- \sum_{\{j|j\neq i\}} \sum_{\{k,\ell|z\leq z_\ell<z+h,\bar\gamma_j(z_k-z_\ell)>0\}} S_{ij}^{k} S_{ij}^{\ell} u\left(z, \tau - (\bar\gamma_i - \bar\gamma_j)(z_\ell - z_k)\right). \end{aligned} \tag{7}$$

In the right member the first term accounts for transmission loss and the other term represents the twice scattered field in mode i. See Figures 2 & 3. All other scattering events are assumed to be negligible. These include those events involving three or more reflections within the interval $(z, z+h)$, events involving a reflection outside the interval $(z, z+h)$, and corrections to terms appearing explicitly.

3. AVERAGING

We shall make the following statistical assumption concerning the S^k_{ij}. Let D be a domain in the (z', z'') plane and ϕ a smooth function with support in D. We shall suppose that

$$\sum_{\{(z_k,\, z_\ell)\in D\}} S^k_{ij} S^\ell_{ij} \phi(z_k, z_\ell) = \int\int_D \sum_k S^k_{ij}\delta(z' - z_k) \sum_\ell S^\ell_{ij}\delta(z'' - z_\ell)\phi(z', z'')dz'dz''. \tag{8}$$

is well approximated by

$$\int\int_D \left[a^{ij}_0 \delta(z' - z'') + a^{ij}(z' - z'')\right] \phi(z', z'')\, dz'dz'', \tag{9}$$

where a^{ij}_0 is a scalar constant and a^{ij} is an even integrable function of one variable $(a^{ij}(-\nu) = a^{ij}(\nu))$. These are smoothed averages of the discrete terms in (8). Among other things, assumption (9) implies the wide-sense stationarity of the ij reflectivity function,

$$\sum_k S^k_{ij}\delta(z - z_k). \tag{10}$$

The quantity in square brackets in (9) is the autocovariance of the reflectivity function, and we may write

$$a^{ij}_0 \delta(z' - z'') + a^{ij}(z' - z'') = \langle \sum_k S^k_{ij}\delta(z' - z_k) \sum_\ell S^\ell_{ij}\delta(z'' - z_\ell)\rangle, \tag{11}$$

the angular brackets standing for average. Although it would be mathematically convenient if this were an ensemble average, it is clear from the analysis in [3] that $\langle\cdot\rangle$ represents a less manageable spatial average.

4. THE INTEGRODIFFERENTIAL EQUATION

Let us now use (8) and (9) in (7) to rewrite the right hand side:

$$\begin{aligned} u(z + h, \tau) - u(z, \tau) = &-\tfrac{1}{2} \sum_{\{j|j\neq i\}} \int_z^{z+h} dz' a^{ij}_0 u(z, \tau) \\ &- \sum_{\{j|j\neq i\}} \int_z^{z+h} dz'' \int_{\{z''|\gamma_j(z'-z'')>0\}} dz'\, a^{ij}(z' - z'')u\left(z, \tau - (\gamma_i - \gamma_j)(z'' - z')\right). \end{aligned} \tag{12}$$

Changing variable of integration in (12) to $\tau' = (\gamma_i - \gamma_j)(z'' - z')$ and dividing by h we get

$$u_z(z, \tau) = -\tfrac{1}{2} \sum_{\{j|j\neq i\}} a^{ij}_0 u(z, \tau) - \sum_{\{j|j\neq i\}} \frac{1}{|\gamma_i - \gamma_j|} \int_{I_{ij}} a^{ij}\left(\tfrac{\tau'}{\gamma_i - \gamma_j}\right) u(z, \tau - \tau')d\tau'. \tag{13}$$

Here the intervals of integration I_{ij} are either $(-\infty, 0)$ or $(0, \infty)$ depending on the signs of γ_j and of $\gamma_i - \gamma_j$. Set

$$a_0^i = \sum_{\{j|j\neq i\}} a_0^{ij}, \quad a^i(\tau) = \sum_{\{j|j\neq i\}} \frac{\chi_{I_{ij}}(\tau)}{|\gamma_i - \gamma_j|} a^{ij}\left(\frac{\tau}{\gamma_i - \gamma_j}\right), \tag{14}$$

where $\chi_{I_{ij}}(\tau)$ is the characteristic function of I_{ij}: $\chi_{I_{ij}}(\tau) = 1$ if $\tau \in I_{ij}$, and 0 otherwise. Then we may rewrite (13) as

$$u_z(z, \tau) = -\tfrac{1}{2}a_0^i u(z, \tau) - \int_{-\infty}^{\infty} a^i(\tau')u(z, \tau - \tau')d\tau'. \tag{15}$$

This integrodifferential equation, which is the main result of this section, governs the evolution of the pulse shape $u(z, \tau)$ as the disturbance travels through the stack of layers.

It is interesting to notice that equation (15) has the same structure as the Kolmogorov-Feller forward equation which arises in the theory of purely discontinuous stochastic processes. See Gnedenko [6] chapter X.

5. COMPARISON OF SIMULATIONS WITH THE LIMITING SOLUTION

For the purpose of testing the usefulness of the integrodifferential equation we used a pseudo-random number generator to generate the layer parameters and a layer matrix code to simulate the pulse propagation. For comparison we also calculated the 'limiting' wave form by solving the integrodifferential equation. The layer thicknesses were exponentially distributed, and the parameters in each layer were uncorrelated with those in any other layer. See Burridge & Chang [3] for further details of the statistical model, the layer code, and the methods used to solve the integrodifferential equation. The comparison of the numerical simulations with the solution of the integrodifferential equation is not only of interest in its own right but, in the absence of a rigorous proof that the actual solution is well approximated by the limiting solution, it provides a corroboration that the integral equation does in fact govern the evolution of the pulse.

In Figure 4, which shows the P wave in the same representation as the S wave of Figure 1, we show the impulse response corresponding to a plane P wave incident at $\theta_P = 45°$ as it travels through a stack of 10,000 layers. The horizontal axis is time t measured in units of P-wave travel time across an average layer from the time of arrival of the directly transmitted impulse. The vertical scale is appropriate to the impulse response in that the pulses have unit area (approximately). The five curves represent the developed pulse after it has traversed 2,000, 4,000, 6,000, 8,000, and 10,000 layers. The further through the stack the pulse travels the lower and broader the pulse becomes.

For comparison we have plotted the limiting solution on the same graph as dashed curves. It is clear that the limiting solution captures the main features of the pulse and that the agreement between the two is very good. The two solutions differ in that the simulation contains some high frequency fluctuations, which are most noticeable

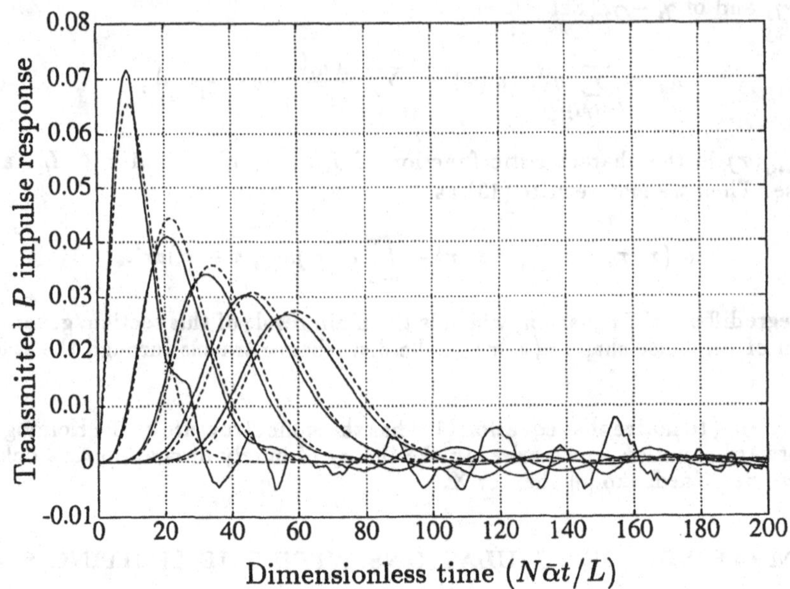

FIGURE 4. This is the P impulse incident at $\theta_P = 45°$ after traveling through 2,000, 4,000, 6,000, 8,000, and 10,000 layers. The layer matrix code simulation is shown in full line and the limiting solution dashed.

at the 2,000th interface. Actually, if it were possible to calculate the exact impulse response it would be found to consist of a discrete distribution of impulses (Dirac delta functions) to which the limiting solution approximates only in the weak sense, i.e. after integration. In Figures 1 and 4 $\theta_S = 25.10°$ and $\theta_P = 45°$ correspond to the same horizontal slowness p.

Note that in these cases, as in others shown in [3], the simulated pulse is very closely approximated by the limiting solution.

6. EXTENSIONS OF THE THEORY

6.1 S waves beyond the critical angle

When an S wave impinges on the stack of layers beyond (but not too near) the critical angle $\sin^{-1}(\bar{\beta}/\bar{\alpha})$ the scattering coefficients $\grave{S}\acute{P}$, $\grave{S}\grave{P}$, $\acute{P}\grave{S}$, and $\grave{P}\grave{S}$ are complex and the P segment in the double-scattering events $\grave{S}\acute{P}\grave{S}$ and $\grave{S}\grave{P}\grave{S}$ tunnels, being evanescent in the z direction. Because the correlation length is small, the tunneling P retains significant amplitude at reconversion to S. The kernel $a(\tau)$ in (14) and (15) is replaced by an analytic extension, which reverts to (14) when all quantities are real. A comparison of simulations with the limiting behavior in this case is shown in Figure 5, which illustrates the propagation of an S wave incident at 50° when the critical angle is 36.86°. For further details see De Hoop, Burridge & Chang [7].

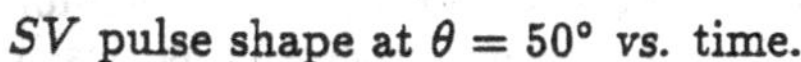

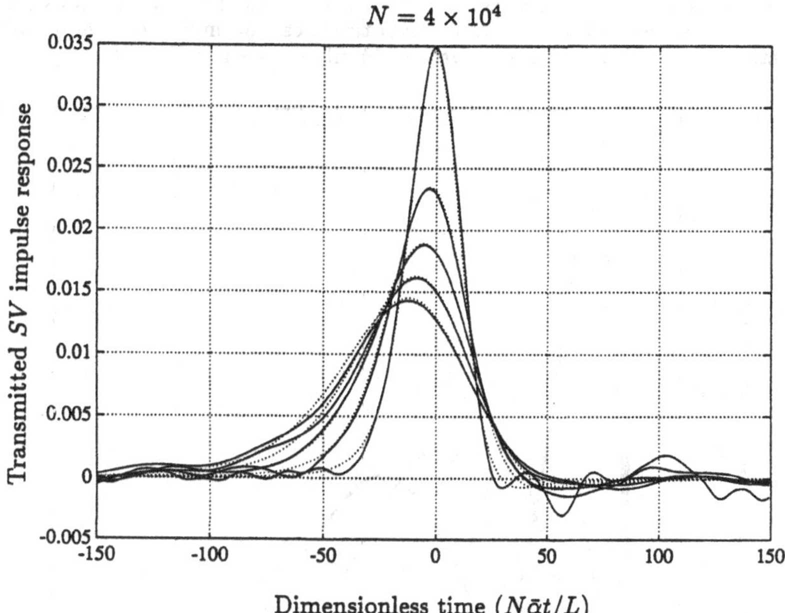

FIGURE 5. Here is shown the SV impulse incident at $\theta_S = 50°$ after traveling through 8,000, 16,000, 24,000, 32,000, and 40,000 layers. The layer matrix code simulation is shown in full line and the limiting solution dotted.

6.2 Point source and the relationship with ray theory

With the extension of the theory just described we are in a position to superpose plane wave components to obtain the response to a point source or other more general wave field. Our results illustrate the fact that amplitudes may be calculated by ray theory in the background medium, while the pulse shapes are predicted by our theory. In the case of curved rays or curved interfaces the kernel $a(\tau)$ of (14) will also depend upon position (arc distance s along the ray): $a(s,\tau)$. However, the theory cannot yet deal with rays which graze or come near to grazing the interfaces. Also the exact arrival time of the pulse is undetermined by the limiting theory, which predicts only the time lag relative to the time of the directly transmitted arrival, which in general differs slightly from the arrival time of the ray in the background medium. For further details see De Hoop, Burridge & Chang [8].

6.3 Synthetic seismograms computed using well-log data

In well logging, measurements of density and wave speeds are made at closely spaced points in oil wells. This data may then be used to compute synthetic seismograms, which are compared with actual seismograms from VSP experiments. Our theory shows how the propagation of the pulses depends upon the covariances of the mechanical properties of the rock. The same theory shows that the computed synthetic seismograms depend upon the covariances of the well-log data. On the basis of the physics of the well-log measurement process it is possible to understand how the covariances of the measurements relate to the covariances of the actual rock parameters. Then corrections to the synthetic travel times can be made for a more detailed interpretation of the VSP data. For further details see Hsu & Burridge [9].

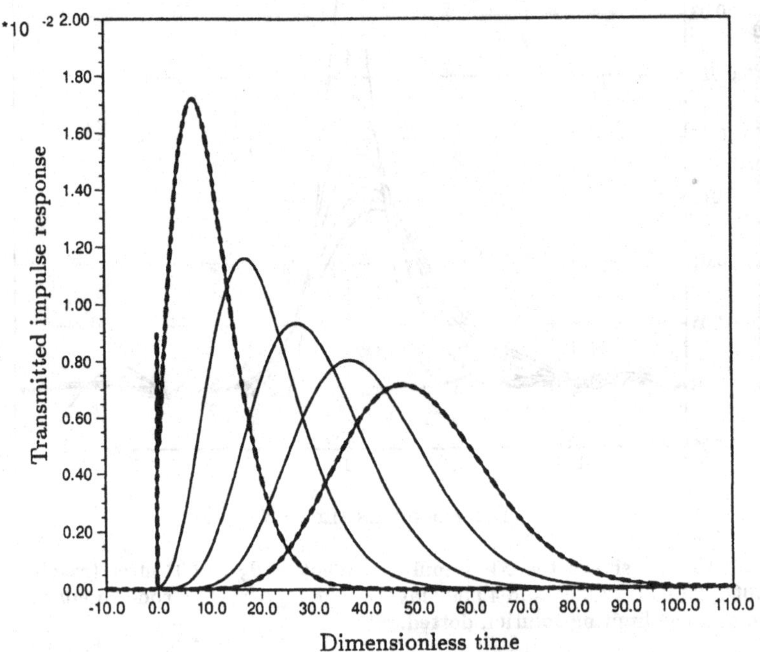

FIGURE 6. Here the solid curves represent the ensemble average pulse shape according to Keller's theory after the pulse has traveled through 2,000, 4,000, 6,000, 8,000, and 10,000 layers when there is variation in the characteristic impedance but no variation in the wave slowness. The dashed curves are the limiting solution according to this paper after traveling through 2,000 and 10,000 layers.

7. RELATIONSHIP WITH OTHER THEORIES

7.1 Comparison with the effective medium theory

Effective medium theory (see e.g. Schoenberg [10]) seeks a homogeneous elastic medium, which is macroscopically 'equivalent' to the finely layered medium. Because wave propagation in a homogeneous medium is non-dispersive, this form of effective medium will not provide the pulse broadening predicted by our theory, and seen in the simulations. However, the time of the sharp arrival in the (anisotropic) effective medium agrees with the arrival time of the centroid of our spreading pulse. Again, this arrival time in both theories is measured relative to the actual arrival time of the directly transmitted impulse. This requires, in the effective medium theory, that the averages be taken over precisely the stack of layers between the source and the receiver. Otherwise the arrival time may be in error by a small amount, which fluctuates from one stack of layers to another while the statistics are held fixed. This small error is comparable to the width of the spreading pulse. See De Hoop, Burridge & Chang [8] for further details.

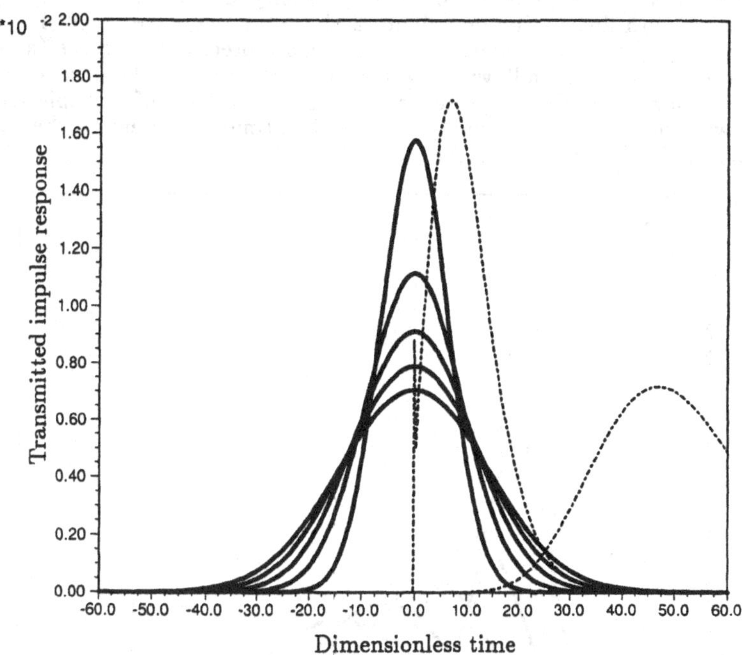

FIGURE 7. Here the solid curves represent the ensemble average pulse shape according to Keller's theory after the pulse has traveled through 2,000, 4,000, 6,000, 8,000, and 10,000 layers. Here there is variation in the slowness but no variation in the characteristic impedance. We also plot for reference the dashed curves for 2,000 and for 10,000 layers when there is no variation in wave slowness.

7.2 Comparison with Keller's theory of wave propagation through random media

Keller [11] and then Karal and Keller [12] investigated the ensemble average of a wave field propagating through a random medium. As in our theory they considered media deviating only sllghtly from a mean background. Recently Stanke and Burridge [13] applied Keller's method to a random stack of finely layered media and studied in particular two special cases of normal incidence in acoustic media: (a) where the impedance fluctuates but the wave speed is constant, and (b) where the wave speed fluctuates while the characteristic impedance remains constant. In (a) there is no fluctuation in arrival time from one realization to another. Since, as our theory predicts, the limiting pulse is a good approximation for each realization, we would expect the ensemble average according to Keller's theory to agree with our (deterministic) results. This is indeed the case. In (b) the reflection coefficients are zero and so there is no scattering. Thus the pulses travel unchanged. The impulse response is itself a single impulse arriving as the directly transmitted signal. Our theory merely predicts this result. The theory of Karal and Keller does not refer to the individual realizations but to the ensemble average, which consists of a broadened (gaussian) pulse whose shape is the probability density of the arrival time. In general, when there are both fluctuating impedance and fluctuating wave speed the ensemble average pulse consists of a convolution of the pulses predicted in cases (a) and (b). Thus our theory sheds the following light on Keller's results: the ensemble average pulse consists of a deterministic pulse, representing the effects of multiple scattering, convolved with the probability density of the arrival time, representing the variability in wave speed. See Figures 6,7, and 8.

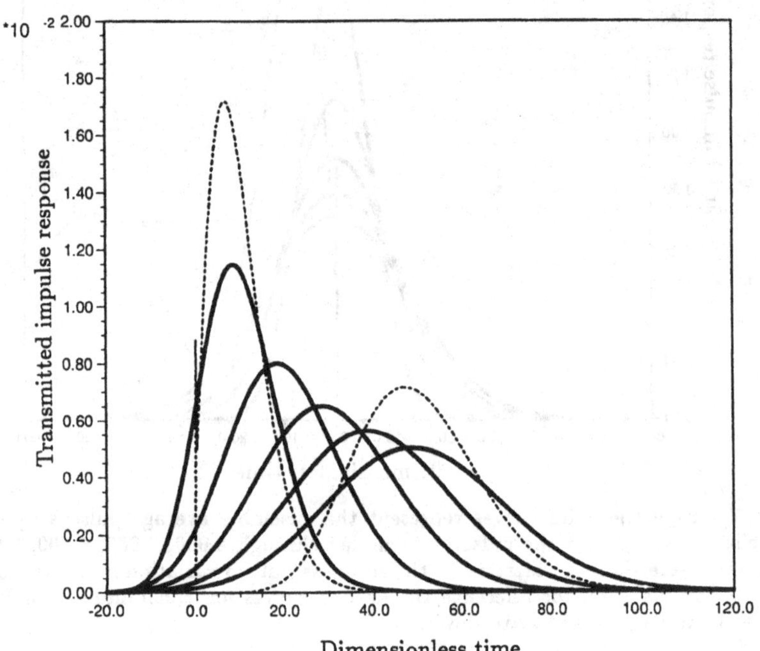

FIGURE 8. Here the solid curves represent the ensemble average pulse shape according to Keller's theory after the pulse has traveled through 2,000, 4,000, 6,000, 8,000, and 10,000 layers when there is an equal (relative) variation in slowness and in characteristic impedance. We also plot for reference the dashed curves for 2,000 and for 10,000 layers when there is no variation in wave slowness. The solid curves are the convolutions of the corresponding curves in Figures 6 & 7.

REFERENCES

[1] R.F. O'Doherty and N.A. Anstey, "Reflections on amplitudes," *Geophysical Prospecting 19,* 430-458 (1971).

[2] R. Burridge, B. White, and G. Papanicolaou, "One-dimensional wave propagation in a highly discontinuous medium", *Wave Motion 10,* 19-44 (1988).

[3] R. Burridge and H.-W. Chang "Multimode, one-dimensional, wave propagation in a highly discontinuous medium", *Wave Motion 11,* 231-249 (1988).

[4] R. Burridge and H.-W. Chang "Pulse evolution in a multimode, one-dimensional, highly discontinuous medium", *in* Elastic Wave Propagation, edited by M.F. McCarthy and M.A. Hayes, Elsevier Science Publishers, Amsterdam, (1989).

[5] K. Aki and P.G. Richards, Quantitative Seismology: Vol. I Theory and Methods, W. H. Freeman, San Francisco, (1980).

[6] B. V. Gnedenko, The Theory of Probability, (translated from the Russian by B. D. Seckler) Chelsea Publishing Company, New York, (1963).

[7] M.V. de Hoop, R. Burridge & H.-W. Chang "Wave propagation with tunneling in a highly discontinuous layered medium", *Wave Motion,* (in press).

[8] M.V. de Hoop, R. Burridge & H.-W. Chang "The pseudo-primary field due to a point source in a finely layered medium", submitted to the *Geophysical Journal International.*

[9] K. Hsu & R. Burridge "Effects of averaging and sampling on the statistics of reflection coefficients", submitted to *Geophysics.*

[10] M. Schoenberg "Reflection of elastic waves from periodically stratified media with interfacial slip", *Geophysical Prospecting 31,* 265-292 (1983).

[11] J.B. Keller "Wave prapagation in random media", in *Proceedings of the 13th Symposium on Applied Mathematics,* American Mathematical Society, New York, (1960).

[12] F.C. Karal & J.B. Keller "Elastic, electromagnetic, and other waves in a random medium", *Journal of Mathematics and Physics 5,* 537-547 (1964).

[13] F.E. Stanke & R. Burridge "Comparison of spatial and ensemble averaging methods applied to wave propagation in finely layered media", in preparation.

Stochastic Geometry and the Intensity of Random Waves.

Benjamin S. White, Exxon Research and Engineering Co.
Route 22 East, Annandale, NJ 08801

Balan Nair, MONY Capital Management
1740 Broadway, New York, NY 10019

Abstract. We outline a theory for the propagation of high frequency waves in the strong fluctuation regime, where random caustics are likely to occur. Using limit theorems for stochastic differential equations, we treat separately the equations of dynamic ray tracing, the parabolic equation, and a regularization of the ray tracing method that is uniformly valid into the caustic boundary layer. This last method is a version of the Gaussian beam summation technique. These three methods give equivalent results for the coherent field, the mutual coherence function and the correlation of intensities at two points separated in space. For the variance of intensity, however, the ray theory method has a singularity which is attributable to the lack of caustic corrections. The regularized ray (Gaussian beam) method, however, gives a finite expression for the variance of intensity at a fixed point in space: it is a universal curve related to the probability of occurence of a caustic. Thus the variance of intensity (scintillation index) is directly related to the statistics of the stochastic geometry of the wavefronts.

Random Ray Theory.

We consider the propagation of waves in a medium whose material properties differ slightly from a constant background. When the random variations are smooth and the wavelength is much smaller than a typical length, l_0, of a random inhomogeneity, geometric methods based on rays and wavefronts are applicable. For simplicity we treat the Helmholtz equation for $x \in R^2$

$$\Delta u + \omega^2 \, C^2(x) \, u = 0 \quad . \tag{1}$$

Here C(x) is the random propagation speed and

R. Spigler (ed.), Applied and Industrial Mathematics, 281–290.

(2) $$C(x) = C_0\ (1 + \sigma\ \hat{C}(x)\ )$$

where σ is a small dimensionless parameter, C_0 is the background speed, and the fluctuation field $\hat{C}$ has mean zero and is homogeneous and isotropic. We non-dimensionalize by taking l_0 as the distance scale, so that l_0 may be set equal to one. Similarly we set C_0=1 by choosing the time scale as the time to traverse l_0. Then the dimensionless frequency ω becomes a large parameter in the small wavelength (high frequency) geometric approximation.

We first review the equations of dynamic ray theory, which produce the geometric picture of wave propagation. We utilize the high frequency Ansatz

(3) $$u \sim e^{i\omega\phi}\ (V + \frac{1}{i\ \omega}\ V^{(1)} + \ .\ .\ .\ ) \qquad .$$

Substitution of (3) into (1) yields, on collecting coefficients of $(i\omega)^{-n}$

(4) $$(\nabla\ \phi)^2 = \frac{1}{C^2}$$

(5) $$2\nabla V\ \cdot\ \nabla\phi + V\ \Delta\phi = 0 \qquad .$$

By definition, the wavefronts are curves of constant ϕ. ϕ can be obtained from the eiconal equation (4) by solving it along its bicharac teristic curves, the rays $X(t,\alpha)$, where α is a parameter that runs along the wavefront and t runs along a ray. For simplicity we consider an initially plane wavefront so that $X(0,\alpha) = X_0(\alpha) = (0,\alpha)^T$. Let $U(t,\alpha)$ be the unit tangent to $X(t,\alpha)$, so that $U(0,\alpha) = U_0 = (1,0)^T$. Let $U^\perp$ be the unit vector orthogonal to U. Then starting from the wavefront ϕ=0, we obtain the wavefront ϕ=t by solving the ray equations

(6) $$\frac{dX}{dt} = C(X)\ U$$
$$\frac{dU}{dt} = -\ (\nabla C\ \cdot\ U^\perp)\ U^\perp \qquad .$$

It can be shown that the rays are orthogonal to the wavefronts, so that the rays and wavefronts form an orthogonal curvilinear coordinate system. Let $A(t,\alpha)$ be the (signed) "raytube area", or the Jacobian of the mapping between wavefronts, i.e., from $X(0,\alpha)$ to $X(t,\alpha)$. Then

$$\frac{\partial X}{\partial \alpha} = A(t,\alpha)\, U^{\perp}(t,\alpha) \tag{7}$$

A can be determined, together with an auxiliary variable $B(t,\alpha)$, by appending to equations (6) two more equations for the propagation of A, B along a ray

$$\frac{dA}{dt} = C\,B$$

(8)

$$\frac{dB}{dt} = -\,(U^{\perp})^{T}\, \nabla\nabla C\, U^{\perp} A + (\nabla C^{T}\, U)\, B$$

Here $\nabla\nabla C$ is the matrix of second derivatives of C at the point $X(t,\alpha)$.

The raytube area can be used to calculate the amplitude V, which is determined by the transport equation (5). It can be shown that

$$V(X(t,\alpha)) = \left\{ \frac{C(X(t,\alpha))\; A(0,\alpha)}{C(X(0,\alpha))\; A(t,\alpha)} \right\}^{1/2} \tag{9}$$

To summarize the ray theory method, equations (6) and (8) are solved starting from each point $(0,\alpha)^{T}$ on the initial wavefront, with initial conditions $X(0,\alpha)$, $U(0,\alpha)$, and $A(0,\alpha)=1$, $B(0,\alpha)=0$. The phase at a point $x=X(t,\alpha)$ is then given by $\phi=t$, the amplitude $V(x)$ is given by equation (9), and the field u can be computed to leading order in ω by equation (3). Thus the solution is known in ray coordinates (t,α), and must be transformed to physical coordinates x by solving $x=X(t,\alpha)$.

We can now consider geometric methods for the solution of the random problem. Early researchers (Chernov [1], Keller [2]) used a regular perturbation series in powers of σ to solve the three-dimensional version of (5) and (6). While this approach gave excellent results for some problems, the regular perturbation series broke down when used over long distances because, it was inferred, of secular terms in the perturbation expansion.

The geometric reason for the limitations of regular perturbations is illustrated in Figure 1. In Figure 1a the vertical line represents a small patch of initially plane wavefront propagating along rays orthogonal to it, which are represented by the horizontal vectors C(X) U. If propagation through a random medium causes positive curvature, the sense shown in Figure 1a, then it can be seen geometrically that deterministic forces will tend to flatten the wavefront again, i.e., an expanding cylindrical wave decreases its curvature with propagation distance. In contrast, if random perturbations cause negative curvature, as in Figure 1b, deterministic forces will tend to collapse the wavefront to a (2-D) point, called a caustic, where the raytube area, pictured as the distance between two "adjacent" rays, becomes zero. Since A=0 is not a small perturbation of A=1, regular perturbations would fail at this point. Note that the breakdown of regular perturbations is generic and relatively sudden: it occurs soon after substantial negative curvatures develop.

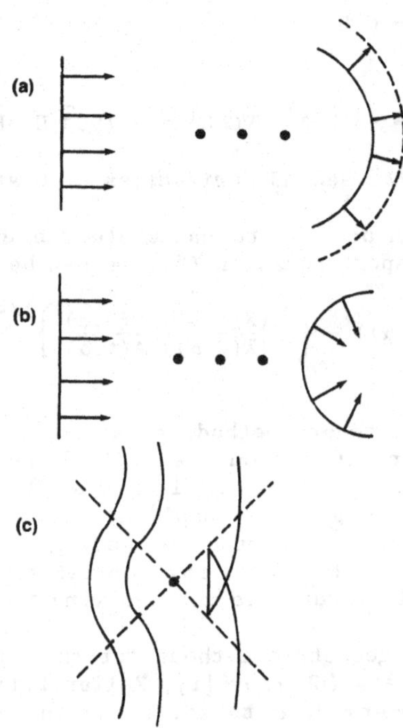

Figure 1. Propagation of a Small Patch of Wavefront. (a)If positive curvature develops, the wavefront tends to flatten again (as an expanding cylindrical wave). (b)If negative curvature develops, the wavefront tends to collapse. (c)Wavefront collapses to a caustic, producing large amplitudes, after which a loop forms.

In fact the occurrence of caustics represents a breakdown of the ray method itself, at least in its simplest form. Substitution of A=0 into equation (9) produces an infinite amplitude. Modifications of ray theory near the caustic show an amplitude of $\omega^{1/6}$ (Ludwig [3]), which is large but not infinite, in a small caustic boundary layer of order $\omega^{-2/3}$ near the caustic itself. Still, such modifications may not be necessary for some aspects of the random problem. For instance, in computing the moments of u(x) at a fixed point x in space, we may heuristically regard $\omega^{-2/3}$ as the order of the probability that a caustic correction will be needed. Then the caustic correction for a moment of order n would be of order $\omega^{-2/3}\,(\omega^{1/6})^n$, which is significant only when $n \geq 4$. In fact, for the careful calculations reported on here (Nair and White [4]), using methods with and without caustic corrections, these corrections are not necessary for the coherent wave (mean of u, n=1), and the mutual coherence function (correlation of u at two points, n=2). They are, however crucial for the calculation of the scintillation index (variance of intensity $I = |u|^2$, n=4). The corrections are not necessary for the correlation of intensities at two distinct points, i.e. $E[|u(x_1)|^2 |u(x_2)|^2]$, which is in agreement with heuristics since we would expect the probability that both x_1 and x_2 are in separate caustic boundary layers to be of order $\omega^{-4/3}$.

Apart from the necessity of amplitude corrections, caustics create a geometric complication, as illustrated in Figure 1c. After passing through a caustic the wavefront develops a loop, as in the letter α, so that a single point in space may be on more than one wavefront, i.e., several rays may go through a fixed point in space. For such points equation (3) must be replaced by a sum of terms of the form of equation (3) corresponding to the rays which go through the point. For the stochastic problem this means that the field at a fixed point x is determined by the random number of randomly chosen rays which pass through x, each ray bearing random amplitudes and phases, as determined from the stochastic differential equations (6) and (8), and equation (9).

Because of these difficulties effective non-geometric methods based on the parabolic equation approximation (Shisov [5], Uscinski [6]) have had more popularity in approaching these problems. The geometric method can, however be made to work in the region of random caustics, and yields results which generally agree with those of the parabolic equation method [4]. In addition, we can compute some aspects of the stochastic geometry itself, and have obtained a new expression for the variance of intensity. It is a universal curve, depending on the statistics of the random medium only through a single distance scale parameter, and is directly related to the probability of caustic formation.

Stochastic Limit Theorems.

To analyze the random problem we use a limit theorem for stochastic ordinary differential equations (Papanicolaou and Kohler [7]). Let $Z^\epsilon(t)$ be a vector-valued random process depending on a small parameter $0<\epsilon<<1$. Let $F(\tau,z)$, $G(\tau,z)$ be random functions such that for fixed nonrandom τ,z, $F(\tau,z)$ has mean zero. Let Z^ϵ satisfy

$$\frac{dZ^\epsilon}{dt} = \frac{1}{\epsilon} F(t/\epsilon^2, Z^\epsilon(t)) + G(t/\epsilon^2, Z^\epsilon(t)) \tag{10}$$

That is, the equation for Z^ϵ has rapidly varying (t/ϵ^2 - scale) and large ($O(1/\epsilon)$) terms. Then under wide hypotheses it can be shown that Z^ϵ has a (weak) limit Z which is a diffusion Markov process. That is, the statistics of Z^ϵ are well approximated by those of a solution, Z, of an equivalent Ito "white noise" equation. Indeed equation (10) is often called the Stratonovich form of the "white noise" equation. The limit process, Z, is characterized in [7]. Its probability density and related statistical quantities can be studied using deterministic partial differential equations, the Kolmogorov, or Fokker-Planck equations.

The ray equations can be brought into the form (10) by the substitution

$$\bar{t} = \sigma^{2/3} t \qquad A = \bar{A}^\sigma$$

$$(11) \qquad X = X_0 + \frac{\bar{t}}{\sigma^{2/3}} U_0 + \bar{X}^\sigma \qquad B = \sigma^{2/3} \bar{B}^\sigma$$

$$U = U_0 + \sigma^{2/3} \bar{U}^\sigma$$

Substitution of (11) into (6), (8) and neglect of terms of order $O(\sigma^{1/3})$ produces a system of equations of the form (10) with $\epsilon = \sigma^{1/3}$. For simplicity we consider here a single ray. It can be shown (Kulkarny and White [8]) that

$$(12) \qquad \bar{X}^\sigma \to \begin{pmatrix} 0 \\ y(\bar{t},\alpha) \end{pmatrix} \qquad \bar{A}^\sigma \to \bar{A}(\bar{t},\alpha)$$

$$\bar{U}^\sigma \to \begin{pmatrix} 0 \\ v(\bar{t},\alpha) \end{pmatrix} \qquad \bar{B}^\sigma \to \bar{B}(\bar{t},\alpha)$$

where (i) (y,v) are independent of $(\bar{A},\bar{B})$, (ii) v is Brownian motion and y is its integral, and (iii) $(\bar{A},\bar{B})$ satisfy the Ito stochastic differential equation

$$(13) \qquad d\bar{A} = \bar{B}\, d\bar{t}$$

$$d\bar{B} = \gamma \bar{A}\, d\beta(\bar{t})$$

where β is a standard Brownian motion and γ is a constant that can be computed from the correlation function of the medium fluctuations, C. γ can be set equal to one by re-scaling $\bar{t}$ and $\bar{B}$. Thus except for scale factors, $\bar{A}$, $\bar{B}$ satisfy a universal equation with no free parameters.

The physical content of this limit is as follows: from (11) the distance scale is large of order $O(\sigma^{-2/3})$; we have small angle scattering since deviations in U are $O(\sigma^{2/3})$; however because of the long distances, rays wander O(1) from their deterministic positions; similarly the raytube area deviates by O(1) from its initial value of unity. From the zeroes in equation (12) it follows that the horizontal distance travelled by the ray, i.e. the distance in the initial direction of propagation is, except for vanishingly small terms, equal to the phase, $\bar{t}/\sigma^{2/3}$. It can be shown that in this limit arclength along a ray is also, except for vanishingly small terms, equal to $\bar{t}/\sigma^{2/3}$.

From the universality of the equations for $\bar{A}$,$\bar{B}$ it follows that the distance along a ray to the first occurrence of a caustic is, except for the scale factor γ, a universal curve with no other free parameters. This curve was derived in Kulkarny and White [8], and generalized to three dimensions in White [9]. For an application in geophysics see White, Nair and Bayliss [10]. The universal curve has been confirmed in the Monte-Carlo simulations of Hesselink and Sturtevant [11].

Using these methods, equations were derived for the joint distribution of an arbitrary number of rays and associated raytube areas in Zwillinger and White [12] with some extension to three dimensions.

In [4] we also use these methods to derive a white noise limit for the parabolic equation approximation to the Helmholtz equation (1). The limit is identified as the Dawson-Papanicolaou process [13]. Thus we do not assume white noise as in the usual derivation of the parabolic moment equations, but derive it as a consequence of a stochastic limit theorem. Consequently we show for the first time that the parabolic equation moment method is valid only on the same $O(\sigma^{-2/3})$ distance scale as our random ray theory.

Transformation to Physical Coordinates. Regularization of Ray Theory.

To compute statistics of the wavefield at a fixed point x in physical space we must transform from ray coordinates. We here outline the main ideas in computing the correlation function of intensities, $I(x_1)$, $I(x_2)$, at two points $x_1=(\bar{t}/\sigma^{2/3}, y_1)$, $x_2=(\bar{t}/\sigma^{2/3}, y_2)$, which are at the same downstream distance $\bar{t}/\sigma^{2/3}$, but separated by a distance $y=y_2-y_1$. For ease of notation we drop overbars on $\bar{t}$, $\bar{A}$.

In the ray theory approximation, $I(x_1)$ can be written in terms of a sum of contributions from all rays passing through x_1.

$$(14) \qquad I(x_1) \sim \sum_{\{\alpha_i :\, y(t,\alpha_i) = y_1\}} \frac{1}{|A(t,\alpha_i)|} .$$

However, since in our approximation

$$(15) \qquad A(t,\alpha) \sim \frac{\partial y(t,\alpha)}{\partial \alpha}$$

equation (14) may be replaced using the δ-function identity by

$$(16) \qquad I(x_1) \sim \int_{-\infty}^{\infty} \delta(y(t,\alpha) - x_1)\, d\alpha .$$

Let $P_\Delta(t,y_1,y_2;\alpha_1,\alpha_2)$ be the joint probability density of $y(t,\alpha_1)$, $y(t,\alpha_2)$. This quantity can be calculated using deterministic partial differential equations corresponding to the white noise limit indicated previously. Then upon multiplying two expressions of the form (16), taking expected values, and interchanging expected values with the double integrals we obtain

$$R(t,y_1-y_2) \equiv E[I(x_1)\, I(x_2)] \sim \int_{\infty}^{\infty} \int_{\infty}^{\infty} P_{\Delta}(t,y_1,y_2;\ \alpha_1,\alpha_2)\ d\alpha_1\ d\alpha_2 \tag{17}$$

In [12] it is shown how R(t,y) can be calculated from a deterministic partial differential equation, and the result is verified by Monte-Carlo simulations. In [4] we show that this PDE can also be obtained from the parabolic moment equation method. It is also shown that in [4], [12] that R(t,y) has a logarithmic singularity as $y \to 0$, a fact attributable to the lack of caustic corrections. Hence this method fails when $y=0$, i.e. $x_1=x_2$.

In [4] we use an extension of this technique to compute other moments of the wavefield. We also use a regularization of ray theory which includes caustic corrections. This method is a modification of the Gaussian beam summation technique of Cerveny, Popov and Psencik [14] (see also White et. al. [15]). Physically this method amounts to decomposing the wavefield into a sum (integral) over beam solutions of the Helmholtz equation. This integral is analogous to the α-integral in equation (16), except that the integrand is a smooth function. Thus many of the ray techniques can be modified for use with this method.

For the second moment of intensity we obtain the regularized expression

$$E[I^2(x_1)] \sim \int_{\infty}^{\infty} \left(\frac{\omega\ \sigma^{2/3}}{2\ \pi} \right)^{1/2} e^{-(\omega\sigma^{2/3})y^2/2}\ R(t,y)\ dy \quad . \tag{18}$$

We assume the Fresnel diffraction limitation that $\omega\sigma^{2/3} >> 1$. Let P(t,A) be the probability density of A(t). Because of equation (13) P(t,A) is universal except for the scale factor γ. Then it can be shown that for t of order O(1)

$$E[I^2(x)] \sim \log(\omega\sigma^{2/3})\ P(t,0) + O(1) \quad . \tag{19}$$

Since in our limit $E[I(x)] \sim 1$, the expression in (19) is asymptotically the variance of intensity. Hence, except for scale factors, the variance of intensity is universal. Since it is essentially the probability that the raytube area is zero, it can be directly related to the (also universal) statistics of random caustic formation.

A plot of P(t,0) versus scaled distance t is shown in Figure 2. It is quite flat for small propagation distances, showing a vanishingly small probability of caustics, it rises to a relatively sharp maximum, and decays with a relatively flat region after the peak. These features are at least in qualitative agreement with numerical calculations based on the parabolic moment equations.

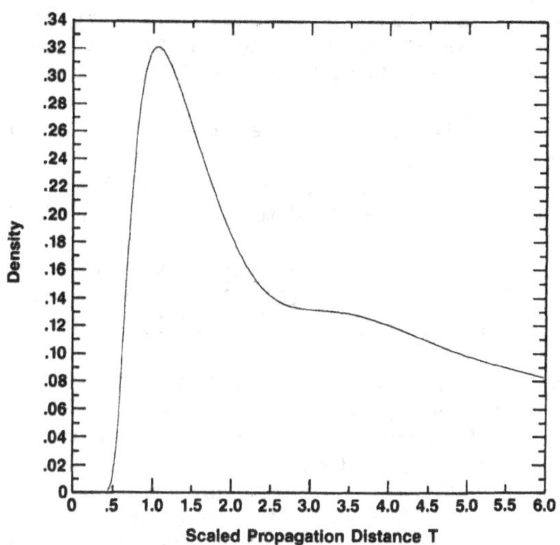

Figure 2. Probability Density of Raytube Area A(t) at A=0 (A Caustic). This universal curve also gives the variance of intensity as a function of propagation distance.

References

1. Chernov, L. A., "Wave Propagation in a Random Medium", trans. R. A. Silverman, McGraw-Hill, New York (1960).
2. Keller, J. B., "Wave Propagation in a Random Medium", Proc. Symp. Appl. Math., Vol. 13, Amer. Math. Soc., Providence (1960).
3. Ludwig, D., "Uniform Asymptotic Expansion at a Caustic", Comm. Pure and Appl. Math. 20, 215-250 (1966).
4. Nair, B., and White, B., "High Frequency Wave Propagation in Random Media - a Unified Approach", SIAM J. App. Math., to appear (1990).
5. Shisov, V. I., Izu. Vysh Uchels Zaved. Radiofizika 11, No.6, 866 (1968).
6. Uscinski, B.J., "The elements of Wave Propagation in Random Media", McGraw-Hill, New York (1977).
7. Papanicolaou, G., and Kohler, W., "Asymptotic Theory of Mixing Ordinary Differential Equations", Comm. Pure and Appl. Math., Vol. 27, No. 5 (1974).
8. Kulkarny, V., and White, B., "Focussing of waves in Turbulent Inhomogeneous Media", Phys. Fluids, 25 (10), 1770-1784 (1982).
9. White, B. S., "The Stochastic Caustic", SIAM J. Appl. Math., vol. 44, No. 1 (1984).
10. White, B., Nair, B., and Bayliss, A., "Random Rays and Seismic Amplitude Anomalies", Geophysics, Vol. 53, No.7, 903-907 (1988).
11. Hesselink, L., and Sturtevant, B., "Propagation of Weak Shocks Through Random Media", J. Fluid Mech. vol. 196, 513-553 (1988).
12. Zwillinger, D., and White, B., "Propagation of Initially Plane Waves in the Region of Random Caustics", Wave Motion 7, 202-227 (1985).
13. Dawson, D., and Papanicolaou, G., "A Random Wave Process", Appl. Math., Optim. 12, 97-114 (1984).
14. Cerveny, V. Popov, M., and Psencik, I., "Computation of Wave Fields in Inhomogeneous Media - Gaussian Beam Approach", Geophys. J. R. Astr. Soc. 70, 109-128 (1982).
15. White, B., Burridge, B., Bayliss, A., and Norris, A., "Some Remarks on the Gaussian Beam Summation Method", Geophys. J. R. Astr. Soc. 89, 579-636 (1987).

PLANE WAVES IN LAYERED RANDOM MEDIA. THE ROLE OF BOUNDARY CONDITIONS

V.I. Klyatskin
Pacific Oceanological Institute
Far Eastern Branch
USSR Academy of Sciences
7 Radio Street, 690032 Vladivostok
USSR

ABSTRACT. It is shown by using the concrete examples that the statistical characteristics of a wave field differ greatly when different boundary conditions are taken. The problem of radiation localization in random media is analyzed in detail.

1. Stationary Boundary Value Wave Problems

At present the propagation of various kinds of waves in natural media is being studied quite vigorously. The mathematical aspects of this propagation are described by a boundary value problem for the wave equation. The application of imbedding techniques permits one to replace the boundary value problems in question by problems with initial conditions and to study statistical characteristics of their solutions [1,2]. The imbedding equations differ, depending on the boundary conditions, and therefore their solutions may have different statistical characteristics. Let us consider some examples.

1.1. Wave Incidence on the Half-Space

Let the layer of the randomly inhomogeneous medium occupy a part of the space, $L_0 \leqslant x \leqslant L$, and suppose the plane wave with amplitude equal to one is incident from the region $x < L$. Then the wave field in the layer is described by the

R. Spigler (ed.), Applied and Industrial Mathematics, 291–299.

boundary value problem

$$\frac{d^2}{dx^2}u(x;L)+\kappa^2[1+\epsilon(x)]u(x;L) = 0, \tag{1}$$

$$\frac{d}{dx}u(L_0;L) = -i\kappa u(L_0;L), \quad \frac{d}{dx}u(x;L)_{x=L} = i\kappa[u(L;L)-2].$$

We suppose here $\epsilon(x) = \epsilon_1(x)+i\gamma$ ($\epsilon_1=\epsilon_1^*$, $\gamma \ll 1$ describes the absorption of the wave), outside the layer $\epsilon = 0$. We introduced the dependence of the solution on the parameter L following the idea of the imbedding method. Then the field on the boundary $x = L$ is $u(L,L) = u_L = I+R_L$, where R_L is the complex reflection coefficient.

The imbedding equations for this boundary value problem are

$$\frac{d}{dL}R_L = 2i\kappa R_L+i\frac{\kappa}{2}\epsilon(L)(1+R_L)^2; \; R_{L_0} = 0, \tag{2}$$

$$\frac{\partial}{\partial L}u(x;L) = i\kappa\left\{1+\frac{1}{2}\epsilon(L)(1+R_L)\right\}u(x;L), \quad u(x;x) = 1+R_x.$$

We are interested in half-space case, i.e. $L_0 \to -\infty$.

For simplicity, let us consider the function $\epsilon_1(x)$ to be the Gaussian δ-correlated random function with parameters $<\epsilon_1(x)> = 0$, $<\epsilon_1(x)\epsilon_1(x')> = 2\sigma^2 l_0\delta(x-x')$. Then there is a 'stationary' probability distribution for the value $W_L = |R_L|^2$:

$$P_\infty(W) = \frac{2\beta}{(1-W)^2}\exp(-2\beta\frac{W}{1-W}), \quad \beta = \frac{\kappa\gamma}{D}, \tag{3}$$

where $D = \kappa^2\sigma^2 l_0/2$ is corresponding diffusion coefficient. If $\beta \to 0$, $P_\infty(W) = \delta(W-1)$ and in this limiting case [1,2]

$$<I(x;L)> = 2, \quad <I^n(x;L)> \sim \exp n(n-1)\xi, \quad \xi = D(L-x), \tag{4}$$

where $I(x;L) = |u(x;L)|^2$ is the intensity of the wave field. The exponential growth of the wave field moments corresponds to the lognormal law of the probability distribution. Physically it means that inside the medium rare but strong intensity jumps of the wave field are present in each realization. The typical realization of the field intensity distribution $I(x;L)$ are given in Fig. 1.

Figure 1. Illustration of a typical realization of the field intensity inside the half-space of random medium.

The intensity statistics of the wave field inside the medium is formed entirely owing to its jumps.

Note that the mean value of the energy flow density

$$S(x;L) = \frac{1}{2i\kappa}\left[u(x;L)\frac{d}{dx}u^*(x;L)-u^*(x;L)\frac{d}{dx}u(x;L)\right]$$

is equal to zero in this case, because $<S(x;L)> = 1-<W_L>$ as $\gamma\to 0$. When finite (however small) damping is present, the exponential growth of the intensity moments has to stop and would change into abrupt attenuation [2,3].

Note that the wave equation (1) gives the equality

$$\kappa\gamma I(x;L) = \frac{d}{dx}S(x;L); \tag{5}$$

integrating (5) over the half-space we obtain

$$\kappa\gamma \int_{-\infty}^{L} dx I(x;L) = S(L;L) = 1-W_L.$$

In the limit for small β we have

$$\int_{-\infty}^{L} dx <I(x;L)> = \frac{2}{D}\ln\frac{1}{\beta} = 2l_\beta,$$

that is, localization linked to the attenuation takes place. The quantity $l_\beta\to\infty$ and localization is absent as $\beta\to 0$.

1.2. Source Location on the Reflective Boundary

If the plane-wave source is placed on the reflective boundary $x = L$ $(d\tilde{u}/dx\,|_{x=L} = 0)$, then the wave field inside the medium is linked by the equality

$$\tilde{u}(x;L) = \frac{2}{1-R_L}u(x;L), \quad \tilde{S}(x;L) = \frac{4}{|1-R_L|^2}S(x;L).$$

The random realization of the field intensity differs in this case from the realization given in Fig. 1 only by multiplication by a constant, and for the limit half-space and a small parameter β we have

$$<\tilde{S}(L;L)> = 1, \quad <\tilde{I}(x;L)> = \frac{8}{\beta}, \quad \int_{-\infty}^{L} dx <\tilde{I}(x;L)> = \frac{4}{\beta D},$$

that is, the mean energy flow and the mean total energy are independent fluctuations of the medium. The intensity inside the medium is described by the equality [4]

$$\lim_{\beta\to 0}\beta <\tilde{I}(x;L)> = 4\varphi_{\text{loc}}(\xi), \quad \xi = D(L-x), \tag{6}$$

$$\varphi_{\text{loc}}(\xi) = 2\pi\int_0^\infty d\tau\,\tau\frac{\text{sh}\pi\tau}{\text{ch}^2\pi\tau}\left[\tau^2+\frac{1}{4}\right]\exp\left\{-\left[\tau^2+\frac{1}{4}\right]\xi\right\}.$$

Note that this formula, obtained on the basis of averaging over fast oscillations is not suitable in the region $\xi\sim 0$, where the correlations, $<R_L R_x^*>$ are significant.

One can see from formula (6) that if $\xi\to 0$, then

$$\varphi_{\text{loc}}(0) = 1, \quad \frac{d}{d\xi}\varphi_{\text{loc}}(0) = -2,$$

and if $\xi \gg 1$ the following asymptotical solution is valid:

$$\varphi_{\text{loc}}(\xi) \approx \frac{\pi^2\sqrt{\pi}}{2}\xi^{-3/2}\exp\left[-\frac{1}{4}\xi\right].$$

Thus localization takes place and the localization curve attenuation rather rapidly when ξ is small ($\varphi_{\text{loc}}(\xi)\approx\exp(-2\xi)$) and the attenuates is much slower if the values of ξ are large. The total area underlying the curve should be equal to 1.

1.3. Source Location in the Unbounded Space

If the source of the plane wave is placed in the point x_0 inside the layer, the wave field is described by the equation

$$\left\{\frac{d^2}{dx^2}+\kappa^2[1+\epsilon(x)]\right\}G(x,x_0) = 2i\kappa\,\delta(x-x_0) \tag{7}$$

with boundary conditions

$$\frac{d}{dx}G(L_0,x_0) = -i\kappa G(L_0,x_0), \quad \frac{d}{dx}G(L,x_0) = i\kappa G(L,x_0).$$

The solution of the boundary value problem (7) has the structure

$$G(x,x_0) = \frac{(1+R_{x_0})(1+\tilde{R}_{x_0})}{1-R_{x_0}\tilde{R}_{x_0}}\exp\left\{i\kappa\int\limits_x^{x_0} d\xi\frac{1-R_\xi}{1+R_\xi}\right\}, \quad x \leqslant x_0, \tag{8}$$

where R_{x_0} is the reflection coefficient from (2) and $\tilde{R}_{x_0}$ is the reflection coefficient for the plane wave which is incident on the layer (x_0,L) from the region $x < x_0$. Note, that the case 1.1. corresponds to $\tilde{R}_{x_0} = 0$, and the case 1.2. corresponds to $\tilde{R}_{x_0} = I$. Thus, the typical realization of the field intensity also has the structure of Fig. 1.

In this case we have

$$<I(x_0,x_0)> = 1+\frac{1}{\beta}, \quad <S(x_0,x_0)> = 1, \quad \int\limits_{-\infty}^{x_0} dx <I(x,x_0)> = \frac{1}{\beta D},$$

as in the case of free space. For small values of β we have

$$<I(x,x_0)> = \frac{1}{\beta}\varphi_{loc}(\xi), \quad \xi = D\,|\,x-x_0\,|, \tag{9}$$

that is, localization of energy also takes place.

1.4. Another Type of Wave Excitation

Let us discuss now a new type of boundary value problem that is, for example, typical for ocean acoustics [2],

$$\frac{d^2}{dx^2}u(x;L)+\kappa^2[1+\epsilon(x)]u(x;L) = 0, \tag{10}$$

$$u(L;L) = 1, \quad \frac{d}{dx}u(L_0;L) = 0.$$

The imbedding equations for this problem have the form

$$\frac{\partial}{\partial L}u(x;L) = -i\kappa\frac{1+r_L}{1-r_L}u(x;L), \quad u(L;L) = 1, \tag{11}$$

$$\frac{d}{dL}r_L = -2i\kappa r_L + \frac{i\kappa}{2}\epsilon(L)(1-r_L)^2, \quad r_{L_0} = -1,$$

where r_L is the analogue of the reflection coefficient. The intensity of the field inside the layer is described by the equation following from (11) $(k=\kappa+i\gamma)$ $(I=|u|^2)$,

$$\frac{\partial}{\partial L}|I(x;L)| = \left[i\kappa\left[\frac{1+r_L^*}{1-r_L^*}-\frac{1+r_L}{1-r_L}\right]+\gamma\left[\frac{1+r_L^*}{1-r_L^*})+\frac{1+r_L}{1-r_L}\right]\right]I(x;L). \tag{12}$$

The influence of fluctuations of means of medium parameters enters into (12) only by the functions r_L and r_L^*. Let us average the rapidly oscillating functions at the right-hand part of (12), using the expression

$$\frac{1}{2\pi}\int_0^{2\pi} d\varphi \frac{1+|r_L|\exp(i\varphi)}{1-|r_L|\exp(i\varphi)} \equiv 1 \ (|r_L|<1).$$

As a result we obtain that the functional $<|u(x;L)|^2>$ does not depend on medium fluctuations ϵ. In view of the linearity of equation (12), this statement is valid for the highest moments of $|u(x;L)|^2$, independently of whether the probability distribution $P_L(R=|r_L|^2)$ becomes a stationary (with respect to L) solution or not. Thus, for the problem under discussion the presence of parameter fluctuations of the medium does not influence the statistical characteristics of the wave intensity, unlike the previous problems where the influence of parameter fluctuations has a cummulative character. The isolated realization of wave intensity in the case also coincides with this one shown in Fig. 1.

2. Non-stationary Boundary Value Problems

Let us now discuss the non-stationary problem of the field of a time-pulse generated in the randomly inhomogeneous medium. The wave field is described in this case by the wave equation

$$\left[\frac{\partial^2}{\partial x^2}-\frac{1}{c^2(x)}\frac{\partial}{\partial t}\left[\frac{\partial}{\partial t}+\tilde{\gamma}\right]\right]u(x,x_0;t) = \frac{2}{c_0}\delta(x-x_0)\frac{\partial}{\partial t}f(t). \tag{13}$$

Its solution can be written as a Fourier integral (the parameter $\tilde{\gamma}$ is supposed to be small):

$$u(x,x_0,t) = \frac{-1}{2\pi}\int_{-\infty}^{\infty} d\omega\, G_\omega(x,x_0)f(\omega)e^{-i\omega t},$$

where the function $G_\omega(x,x_0)$ is the solution of the stationary problem (7) $(c^{-2}(x)=c_0^{-2}[1+\epsilon(x)]$, $\kappa=\omega/c_0$, $\gamma=\tilde{\gamma}/\omega)$.

For the mean of the field intensity $I(x,X_0,t)=u^2(x,x_0,t)$ we have

$$<I(x,x_0;t)> = \frac{1}{(2\pi)^2}\int\int_{-\infty}^{\infty} d\omega d\psi <I_{\omega,\psi}(x,x_0)> f\left[\omega+\frac{\psi}{r}\right] f^*\left[\omega-\frac{\psi}{r}\right] e^{-i\psi t},$$

where $<I_{\omega,\psi} = <G_{\omega+\psi/r} G^*_{\omega-\psi/r}>$.

When $t\to\infty$ the value of this integral is determined by the integrand's behaviour for small values of ψ, i.e. by the quantity

$$<I(x,x_0;t)> = \frac{1}{(2\pi)^2}\int_{-\infty}^{\infty} d\omega |f(\omega)|^2 \int_{-\infty}^{\infty} d\psi <I_{\omega,\psi}(x,x_0)> e^{-i\psi t}. \tag{14}$$

Analogously, for the energy enclosed in the half-space $(-\infty,x_0)$ we have

$$\epsilon(t) = \int_{x_0}^{\infty} dx <I(x,x_0;t)> = \frac{c_0}{(2\pi)^2}\int_{-\infty}^{\infty} d\omega |f(\omega)|^2 \int_{-\infty}^{\infty} \frac{d\psi}{\tilde{\gamma}-i\psi} <S_{\omega,\psi}(x,x_0)> e^{-i\psi t}. \tag{15}$$

Here $S_{\omega,\psi}$ is the two-frequency analogue of the energy flow density. All these two-frequency characteristics may be calculated by analytic continuation into the complex region with respect to the parameter β for the corresponding solutions of stationary problems [4,5]. This means that when solving the stationary problems, the parameter β should be replaced by

$$\beta(\omega,\psi) = \frac{1}{c_0 D(\omega)}(\tilde{\gamma}-i\psi).$$

By calculating the integrals over ψ, we obtain the results $(\tilde{\gamma}\to 0)$.

2.1. *Source of a Time-Pulse inside the Medium*

In this case there is a constant limit

$$<I(x,x_0;\infty)> = \frac{c_0}{2\pi}\int_{-\infty}^{\infty} d\omega D(\omega) |f(\omega)|^2 \varphi_{\text{loc}}(\xi), \quad \xi = D(\omega)|x-x_0|,$$

where $\varphi_{\text{loc}}(\xi)$ is the localization curve (6). Analogously,

$$\epsilon(\infty) = \frac{c_0}{2\pi}\int_{-\infty}^{\infty} d\omega |f(\omega)|^2.$$

For the model $\epsilon_1(x)$ of white noise, when $D(\omega) = \sigma^2\omega^2 l_0/2c_0^2$, we have the following law for sufficiently large values of $|x-x_0|$,

$$<I(x,x_0;\infty)> \sim |x-x_0|^{-3/2}.$$

Similar behaviour of the mean intensity exists in the problem when the source of the time pulsing lies on the reflective boundary.

2.2. *Time-Pulse Incidence on the Half-Space*

In this case we have

$$<I(L;L;t)> = \frac{1}{(2\pi)^2}\int_{-\infty}^{\infty} d\omega |f(\omega)|^2 \int_{-\infty}^{\infty} d\psi \{1+W^{(1)}_{\omega,\psi}\} e^{-i\psi t},$$

$$\epsilon(t) = \frac{c_0}{(2\pi)^2}\int_{-\infty}^{\infty} d\omega |f(\omega)|^2 \int_{-\infty}^{\infty} \frac{d\psi}{0-i\psi}\{1-W^{(1)}_{\omega,\psi}\} e^{-i\psi t},$$

where

$$W^{(1)}_{\omega,\psi} = \beta(\omega,\psi)\int_0^{\infty} du \frac{u}{u+2}\exp(-\beta(\omega,\psi)u)$$

is the analytic continuation of the corresponding expression for $<|R_L|^2>$ with respect to the parameter β.

Accomplishing the integration with respect to ψ and u, we find asymptotic expressions for t large:

$$<I(L,L;t)> = \frac{c_0}{\pi}\int_{-\infty}^{\infty} d\omega |f(\omega)|^2 \frac{D(\omega)}{[2+D(\omega)c_0 t]^2},$$

$$\epsilon(t) = \frac{c_0}{\pi}\int_{-\infty}^{\infty} d\omega |f(\omega)|^2 \frac{1}{2+D(\omega)c_0 t}.$$

For the model of white noise we have asymptotic behaviour of following type as $t\to\infty$, if $|f(\omega)|_{\omega\to 0} \neq 0$,

$$<I(L,L;t)> \sim t^{-3/2}, \quad \epsilon(t) \sim t^{-1/2}.$$

An additional exponential attenuation $\exp(-\tilde{\gamma}t)$ will be put over all the above dependencies if there is a small attenuation $\tilde{\gamma}$.

Note that some of the expressions were obtained before in [6] in another way.

References

[1] Klyatskin V.I. (1980), *Stochastic equations and waves in randomly inhomogeneous media,* Nauka, Moscow (in Russian); (1986), Editions de Physique (in French).

[2] Klyatskin V.I. (1986), *The imbedding method in the theory of wave propagation,* Nauka, Moscow (in Russian).

[3] Babkin G.I., Klyatskin V.I. (1982), *Statistical theory of radiative transfer layered media,* Wave Motion, v. 4, p. 327.

[4] Klyatskin V.I. (1990), *Localization and waves in the layered randomly inhomogeneous media,* Wave Motion (in press).

[5] Shevtsov B.M. (1982), *Three dimensional problem of inverse scattering in layered randomly inhomogeneous media,* Izv. Vuzov, Radiofizica, v. 25, p. 1032 (in Russian).

[6] Burridge R., Papanicolaou G., White B., Sheng P. (1989), *Probing a random medium with a pulse,* SIAM Appl. Math., v. 49, p. 582.

QUANTUM TRANSPORT MODELS FOR SEMICONDUCTORS

Anton ARNOLD[1] and Peter A. MARKOWICH[1]
Technische Universität Berlin
Fachbereich Mathematik
Straße des 17. Juni 136
D-1000 Berlin 12
FRG

1 Introduction

During the last decade the simulation of semiconductor devices became a valuable and indispensable tool for the design of new devices. In this paper we present and analyze new simulation models which also account for quantum effects.

The recent need for quantum models in semiconductor device simulations was prompted by two different technological developments: Due to the ongoing miniaturization of devices, extremely high electric fields (up to $10^6 V/cm$) appear in the active region, which makes quantum effects important for the charge carrier transport. On the other hand, several devices (e.g. the resonant tunneling diode) whose performance depend crucially on quantum effects (like tunneling) are in use. It is obvious that these new devices cannot be correctly described by conventional drift-diffusion, hydrodynamic or (semi) classical Boltzmann models. Therefore quantum kinetic equations have to be employed. Today's supercomputer capacities also give a possibility to numerically handle these kinetic equations, even for complicated geometries in two (spatial) dimensions.

2 The Wigner Equation - Models and Analysis

2.1 LINEAR MODELS

The Wigner distribution function represents a quantum mechanical description of the elctron's motion in the position-velocity (x, v)-phase space, which is equivalent to the conventional wave function or density matrix formalism ([1]). From the Schrödinger equation

$$i\hbar\psi_t = -\frac{\hbar^2}{2m}\Delta_x\psi + qV(x,t)\psi, x \in \mathbb{R}^d, \tag{2.1}$$

for the wave-function $\psi = \psi(x,t)$ one obtains the Wigner (or Quantum-Liouville) equation, which governs the evolution of the (real valued) Wigner function

$$w(x,v,t) = (2\pi)^{-\frac{d}{2}} \int_{\mathbb{R}^d} \overline{\psi}(x + \frac{\hbar}{2m}\eta, t)\psi(x - \frac{\hbar}{2m}\eta, t)e^{iv\cdot\eta}d\eta; x, v \in \mathbb{R}^d. \tag{2.2}$$

The equation reads

$$w_t + v \cdot \nabla_x w - \frac{q}{m}\theta_\hbar[V]w = 0, \tag{2.3}$$

[1]The authors acknowledge support from the Austrian "Fonds zur Förderung der wissenschaftlichen Forschung" under Grant No. P6771.

R. Spigler (ed.), Applied and Industrial Mathematics, 301–307.

with the pseudo-differential operator

$$\theta_\hbar[V]w = i\frac{\hbar}{m}[V(x+\frac{\hbar}{2mi}\nabla_v,t) - V(x-\frac{\hbar}{2mi}\nabla_v,t)]w \tag{2.4}$$
$$= \frac{\hbar}{m}\frac{i}{(2\pi)^d}\int_{\mathbb{R}^d_\eta}\int_{\mathbb{R}^d_{v'}}[V(x+\frac{\hbar}{2m}\eta,t) - V(x-\frac{\hbar}{2m}\eta,t)]w(x,v',t)e^{i(v-v')\eta}dv'd\eta.$$

m and q denote the electrons mass and charge, resp., and $\hbar$ is the reduced Planck constant. (2.3) describes the motion of non-interacting electrons in vacuum under the action of the prescribed electrostatic potential V. The Wigner formalism is the quantum mechanical analogue of the classical phase space setting (Liouville equation). For smooth potentials and a small parameter $\hbar$ (in an appropriate scaling) the Wigner solution is found to be a perturbation of the corresponding classical phase space distribution function. Although the Wigner function may assume negative values, the particle density

$$n(x,t) = \int_{\mathbb{R}^d} w(x,v,t)dv \tag{2.5}$$

can be shown to stay non-negative (under reasonable assumptions on V and on the initial data).

Any L^2-solution of the Wigner equation can be expanded into a series of pure states ([2]):

$$w = \sum_{j=1}^{\infty}\lambda_j w_j,\ \lambda_j \geq 0,\ \sum_{j=1}^{\infty}\lambda_j = 1, \tag{2.6}$$

where each w_j corresponds via (2.2) to a pure wave function ψ_j, whose time evolution is governed by (2.1). The coefficients λ_j, constant in time, give the probability to find the electron ensemble in the state ψ_j. By this expansion the Wigner equation is equivalent to countably many Schrödinger equations.

The Wigner operator (2.3) generates a semigroup on $L^2(\mathbb{R}^d_x \times \mathbb{R}^d_v)$, which implies the global existence and uniqueness of a classical solution in the whole space case ([3]). Also the bounded domain problem with inflow boundary conditions has been analyzed in [3]. Even in this case the potential V has to be known on all of $\mathbb{R}^d$, due to the Fourier transform involved in $\theta_\hbar[V]$.

In an external electromagnetic field the electron's spin interacts with the magnetic field. Existence and uniqueness results for the resulting 2×2 matrix-valued Wigner function are given in [4].

2.2 SELF-CONSISTENT WIGNER-POISSON MODEL

Since electrons are negatively charged particles, they interact by a repulsive Coulomb force. In a self-consistent model the electrostatic potential V is assumed to result from the electron density $n(x,t)$ and an external charge density $C(x)$, which in our case represents the doping profile of the considered semiconductor. The potential is then given by the Newtonian solution of the Poisson equation:

$$\Delta V(x,t) = \frac{q}{\varepsilon_s}(n(x,t) - C(x)), \tag{2.7}$$

where ε_s is the permittivity of the semiconductor, which is nonlinearly coupled to the Wigner equation (2.3). The derivation of this self-consistent Wigner-Poisson (or Quantum Vlasov) model by a BBGKY-hierarchy from the many body problem is analogous to the Vlasov equation.

In contrast to the Vlasov equation which preserves all L^p-norms of solutions, no L^1-theory is yet available for the Wigner equation. So it is not *a priori* clear, how to mathematically define the density n. Therefore it seems easier for the mathematical analysis to exploit the equivalence of the Wigner equation to countably many Schrödinger equations. This idea was pursued in the proofs of global existence and uniqueness of classical solutions in the one, two and three dimensional cases ([5] ,[6] ,[7]). The scaled self-consistent problem for the sequence of wavefunctions $\Psi = (\psi_j)_{j\in \mathbb{N}}$ then reads:

$$\Psi_t = \frac{i}{2}\Delta_x\Psi - iV(x,t)\Psi, x \in \mathbb{R}^d, \quad (2.8)$$

$$\Delta V(x,t) = C(x) - n(x,t), \quad (2.9)$$

$$n[\Psi](x,t) = \sum_{j=1}^{\infty}\lambda_j|\psi_j(x,t)|^2, \quad (2.10)$$

$$\Psi(t=0) = \Psi^I. \quad (2.11)$$

We remark that the proofs for $d = 1,2$ and 3 are very different, since the Newtonian potential solution of (2 . 9) behaves substantially different in the three cases.

Next we state the existence and uniqueness results for the two and three dimensional Schrödinger-Poisson problem (2 . 8)-(2 . 11) and briefly sketch the idea of the proofs. The corresponding result for the Wigner-Poisson problem then follows by equivalence of the two formulations.

First we define the Banach spaces

$$X := \{ \Psi \mid \psi_j \in L^2(\mathbb{R}^2), j \in \mathbb{N}, \| \Psi \|_X^2 := \sum_{j=1}^{\infty} \lambda_j \| \psi_j \|^2_{L^2(\mathbb{R}^2)} < \infty\}, \quad (2.12)$$

$$Y := \{ \Psi \mid \psi_j \in H^2(\mathbb{R}^2), |x|\psi_j \in L^2(\mathbb{R}^2), j \in \mathbb{N}, \quad (2.13)$$
$$\| \Psi \|_Y^2 := \sum_{j=1}^{\infty} \lambda_j(\| \psi_j \|^2_{H^2(\mathbb{R}^2)} + \| |x|\psi_j \|^2_{L^2(\mathbb{R}^2)}) < \infty\},$$

with the (closed) subset of charge neutral wave functions

$$\tilde{Y} := \{\Psi \in Y | \int_{\mathbb{R}^2}(C - n[\Psi])dx = 0. \quad (2.14)$$

Theorem 2.1 *Let $C \in L^\infty(\mathbb{R}^2)$ and $(1+|x|)C \in L^1(\mathbb{R}^2) \cap L^2(\mathbb{R}^2)$. If the initial (charge neutral) wavefunction satisfies $\Psi^I \in \tilde{Y}$ and $\nabla(|x|\Psi^I) \in X$, then the two dimensional Schrödinger-Poisson problem (2 . 8)-(2 . 11) has a unique global, classical solution.*

Proof (Sketch): The local-in-time-part consists of a fixed point iteration for Ψ, where the (linear) Schrödinger equation (2 . 8) and the Poisson equation (2 . 9) are solved consecutively. The assumed charge neutrality of the system assures (in $\mathbb{R}^2$) that the Newtonian potential $V = V[\Psi(t)]$ is $L^\infty(\mathbb{R}^2)$ and a bounded perturbation of $\frac{i}{2}\Delta_x$. The contraction argument is then applied in $C([0,T],\tilde{Y})$.

The global in time result is obtained by successive *a priori* estimates of $\| \Psi \|_X$, $\| \nabla\Psi \|_X$, $\| |x|\Psi \|_X$ and $\| \Delta\Psi \|_X$. We remark that the potential energy of an electrostatic system is always positive. Therefore the energy estimate

$$\| \nabla\Psi(t) \|_X^2 + \| \nabla V(t) \|^2_{L^2(\mathbb{R}^2)} = \text{ const } , t \geq 0 \quad (2.15)$$

gives the *a priori* estimate for $\nabla\Psi$. □

For the three dimensional problem we first define the Banach space

$$Z_\Omega := \{\Psi|\psi_j \in H^2(\Omega)\cap H_0^1(\Omega), j\in I\!N, \| \Psi \|^2_{Z_\Omega} := \sum_{j=1}^{\infty} \lambda_j \| \Psi_j \|^2_{H^2(\Omega)} < \infty\}. \tag{2.16}$$

Theorem 2.2 *Let $\Psi^I \in Z_{I\!R^3}$ (see (2 . 16)) hold.*

Then the three dimensional Schrödinger-Poisson problem (2 . 8)-(2 . 11) with $C \equiv 0$ has a unique global, classical solution.

Proof (Sketch): We remark that the Newtonian potential $V = V[\Psi]$ is bounded in $I\!R^3$. The problem (2 . 8)-(2 . 11) is first approximated by Schrödinger-Poisson problems posed on balls $B_R \subseteq I\!R^3$, where only finitely many Schrödinger equations are coupled to the Poisson equation. For these problems the nonlinearity $-iV[\Psi]\Psi$ in (2 . 8) is a local Lipschitz perturbation of $\frac{i}{2}\Delta_x\Psi$ in Z_{B_R}.

This and *a priori* estimates imply the global existence of unique solutions for the problems on B_R. Uniform estimates in $R > 1$ of these solutions are obtained from the key estimate

$$\| u \|_{L^6(B_R)} \leq C \| \nabla u \|_{L^2(B_R)}, u \in H_0^1(B_R), \tag{2.17}$$

where C is independent of $R > 0$ (which, e.g., allows to control the potential in $L^6(B_R)$ independently of R). Then a compactness argument gives convergence of the approximating problems to the classical solution of (2 . 8)-(2 . 11) on $I\!R^3$. □

2.3 AN ENERGY BAND MODEL

The previous models do not account for the influence of the semiconductor crystal on the charge transport. Therefore we extend the model by incorporating the quantum effects of the periodic crystal lattice potential via the band-diagram of the considered semiconductor ([8]). In the single band approximation, where no band transitions of the electrons are considered, the semi-classical evolution of an electron ensemble in the n-th energy band ($\varepsilon_n(k)$) is governed by the Hamilton function

$$H(x,k,t) = \varepsilon_n(k) + qV(x,t), \tag{2.18}$$

with the space variable $x \in I\!R^3$ and the canonically conjugate crystal momentum $\hbar k$. By the quantum mechanical correspondence principle ($x \to i\nabla_k$) one obtains the Wigner equation in a crystal:

$$w_t + \frac{i}{\hbar}[\varepsilon_n(k+\frac{1}{2i}\nabla_x) - \varepsilon_n(k-\frac{1}{2i}\nabla_x) - qV(x+\frac{1}{2i}\nabla_k) + qV(x-\frac{1}{2i}\nabla_k)]w = 0 \tag{2.19}$$

for the Wigner function $w = w(x,k,t)$. Since $\varepsilon_n(k)$ is periodic with period domain B (first Brillouin zone), x has to be restricted to $\frac{1}{2}\cdot L^\#$, where $L^\#$ is the set of crystal lattice vectors. In contrast to the vacuum Wigner equation, x is a discrete variable and k varies in a bounded domain in the crystal case. Moreover, (2 . 19) is nonlocal in k and x because of the non-quadraticity of $\varepsilon_n(k)$. The pseudo-differential operators in (2 . 19) are discrete analogues of $\theta_\hbar[V]$, defined in (2 . 4).

In a selfconsistent model (2 . 19) is coupled to a 'discrete' Poisson equation, with the electron density

$$n(x,t) = \int_B w(x,k,t)dk, x \in \frac{1}{2}\cdot L^\#. \tag{2.20}$$

This nonlinear model has been analysed in one and three dimensions ([9], [10]).

In order to understand the relationship between the semi-classical and the Wigner-Poisson model in a crystal one introduces the scaling $x = x_o x_s$ and $k = \frac{k_s}{\ell}$, where x_o is a typical length-scale of the semiconductor and $L = \ell^3$ is the volume of a primitive crystal cell. The limit $\alpha = \frac{\ell}{x_o} \to 0$ represents the (formal) transition from (2 . 19) to the semi-classical model:

$$w_t + [a\nabla_k \varepsilon_n(k)\nabla_x - b\nabla_x V(x)\nabla_k]w = 0, x \in \mathbb{R}^3, k \in \ell \cdot B, \tag{2.21}$$

coupled to the (continuous) Poisson equation

$$\Delta V(x,t) = c(C(x) - n(x,t)), x \in \mathbb{R}^3 \tag{2.22}$$

(here we suppressed the index 's' for the scaled variables). The dimensionless constants a, b, c are of order 0.1 to 1 in the example of a typical tunneling diode (see [8] for details).

Due to the nonlocality of (2 . 19) in x and k, this Wigner-Poisson model is very difficult to be handled numerically. Replacing the ε_n-pseudo-differential operators by

$$\frac{1}{\hbar}\nabla_k \varepsilon_n(k) \cdot \nabla_x w(x,k,t), x \in \mathbb{R}^3, k \in B, \tag{2.23}$$

coupled to the (continuous) Poisson equation gives an approximating (local in k) model, which is still capable of simulating quantum effects like tunneling. Existence and uniqueness results of this equation, posed on a bounded crystal are presented in [11].

3 Discretisation

During the last few years different numerical methods for the Wigner(-Poisson) equation have been used to simulate quantum effects in ultra-integrated semiconductor devices and plasmas: finite difference schemes ([12], [13]), spectral collocation methods ([14]) an a fractional step method ([15]).

Recently also a deterministic particle method has been applied. For the application of this method the Wigner equation is at first reformulated as

$$w_t + v \cdot \nabla_x w + \frac{q}{m} E_1 \nabla_v w = \int_{\mathbb{R}^d_{v'}} \varphi(x, v - v')w(x, v', t)dv', \tag{3 . 1}$$

with a constant electric field E_1 and $\varphi(x,v)$ resulting from the (possibly nonsmooth) potential $V_2(x)$:

$$\varphi(x,y) = \frac{q}{\hbar} \cdot \frac{i}{(2\pi)^d} \int_{\mathbb{R}^d_\eta} [V_2(x + \frac{\hbar}{2m}\eta) - V_2(x - \frac{\hbar}{2m}\eta)]d\eta. \tag{3 . 2}$$

The idea of this method relies on the definition of a discrete set of "particles" located at $(x_i(t), v_i(t))$, $i \in \mathbb{Z}^{2d}$, with constant control volumes ω_i and weights $w_i(t)$. Then the Wigner function w is approximated in the sense of measures as a linear combination of delta distributions:

$$\tilde{w}(x,v,t) = \sum_i \omega_i w_i(t)\delta(x - x_i(t)) \otimes \delta(v - v_i(t)). \tag{3 . 3}$$

The particles are moved along the characteristics of the left hand side of (3 . 1) and the time evolution of the weights acccounts for right hand integral operator. Since the integral kernel is singular in x-direction, it has to be smoothed for the discrete particle approximation. Stability and convergence results for this method, also for the physically relevant case of nonsmooth potential barriers, are given in [16].

4 Open problems

The question of adequate boundary conditions for the contacts and the insulating boundaries of a semiconductor are not yet solved satisfactorily. Moreover it is not known yet how to appropriately extend the electrostatic potential V outside of the device, which is necessary for the evaluation of the pseudo-differential operator $\Theta_\hbar[V]$ in the bounded domain case.

In the physics literature quantum scattering terms for the Wigner formalism can only be found for very particular casses. So a general, mathematically and numerically tractable quantum analogue of the Boltzmann equation is not available yet. However, a relaxation time model is a reasonable physical approximation in some situations ([17]).

In the application to semiconductor devices the crystal lattice has a significant influence on the electron transport. The presented crystal model (2 . 19), (2 . 23) for the Wigner equation is a one band approximation. Hence an extension to a full energy band model, which also incorporates transitions is desirable.

References

[1] V.I. Tatarskii, The Wigner representation of quantum mechanics, Sov. Phys. Usp. 26, 311-327 (1983).

[2] P.A. Markowich, On the equivalence of the Schrödinger and the quantum Liouville equation, Math. Meth. Appl. Sci. 11, 459-469 (1989).

[3] P.A. Markowich and C. Ringhofer, An analysis of the quantum Liouville equation, ZAMM 69, 121-127 (1989).

[4] A. Arnold and H. Steinrück, The 'electromagnetic' Wigner equation for an electron with spin, ZAMP 40, No. 6, 793-815 (1989).

[5] H. Steinrück, The one dimensional Wigner-Poisson problem and its relation to the Schrödinger-Poisson problem, to appear in SIAM (1990).

[6] A. Arnold and F. Nier, The two-dimensional Wigner-Poisson problem for an electron gas in the charge neutral case, manuscript (1990).

[7] F. Brezzi and P.A. Markowich, The three-dimensional Wigner-Poisson problem: Existence, uniqueness and approximation, submitted (1989).

[8] A. Arnold, P. Degond, P.A. Markowich and H. Steinrück, The Wigner-Poisson problem in a crystal, Appl. Math. Lett. 2, No. 2, 187-191 (1989).

[9] H. Steinrück, The Wigner-Poisson problem in a crystal: Existence, uniqueness, semiclassical limit in the one dimensional case, to appear in ZAMM (1990).

[10] P. Degond and P.A. Markowich, A mathematical analysis of quantum transport in three dimensional crystals, to appear in Annali di Mat. Pura ed Appl. (1990).

[11] P. Degond and P.A. Markowich, A quantum-transport model for semiconductors: the Wigner-Poisson problem on a bounded Brillouin zone, to appear in M^2AN (1990).

[12] U. Ravaioli, M.A. Osman, W. Pötz, N. Kluksdahl and D.K. Ferry, Investigation of ballistic transport through resonant-tunneling quantum wells using Wigner function approach, Physica 134B, 36-40 (1985).

[13] W.R. Frensley, Wigner function model of a resonant-tunneling semiconductor device, Phys. Rev. B 36, 1570-1580 (1987).

[14] C. Ringhofer, A spectral method for the numerical simulation of quantum tunneling phenomena, Technical Report No. 115, Arizona State University (1988).

[15] N. D. SUH, L'étude des structures cohérentes dans l'éspace des phases du plasma unidimensionnel électrostatique: aspect classique et quantique, thesis at University of Orleans (1989).

[16] A. Arnold and F. Nier, Numerical analysis of the deterministic particle method applied to the Wigner equation, manuscript (1990).

[17] G. Grimvall, *The electron-phonon interaction in metals*, North Holland, Amsterdam - New York - Oxford (1981).

[11] P. Degond and P.A. Markowich, A quantum transport model for semiconductors: the Wigner-Poisson problem on a bounded Brillouin zone, to appear in M2AN (1990).

[12] U. Ravaioli, M.A. Osman, W. Pötz, N. Kluksdahl and D.K. Ferry, Investigation of ballistic transport through resonant-tunnelling quantum wells using Wigner function approach, Physica 134B, 36-40 (1985).

[13] W.R. Frensley, Wigner-function model of a resonant-tunneling semiconductor device, Phys. Rev. B 36, 1570-1580 (1987).

[14] C. Ringhofer, A spectral method for the numerical simulation of quantum tunneling phenomena, Technical Report No. [illegible], Arizona State University (1988).

[15] [illegible] Étude des caractéristiques [illegible] dans l'espace des phases du [illegible] électrostatique [illegible], thèse, University of [illegible] (1989).

[16] A. Arnold and F. Nier, Numerical analysis of the deterministic particle method applied to the Wigner equation, Math. Comp., to appear (1990).

[17] [illegible] North Holland, Amsterdam New York Oxford (1984).

PARTICLE TRANSPORT IN RANDOM MEDIA

G. C. POMRANING*
Department of Mathematics
University of Bologna
Piazza di Porta S. Donato, 5
40127 Bologna, Italy

*Permanent address: *School of Engineering and Applied Science, University of California, Los Angeles, Los Angeles, CA 90024-1597, U.S.A.*

ABSTRACT. A model is discussed which describes the ensemble-averaged intensity for linear transport of neutral particles through a turbulent (stochastic) mixture of two immiscible fluids. The mixing of the fluids is assumed to be Markovian. The Liouville master equation approach is used to develop this model, and in the absence of time dependence and particle-fluid scattering interactions, this approximate model becomes exact and coincides with a result obtained by the method-of-smoothing. Two asymptotic limits of this model are obtained, corresponding to: (1) a small amount of an opaque fluid admixed with a large amount of a transparent fluid (a renormalized transport formulation); and (2) a nearly isotropic intensity (a Hilbert-like expansion).

1. Introduction

An interesting applied/industrial mathematical problem arose recently in the context of the inertially confined fusion (ICF) concept for energy production. This concept involves the implosion of a microsphere of deuterium-tritium gas. This implosion is accomplished by surrounding this gas mixture with a spherical annulus of some heavy material, such as lead, and depositing energy into this heavy material by exposing the lead-gas configuration to a uniform bath of laser light or energetic ions. The energy required to obtain a useful implosion is such that very high temperatures are achieved, and the implosion and subsequent explosion are described by radiation-hydrodynamics.[1] Thus a proper treatment of radiative transfer is crucial for the successful numerical simulation of the performance of any ICF capsule. It is also well known that during the implosion process, various Rayleigh-Taylor instabilities are likely to be excited. The result of such an instability is the development of a turbulent mixture of two materials. The applied/industrial mathematics problem can then be stated as follows: "How does one model radiative transfer in a turbulent (random) mixture of two immiscible fluids?" The challenge is to derive a model for this transport of energy in a stochastic medium which describes to a reasonable approximation the underlying physics, but is simple enough to be used in a design context of ICF capsules. This paper

R. Spigler (ed.), Applied and Industrial Mathematics, 309–315.

is concerned with the development of such a model, under the assumption of Markovian mixing of the two fluids.

2. Preliminary Remarks

The generic linear transport equation we will be concerned with which describes neutral particle flow is written[1]

$$\frac{1}{v}\frac{\partial \psi}{\partial t} + \mathbf{\Omega}\cdot\nabla\psi + \sigma\psi = \frac{\sigma_s}{4\pi}\varphi + S, \tag{1}$$

where

$$\varphi = \int_{4\pi} d\mathbf{\Omega}\,\psi(\mathbf{\Omega}). \tag{2}$$

The dependent variable in Eq.(1) is the particle intensity in phase space $\psi(\mathbf{r},\mathbf{\Omega},t)$, with $\mathbf{r},\mathbf{\Omega}$, and t denoting the spatial, angular (particle flight direction), and time variables, respectively. The quantity v is the particle speed, $\sigma(\mathbf{r},t)$ is the macroscopic total cross section, $\sigma_s(\mathbf{r},t)$ is the macroscopic scattering cross section, and $S(\mathbf{r},\mathbf{\Omega},t)$ denotes any source of particles. We have assumed isotropic and coherent (no energy exchange) scattering in Eq.(1), but this simplification is not necessary for the essentials of the considerations to follow. Thus Eq.(1) is a monoenergetic (one group) transport equation, and there is no need to display the independent energy variable which is simply a parameter. In the radiative transfer context, the particles of interest are photons, and ψ is the specific intensity of radiation; v is the vacuum speed of light; $\sigma = \rho\kappa$ and $\sigma_s = \rho\kappa_s$, with ρ denoting the fluid density, κ the total opacity, and κ_s the scattering opacity; $S = (\sigma - \sigma_s)B$, with B denoting the Planck function; and $\varphi = vE$, with E denoting the radiative energy density per unit frequency.

To treat the case of a binary statistical mixture, the quantities σ, σ_s, and S in Eq.(1) are considered as discrete random variables, each of which assumes, at any $\mathbf{r},\mathbf{\Omega},t$, one of two sets of values characteristic of the two fluids constituting the mixture. We denote the two fluids by an index i, with $i = 0, 1$, and in the ith fluid these three quantities are denoted by $\sigma_i(\mathbf{r},t), \sigma_{si}(\mathbf{r},t)$, and $S_i(\mathbf{r},\mathbf{\Omega},t)$. That is, as a particle (photon in the radiative transfer context) traverses the mixture along any path, it encounters alternating segments of the two fluids, each of which has known deterministic values of σ, σ_s, and S. The statistical nature of the problem enters through the statistics of the fluid mixing, i.e., through the statistical knowledge as to what fluid is present in the mixture at space point $\mathbf{r}$ and time t. Since σ, σ_s, and S in Eq.(1) are (two state, discrete) random variables, the solution of Eqs.(1) and (2) for ψ and φ is stochastic, and we let $\bar{\psi}$ and $\bar{\varphi}$ denote the corresponding ensemble-averaged quantities. The goal is to develop accurate and, hopefully, simple models for these ensemble averages.

To discuss this problem in a qualitative way, it is convenient to consider the simple case of time-independent transport in a purely absorbing ($\sigma_s = 0$) medium. Then Eq.(1) reduces to

$$\frac{d\psi}{ds} + \sigma\psi = S, \tag{3}$$

where s denotes the spatial variable in the direction of propagation. If we write any random variable Q as the sum of its ensemble average $\bar{Q}$ and the deviation from this average $\tilde{Q}$, an ensemble averaging of Eq.(3) gives

$$\frac{d\bar{\psi}}{ds} + \bar{\sigma}\bar{\psi} + (\overline{\tilde{\sigma}\tilde{\psi}}) = \bar{S}. \tag{4}$$

It is clear from Eq.(4) that one needs to calculate (or approximate) the cross correlation term involving the ensemble average of the product $\tilde{\sigma}$ and $\tilde{\psi}$ to obtain a formulation for the quantity of interest, namely $\bar{\psi}$.

On physical grounds, and it can also be shown mathematically by using asymptotics,[2] it is clear that this term is very small and can be neglected when $\sigma_i l_i << 1$, where l_i is a characteristic chord length for the fluid packets of fluid i. However, when this inequality is not satisfied, the neglect of the cross correlation term in Eq. (4) can lead to large errors in $\bar{\psi}$. To see this, consider Eqs.(3) and (4) in the absence of a source ($S = 0$). Let fluid 0 be composed of optically thin packets ($\sigma_0 l_0 << 1$), and let fluid 1 be composed of optically thick packets ($\sigma_1 l_1 >> 1$). Further, assume that fluid 1 is sparse, i.e., present in only a very small amount. Physically, then, we have a near vacuum interspersed with very few fluid packets of essentially infinite optical thickness. Particles incident upon a piece of this mixture will, on the average, have a good chance of passing through the mixture. They will not often encounter the sparse absorbing packets, but will pass between these packets and hence will be transmitted through the mixture. On the other hand, the neglect of the cross correlation term in Eq.(4) will lead to exponential attenuation for $\bar{\psi}$ with a scale length $1/\bar{\sigma}$, and $\bar{\sigma}$ will be very large, essentially infinite, since σ_1 is very large. Thus the neglect of the cross correlation term in Eq.(4) will lead to virtually no transmission through the mixture. This simple gedanken experiment suggests that the neglect of the cross correlation term in Eq.(4) will always underestimate the particle transmission through a random mixture. Simply stated then, the challenge is to compute or approximate the cross correlation term in Eq.(4). To do this, one must specify the statistics of the fluid mixing, and we assume that this mixing is Markovian.

3. The Transport Model

We consider the simple transport equation given by Eq. (3), under the assumption of Markovian mixing. These mixing statistics are defined by a no-memory statement. Given that point s is in fluid i, the probability of point $s + ds$ being in fluid $j \neq i$ is given by

$$Prob(i \rightarrow j) = \frac{ds}{\lambda_i(s)},\ j \neq i, \tag{5}$$

where the $\lambda_i(s)$ are the prescribed Markovian transition probabilities. The probability $p_i(s)$ of finding fluid i at position s, and the correlation length $\lambda_c(s)$ for the Markov mixture are uniquely and simply related to the $\lambda_i(s)$.[3] With the mixing statistics defined by Eq. (5), one can obtain an exact description for $\bar{\psi}$ corresponding to the transport problem given by Eq. (3). This is given by the two coupled equations [4-7]

$$\frac{d}{ds}\begin{bmatrix} \bar{\psi} \\ \chi \end{bmatrix} + \begin{bmatrix} \bar{\sigma} & \nu \\ \nu & \hat{\sigma} \end{bmatrix}\begin{bmatrix} \bar{\psi} \\ \chi \end{bmatrix} = \begin{bmatrix} \bar{S} \\ T \end{bmatrix}, \tag{6}$$

where

$$\nu\chi = (\overline{\tilde{\sigma}\tilde{\psi}}), \quad \nu = (p_0 p_1)^{1/2}(\sigma_0 - \sigma_1), \tag{7}$$

$$\bar{\sigma} = p_0\sigma_0 + p_1\sigma_1, \quad \hat{\sigma} = p_1\sigma_0 + p_0\sigma_1 + \frac{1}{\lambda_c}, \tag{8}$$

$$\bar{S} = p_0 S_0 + p_1 S_1, \quad T = (p_0 p_1)^{1/2}(S_0 - S_1). \tag{9}$$

Two derivations of Eq. (6) can be given. The first is based upon the so-called method of smoothing.[4,5,7] Subtraction of Eq. (4) from Eq. (3) gives

$$\frac{d\tilde{\psi}}{ds} + \bar{\sigma}\tilde{\psi} + [\tilde{\sigma}\tilde{\psi} - (\overline{\tilde{\sigma}\tilde{\psi}})] = \tilde{S} - \tilde{\sigma}\bar{\psi}. \tag{10}$$

Inserting a smallness parameter in front of the bracketted term in Eq. (10) and seeking a solution as a power series in this smallness parameter gives the Neumann series

$$\tilde{\psi} = \sum_{n=0}^{\infty} (-1)^n [G(I-P)\tilde{\sigma}]^n G(\tilde{S} - \tilde{\sigma}\bar{\psi}), \tag{11}$$

where I is the identity and G and P are Green's and projection operators defined by

$$Gf = \int_0^s ds' \left[exp - \int_{s'}^s d\xi\bar{\sigma}(\xi)\right] f(s'), \quad Pf = \bar{f}. \tag{12}$$

For Markovian statistics, all of the terms in the sum of Eq. (11) can be evaluated and the infinite summation performed. The result of these algebraic manipulations is just Eq. (6).

The second derivation of Eq. (6) is based upon the observation that the transport problem given by Eq. (3) describes a Markov process; it is an initial value problem with the spatial variable s playing the role of time. Thus, together with the assumed Markovian mixing, we have a joint Markov process and we can write a Liouville master equation for the joint probability $P_i(s,\psi)$, defined such that $P_i d\psi$ is the probability of finding fluid i and intensity ψ lying between ψ and $\psi + d\psi$ at position s.[6,7] This master equation is given by

$$\frac{\partial P_i}{\partial s} - \frac{\partial}{\partial\psi}(\psi\sigma_i P_i - S_i P_i) = \frac{P_j}{\lambda_j} - \frac{P_i}{\lambda_i}, \; j \neq i. \tag{13}$$

The probability $p_i(s)$ is just the integral of $P_i(s,\psi)$ over ψ, and the conditional ensemble-averaged intensity given that position s is in fluid i, $\psi_i(s)$, is the integral of $\psi P_i(s,\psi)$ over ψ, divided by p_i. Multiplying Eq. (13) by ψ and integrating over ψ gives two coupled equations for the ψ_i. Making the change of variables

$$\bar{\psi} = p_0\psi_0 + p_1\psi_1, \quad \chi = (p_0 p_1)^{1/2}(\psi_0 - \psi_1) \tag{14}$$

once again leads to Eq. (6).

No such exact results are available when time-dependence and scattering are present in the transport problem, i.e., when Eq. (1) is the underlying stochastic transport equation. In this case, the algebra of the method of smoothing cannot be carried through. Further, with scattering the transport problem is not a Markov process (it is a boundary value problem) and hence the master equation approach cannot rigorously be used. Nevertheless, it has been suggested[7] that the use of the master equation might lead to a useful, albeit approximate, model. This model is

$$\left(\frac{1}{v}\frac{\partial}{\partial t}+\boldsymbol{\Omega}\cdot\nabla+\boldsymbol{\Sigma}\right)\begin{bmatrix}\bar{\psi}\\ \chi\end{bmatrix} = \frac{1}{4\pi}\boldsymbol{\Sigma}_s\begin{bmatrix}\bar{\varphi}\\ \eta\end{bmatrix}+\begin{bmatrix}\bar{S}\\ T\end{bmatrix}, \tag{15}$$

where

$$\boldsymbol{\Sigma} = \begin{bmatrix}\bar{\sigma} & \nu\\ \nu & \hat{\sigma}\end{bmatrix}, \quad \boldsymbol{\Sigma}_s = \begin{bmatrix}\bar{\sigma}_s & \nu_s\\ \nu_s & \hat{\sigma}_s\end{bmatrix}, \tag{16}$$

and, in addition to Eqs. (7) through (9), we have

$$\bar{\varphi} = \int_{4\pi} d\Omega\bar{\psi}, \quad \eta = \int_{4\pi} d\Omega\chi, \quad \nu_s = (p_0p_1)^{1/2}(\sigma_{s0}-\sigma_{s1}), \tag{17}$$

$$\bar{\sigma}_s = p_0\sigma_{s0}+p_1\sigma_{s1}, \quad \hat{\sigma}_s = p_1\sigma_{s0}+p_0\sigma_{s1}. \tag{18}$$

Several other derivations of this model are available. Sahni[8] has shown that the assumption of independent particle flight paths leads to Eq. (15), and he has also shown[9] that nuclear reactor noise techniques can be used to obtain this model. It has also been demonstrated[10] that one can derive exact particle balance equations in this stochastic transport setting, but involving two ensemble averages conditioned upon being in fluid i, namely a volumetric average ψ_i and a surface (interface between fluids) average ψ_i^s. Making the approximation that these two ensemble averages are equal again yields the model given by Eq. (15).[10] Numerical comparisons with exact Monte Carlo results[10] shows the accuracy of this model to be of the order of ten percent or so, with the error being the largest, as expected, for problems with multiple scattering interactions.

If one accepts Eq. (15) as a reasonable model of particle transport in a binary Markovian mixture, two simplifications involving certain asymptotic limits have been reported.[2] The first of these corresponds to a small amount, of $0(\epsilon^2)$, of one fluid admixed with an $0(1)$ amount of the second fluid. Further, the first fluid has large, of $0(1/\epsilon^2)$, source and cross sections compared to $0(1)$ quantities for the second fluid. In this asymptotic limit, one finds a renormalized transport equation of the standard form given by[2]

$$\left(\frac{1}{v}\frac{\partial}{\partial t}+\boldsymbol{\Omega}\cdot\nabla+\sigma_{eff}\right)\bar{\psi} = \frac{\sigma_{s,eff}}{4\pi}\bar{\varphi}+S_{eff}+0(\epsilon^2), \tag{19}$$

with

$$\sigma_{eff} = \bar{\sigma}-\frac{\nu^2}{\hat{\sigma}}, \quad \sigma_{s,eff} = \sigma_{eff}-\sigma_{a,eff}, \tag{20}$$

and

$$\sigma_{a,eff} = (\bar{\sigma}-\bar{\sigma}_s)-\frac{(\nu-\nu_s)^2}{(\hat{\sigma}-\hat{\sigma}_s)}, \quad S_{eff} = \bar{S}-\frac{(\nu-\nu_s)}{(\hat{\sigma}-\hat{\sigma}_s)}T. \tag{21}$$

Equation (21) for S_{eff} assumes that the sources are isotropic, i.e., $S_i \neq S_i(\boldsymbol{\Omega})$. The corresponding expression for anisotropic sources is more complex.[2] The result given by Eqs. (19) through (21) is robust in that all of these effective cross sections and source are always nonnegative, even when one is far from the asymptotic limit under consideration. This derivation of a renormalized transport equation as an asymptotic limit has been generalized to a mixture containing an arbitrary number of fluid components, and leads to an entire class of reduced (in complexity) transport descriptions.[11]

The second asymptotic limit of Eq. (15) is one in which the scattering interaction is dominant, of 0(1). Absorption, sources, and all derivatives are assumed small, of $0(\epsilon^2)$, except for the spatial derivative which is scaled to be $0(\epsilon)$. This scaling eliminates the angular variable in the transport problem, and is analogous to the Hilbert expansion which gives the Euler fluid equations from the Boltzmann equation. If we restrict ourselves to isotropic sources as well as an isotropic correlation length, one finds in this asymptotic limit two coupled diffusion equations given by[2]

$$\left(\frac{1}{v}\frac{\partial}{\partial t} - \nabla\cdot\mathbf{D}\nabla + \Sigma_a\right)\begin{bmatrix}\bar{\varphi}\\ \eta\end{bmatrix} = 4\pi\begin{bmatrix}\bar{S}\\ T\end{bmatrix} + 0(\epsilon^2), \tag{22}$$

where $\Sigma_a = \Sigma - \Sigma_s$ and

$$\mathbf{D} = \frac{1}{3}\Sigma^{-1} = \frac{1}{3(\bar{\sigma}\hat{\sigma} - \nu^2)}\begin{bmatrix}\hat{\sigma} & -\nu\\ -\nu & \bar{\sigma}\end{bmatrix}. \tag{23}$$

It can easily be shown that the diffusion coefficient matrix $\mathbf{D}$ is positive definite under all choices of the physical parameters, thus assuring robustness in this diffusion approximation. Since the scaling which led to Eq. (22) corresponds to a singular perturbation problem, both boundary and temporal layers exist. Performing the appropriate asymptotic matching analyses, one obtains boundary and initial conditions for Eq. (22). The boundary condition is of the mixed (Robbin) type, involving the number 0.710446... , the classic linear extrapolation distance for the Milne problem.[2]

4. Concluding Remarks

Under the assumption of Markovian mixing as defined by Eq.(5), we have described a model giving the ensemble-averaged intensity for the linear transport of particles through a two component stochastic medium. This model consists of two coupled transport equations, written in matrix form as Eq.(15). Numerical comparisons of the predictions of this model with essentially exact results indicate that this model is accurate to about ten percent or so, with the accuracy best for problems with a minimum of multiple scattering interactions. Two simplifications of this model, corresponding to two different asymptotic limits, have been obtained. The first of these leads to a single renormalized transport equation given by Eq.(19), involving an effective source and cross sections to account for the stochastic nature of the problem. The second asymptotic limit, given by Eq.(22), reduces the dimensionality of the model by eliminating the angular variable from the problem, and has the form of two coupled diffusion equations.

Certain exact results in the case of Markovian mixing and including the scattering interaction, distinct from the model given by Eq.(15), have been reported in the literature. Further, some results are available for non-Markovian mixtures. In this more general case, the description of particle transport in stochastic mixtures generally involves integral equations, resulting from the application of the theory of alternating renewal processes. Space limitations prevent us from discussing these items here, but a short description of and references to these works are available in a recent paper.[12]

Acknowledgements

The author wishes to express his gratitude to the Italian National Research Council (C.N.R.) for the opportunity to visit the University of Bologna under C.N.R.'s Visiting Professor Program. The preparation of this paper was also partially supported by the U.S. Department of Energy under grant DE-FG03-89ER14016.

References

1. Pomraning, G.C. (1973) *The Equations of Radiation-Hydrodynamics*, Pergamon Press, Oxford.

2. Malvagi, F., Levermore, C. D., and Pomraning, G. C. (1989) "Asymptotic Limits of a Statistical Transport Description," *Transport Th. Statis. Phys.*, **18**, 287-311.

3. Pomraning, G. C. (1989) "Statistics, Renewal Theory, and Particle Transport," *J. Quant. Spectros. Radiat. Transfer* **42**, 279-293.

4. Pomraning, G. C. (1986) "Transport and Diffusion in a Statistical Medium," *Transport Th. Statis. Phys.* **15**, 773-802.

5. Levermore, C. D., Pomraning, G. C., Sanzo, D. L., and Wong, J. (1986) "Linear Transport Theory in a Random Medium," *J. Math. Phys.* **27**, 2526-2536.

6. Vanderhaegen, D. (1986) "Radiative Transfer in Statistical Heterogeneous Mixtures," *J. Quant. Spectros. Radiat. Transfer* **36**, 557-561.

7. Pomraning, G. C., Levermore, C. D. and Wong, J. (1989) "Transport Theory in Binary Statistical Mixtures," in P. Nelson et al. (eds.), *Lecture Notes in Pure and Applied Mathematics* **Vol. 115**, Marcel Dekker, Inc., New York, pp. 1-35.

8. Sahni, D. C. (1989) "Equivalence of Generic Equation Method and the Phenomenological Model for Linear Transport Problems in a Two State Random Scattering Medium," *J. Math. Phys.* **30**, 1554-1559.

9. Sahni, D. C. (1989) "An Application of Reactor Noise Techniques to Neutron Transport Problems in a Random Medium," *Annals Nucl. Energy* **16**, 397-408.

10. Adams, M. L., Larsen, E. W., and Pomraning, G. C. (1989) "Benchmark Results for Particle Transport in a Binary Markov Statistical Medium," *J. Quant. Spectros. Radiat. Transfer* **42**, 253-266.

11. Malvagi, F. and Pomraning, G. C. (1990) "Renormalized Equations for Linear Transport in Stochastic Media," *J. Math. Phys.*, in press.

12. Pomraning, G. C. (1990) "Radiative Transfer in Rayleigh-Taylor Unstable ICF Pellets," *Laser and Particle Beams*, in press.

DETERMINING CONDUCTIVITY BY BOUNDARY MEASUREMENTS, THE STABILITY ISSUE

GIOVANNI ALESSANDRINI
Dipartimento di Matematica "Vito Volterra"
Università di Ancona
Via delle Brecce Bianche, ANCONA, ITALY

ABSTRACT. We consider the inverse boundary value problem of determining the coefficient γ in the elliptic equation $div(\gamma\, grad\, u)=0$ in Ω, when all possible pairs of Dirichlet and Neumann data on $\partial\Omega$ are known. We dicuss the problem of continuous dependence of γ on the data, we state a Theorem, obtained in [2], which gives sufficient conditions under which such continuity holds, and we sketch a new simplified proof.

1.Introduction

There are various applications, like medical imaging and geophisical prospection, in which the following problem arises. Given an electrically conducting body Ω satisfying Ohm's law with isotropic conductivity density γ, to determine γ from the knowledge of all possible currents and potentials measured at the boundary.

Mathematically the problem can be formulated as follows. To find the positive coefficient γ in the equation

$$div(\ \gamma\ grad\ u\ ) = 0 \qquad \text{in } \Omega, \tag{1.1}$$

when, for every solution u, the pair

$$u|_{\partial\Omega}(potential),\ \gamma\frac{\partial u}{\partial\nu}\Big|_{\partial\Omega}(current)$$

is known, here Ω denotes a bounded domain in $\mathbb{R}^n$, $n\geq 2$, with sufficiently smooth boundary $\partial\Omega$, and ν denotes the exterior unit normal to $\partial\Omega$. In other words, we want to determine γ assuming that we know Λ_γ, where we denote by $\Lambda_\gamma: H^{1/2}(\partial\Omega)\rightarrow H^{-1/2}(\partial\Omega)$ the so-called Dirichet-to-Neumann map, that is the linear operator such that, for any $\phi\in H^{1/2}(\partial\Omega)$

$$\Lambda_\gamma\phi = \gamma\frac{\partial u}{\partial\nu}\Big|_{\partial\Omega}$$

where $u\in H^1(\Omega)$ is the (weak) solution to (1.1) with Dirichlet data ϕ, this setting goes back to A.P.Calderon [4].

This problem can be investigated from three main viewpoints:

R. Spigler (ed.), Applied and Industrial Mathematics, 317–324.

(I) UNIQUENESS

This is the aspect for which principal progress has been achieved in the latest years: R.Kohn and M.Vogelius proved that Λ_γ uniquely determines γ when the space dimension $n \geq 2$ and γ is taken in the class of analytic functions, [6], J.Sylvester and G.Uhlmann proved uniqueness when $n \geq 3$ and γ is C^∞, [13]. Further refinements of these two main results can be found in [7], [12], [10], [9], [11], [2]. Let us mention that, in the author's knowledge, the question of uniqueness when $n=2$ and γ is assumed smooth but not necessarily analytic is still open.

(II) STABILITY

That is the continuous dependence of the unknown γ from the data Λ_γ . This matter has been studied by the author, [1], [2], and will be the subject of this note. By an example we will show that, in general, instability indeed occurs and we will state sufficient hypotheses under which stability holds, Theorem 1.1.

(III) ALGORITHMS

It is obvious that this aspect is the most important for the point of view of applications. In practice, the available boundary measurements will be finite in number, only on a discrete mesh, and affacted by errors. In one word, only an approximate knowledge of Λ_γ will be at our disposal. This shows that a preliminary treatment of step (II) is necessary.

For this reason we will focus on (II) and we will not continue the discussion of point (III) any further.

Let us just mention that, recently, a lot of research has been carried on it: see for instance [8], [14], [3] and [5] where the reader can find a very detailed list of references.

Now we show by an example that, in general, γ does not depend continuously on Λ_γ . Just for simplicity, we can consider the case when $n=2$.

Here and in the sequel, $\| \ \|_*$ will denote the operator norm on $L(H^{1/2}(\partial\Omega),H^{-1/2}(\partial\Omega))$, this is in fact the natural norm for such Dirichlet-to-Neumann operators.

Example 1.1 We suppose Ω to be the unit disk and we let r, θ be the polar coordinates, $x=r(\cos\theta,\sin\theta)$. Given numbers A, r_0, $A>0$, $0<r_0<1$, we consider the following two coefficients,

$$\gamma_1 \equiv 1 \text{ in } \Omega \ , \ \gamma_2 = \begin{cases} 1+A \text{ if } r \leq r_0 \ , \\ \\ 1 \text{ if } r_0 < r \leq 1 \ . \end{cases} \tag{1.2}$$

Let us denote by Λ_1,Λ_2 the Dirichlet-to-Neumann maps Λ_γ when $\gamma=\gamma_1,\gamma_2$ respectively. By separation of variables, the expressions of Λ_1 and Λ_2 are easily found in terms of Fourier series. We see that

$$\|\Lambda_1 - \Lambda_2\|_* \leq A r_0 \rightarrow 0 \ , \text{ as } r_0 \rightarrow 0+ \ , \tag{1.3}$$

while

$$\|\gamma_1-\gamma_2\|_{L^\infty(\Omega)}=A=const.\ \text{as}\ r_0\to 0+\ . \tag{1.4}$$

Therefore we need a remedy to the ill-posedness of this problem and, in such cases, the standard strategy consists in assuming that additional prior information is available on the unknown of the problem. In Theorem 1.1 below we will show that if prior information on γ is assumed in the form of the following on γ yield then stability holds. Namely we will assume that for a given number $E>0$ the unknown γ satisfies

$$E^{-1}\leq\gamma(x)\ ,\ \text{for every}\ x\in\Omega, \tag{1.5}$$

$$\|\gamma\|_{W^{2,\infty}(\Omega)}=\|\gamma\|_{L^\infty(\Omega)}+\|D\gamma\|_{L^\infty(\Omega)}+\|D^2\gamma\|_{L^\infty(\Omega)}\leq E\ . \tag{1.6}$$

Let us notice here that (1.5) consists just of a non-degeneracy assumption on the equation (1.1), while (1.6) is a smoothness requirement on γ . It is remarkable that at the present time all the techniques used in treating uniqueness or stability for this problem always require the existence of second derivatives of γ (at least in a piecewise sense, see for instance [7], [9], [2]).

Theorem 1.1 *Let Ω be a bounded domain in $\mathbf{R}^n$ with Lipschitz boundary $\partial\Omega$, and let $n\geq 3$. Let γ_1,γ_2 satisfy (1.5),(1.6) and let Λ_1,Λ_2 be the corresponding Dirichlet-to-Neumann maps. Then we have the following stability estimate*

$$\|\gamma_1-\gamma_2\|_{L^\infty(\Omega)}\leq\omega(\|\Lambda_1-\Lambda_2\|_*)\ , \tag{1.7}$$

where, the function ω is such that

$$\omega(t)\leq C|\log t|^{-\delta}\ ,\ \text{for every}\ t\ ,\ 0<t<1\ , \tag{1.8}$$

and C , δ are positive numbers, the former depends only on Ω , E and n while the latter depends only on n .

2. Proof of Theorem 1.1

We will just indicate the main steps of the proof giving only the details of one of them, Lemma 2.3 below, using some new arguments which simplify the previous proof in [2]. It is worthwile to mention that the main stream of the proof follows the ideas used by Sylvester and Uhlmann [12] for the uniqueness result, on the other hand, the proof of stability requires more care on the quantitative side.

The first thing to do is a change of variables in (1.1). For a given γ and for any solution u to (1.1) we define

$$v=\sqrt{\gamma}\,u\ , \tag{2.1a}$$

$$q=\frac{\Delta\sqrt{\gamma}}{\sqrt{\gamma}}\ , \tag{2.1b}$$

it is easily seen that v satisfies the elliptic equation

$$-\Delta v+qv=0 \text{ in } \Omega \,, \tag{2.2}$$

thus we can introduce an appropriate Dirichlet-to-Neumann map for this new equation, as follows

$$\widetilde{\Lambda}_q v|_{\partial\Omega}=\frac{\partial v}{\partial\nu}\Big|_{\partial\Omega} \,, \tag{2.3}$$

and it is easily noticed that $\widetilde{\Lambda}_q$ is linked to $\gamma|_{\partial\Omega}$, $\frac{\partial\gamma}{\partial\nu}\Big|_{\partial\Omega}$ and Λ_γ by the formula

$$\widetilde{\Lambda}_q=\frac{1}{\sqrt{\gamma}}\Lambda_\gamma\Big(\frac{1}{\sqrt{\gamma}}(\cdot)\Big)+\frac{1}{2\sqrt{\gamma}}\frac{\partial\gamma}{\partial\nu}\Big|_{\partial\Omega}(\cdot)\,. \tag{2.4}$$

Therefore we can split the problem of estimating the dependence of γ on Λ_γ in the following four sub-problems.

(*i*) To estimate the dependence of $\gamma|_{\partial\Omega}$ and $\frac{\partial\gamma}{\partial\nu}\Big|_{\partial\Omega}$ on Λ_γ.

(*ii*) To estimate the dependence of $\widetilde{\Lambda}_q$ on Λ_γ, $\gamma|_{\partial\Omega}$ and $\frac{\partial\gamma}{\partial\nu}\Big|_{\partial\Omega}$.

(*iii*) To estimate the dependence of q on $\widetilde{\Lambda}_q$.

(*iv*) To estimate the dependence of γ on q and $\gamma|_{\partial\Omega}$.

Let us state here four Lemmas which summarize the answer to the above sub-problems and whose combination yields immediately the proof of Theorem 1.1.

We will denote by $q_1,q_2,\widetilde{\Lambda}_1,\widetilde{\Lambda}_2$ the coefficients given by (2.1b), and the Dirichlet-to-Neumann maps associated to equation (2.2) when $\gamma=\gamma_1,\gamma_2$ respectively. The hypotheses of Theorem 1.1 will be assumed throghout.

Lemma 2.1 *For every β, $0<\beta<1/2$, the following estimates hold*

$$\|\gamma_1-\gamma_2\|_{L^\infty(\partial\Omega)}\leq C_1\|\Lambda_1-\Lambda_2\|_* \,, \tag{2.5a}$$

$$\|D(\gamma_1-\gamma_2)\|_{L^\infty(\partial\Omega)}\leq C_2\|\Lambda_1-\Lambda_2\|_*^\beta \,. \tag{2.5b}$$

A proof of this Lemma can be found in [2], it is based on the construction of solutions to (1.1) having isolated singularities with prescribed asymptotics of arbitrarily high order on points outside, but near, the boundary. Let us just mention here that different techniques could be applied: for instance, in [1] a proof was given by an adaptation of a device due to Kohn and Vogelius [6], but much stronger a-priori assumptions were needed, see also related estimates in [13] where methods of microlocal analysis are used.

Lemma 2.2 *The following estimate holds*

$$\|\widetilde{\Lambda}_1-\widetilde{\Lambda}_2\|_* \leq$$

$$\leq C\left[\|\Lambda_1-\Lambda_2\|_* + \|\gamma_1-\gamma_2\|_{W^{1,\infty}(\partial\Omega)} + \left\|\frac{\partial\gamma_1}{\partial\nu}-\frac{\partial\gamma_2}{\partial\nu}\right\|_{L^\infty(\partial\Omega)}\right]. \tag{2.6}$$

Proof Straightforward consequence of (2.4). □

Lemma 2.3 *The following estimate holds*

$$\|q_1-q_2\|_{H^{-1}(\Omega)} \leq \omega(\|\widetilde{\Lambda}_1-\widetilde{\Lambda}_2\|_*) , \tag{2.7}$$

where ω is as in (1.8).

See Section 3 below for a proof.

Lemma 2.4 *There exists α, $0<\alpha<1$ depending only on n , such that*

$$\|\gamma_1-\gamma_2\|_{L^\infty(\Omega)} \leq C\left[\|q_1-q_2\|_{H^{-1}(\Omega)} + \|\gamma_1-\gamma_2\|_{W^{1,\infty}(\partial\Omega)}\right]^\alpha . \tag{2.8}$$

Proof Starting from (2.1b) we can see that

$$\phi = log\frac{\gamma_1}{\gamma_2}$$

satisfies the following elliptic equation

$$div((\gamma_1\gamma_2)^{1/2} grad\phi) = 2(\gamma_1\gamma_2)^{1/2}(q_1-q_2) \quad \text{in } \Omega ,$$

then, by the use of an H^1 estimate and, by interpolation with (1.6), we readily obtain (2.8). □

3. Proof of Lemma 2.3

The key tool is a Theorem due to Sylvester and Uhlmann [12], which, for the present purposes, can be stated as follows.

Lemma 3.1 *If γ satisfies* (1.5),(1.6) *and if q is given by* (2.1*b*), *then, for every $\xi \in \mathbb{C}^n$ such that $\xi \neq 0$ and $\xi \cdot \xi = 0$, there exists a solution v of the equation*

$$-\Delta v + q v = 0 \quad in\ \Omega , \tag{3.1}$$

which can be represented as

$$v(x) = e^{\xi \cdot x}(1+R(\xi,x)) , \quad x \in \Omega , \tag{3.2}$$

where

$$\|R(\xi,\cdot)\|_{L^2(\Omega)} \le C(\Omega)|\xi|^{-1}\|q\|_{L^\infty(\Omega)} \ . \tag{3.3}$$

Furthemore, v satisfies

$$\|v\|_{H^1(\Omega)} \le C(\Omega,E)e^{C(\Omega)|\xi|} \ . \tag{3.4}$$

Concerning the proof, let us just mention that the hard part is when

$$|\xi| \ge const.\|q\|_{L^\infty(\Omega)}$$

and it can be found in [10], on the other hand, when

$$0 < |\xi| \le const.\|q\|_{L^\infty(\Omega)},$$

it easily seen that $v=\sqrt{\gamma}$ satisfies (3.1)—(3.4).

Now, from Green's formula, it easily follows that

$$\int_\Omega (q_1-q_2)v_1v_2dx = \int_{\partial\Omega} v_1(\widetilde{\Lambda}_1-\widetilde{\Lambda}_2)v_2dS \ , \tag{3.5}$$

where v_1,v_2 are any two solutions to (3.1) when $q=q_1,q_2$ respectively, and dS denotes the surface measure on $\partial\Omega$. Let us notice incidentally, that identity (3.5), and close relatives of it, have been used by all authors, in a more or less hidden form, in the treatment of this problem. An explicit use in the above form seems to make things simpler.

We will use (3.5) taking as v_1,v_2 the special solutions obtained in Lemma 3.1 with $\xi=\xi_1,\xi_2$ chosen as follows. By a simple construction, we can see that, for any $k\in\mathbf{R}^n$ and any $r>0$, there exist $\xi_1,\xi_2\in\mathbb{C}^n$ satisfying the following properties

$$\xi_1\cdot\xi_1=\xi_2\cdot\xi_2=0 \ , \ \xi_1+\xi_2=ik \ , \ |\xi_1|^2=|\xi_2|^2=\frac{|k|^2}{2}+2r^2 \ .$$

Notice that this tricky observation, which is due to Sylvester and Uhlmann [12], applies only when the space dimension n is ≥ 3 . With this choice, we deduce from (3.2), (3.3), (3.5),

$$\left|((q_1-q_2)\chi_\Omega)^\wedge(k) - \int_{\partial\Omega} v_1(\widetilde{\Lambda}_1-\widetilde{\Lambda}_2)v_2dS\right| \le \frac{C(\Omega,E)}{|k|+r} \ , \tag{3.6}$$

where $(\cdot)^\wedge$ denotes the Fourier transform. Furthermore, from (3.4), we obtain

$$|((q_1-q_2)\chi_\Omega)^\wedge(k)| \le C(\Omega,E)\left[e^{C(\Omega)(|k|+r)}\|\widetilde{\Lambda}_1-\widetilde{\Lambda}_2\|_* + \frac{1}{|k|+r}\right] , \tag{3.7}$$

on the other hand, by Parseval identity and (1.6), we have

$$\|((q_1-q_2)\chi_\Omega)^\wedge(k)\|^2_{L^2(\mathbf{R}^n)} \le C(\Omega,E) \ , \tag{3.8}$$

hence, fixing any $\rho>0$, we deduce

$$\|q_1-q_2\|^2_{H^{-1}(\Omega)} \leq \|(q_1-q_2)\chi_\Omega\|^2_{H^{-1}(\mathbf{R}^n)} =$$

$$= \int_{\mathbf{R}^n} |((q_1-q_2)\chi_\Omega)^{\widehat{}}(k)|^2 (1+|k|^2)^{-1} dk \leq$$

$$\leq \int_{|k|<\rho} |((q_1-q_2)\chi_\Omega)^{\widehat{}}(k)|^2 dk + \int_{|k|>\rho} |((q_1-q_2)\chi_\Omega)^{\widehat{}}(k)|^2 dk (1+\rho^2)^{-1} \leq$$

$$\leq C(\Omega,E)\left[e^{C(\Omega)(\rho+r)}\|\widetilde{\Lambda}_1-\widetilde{\Lambda}_2\|_* + \frac{\rho^n}{r} + \frac{1}{1+\rho^2}\right], \tag{3.9}$$

finally, taking the minimum of the right hand side as $r,\rho>0$, we obtain the desired estimate.

References

[1] Alessandrini, G. (1988) 'Stable determination of conductivity by boundary measurements', Appl. Anal. 27, 153-172.

[2] Alessandrini, G. (1989) 'Singular solutions of elliptic equations and the determination of conductivity by boundary measurements' to appear on J. Differential Equations.

[3] Breckon, W. and Pidcock, M. (1989) 'Progress in electrical impedance tomography' in P. Sabatier (ed.), *Some Topics in Inverse Problems*, World Scientific.

[4] Calderon, A.P. (1980) 'On an inverse boundary value problem' Seminar on Numerical Analysis and its Applications to Continuum Phisics Soc. Brasileira de Matematica, Rio de Janeiro 65-73.

[5] Kohn, R. and McKenney, A. (1989) 'Numerical implementation of a variational method for electrical impedance tomography' preprint.

[6] Kohn, R. and Vogelius, M. (1984) 'Determining conductivity by boundary measurements' Comm. Pure Appl. Math. 37, 113-123.

[7] Kohn, R. and Vogelius, M. (1985) 'Determining conductivity by boundary measurements II. Interior results' Comm. Pure Appl. Math. 38, 643-667.

[8] Kohn, R. and Vogelius, M. (1987) 'Relaxation of a variational method for impedance computed tomography' Comm. Pure Appl. Math. 40, 745-777.

[9] Isakov, V. (1988) 'On uniqueness of recovery of a discontinuous conductivity coefficient' Comm. Pure Appl. Math. 41, 865-877.

[10] Nachman, A., Sylvester, J. and Uhlmann, G. (1988) 'An n—dimensional Borg—Levinson theorem' Comm. Math. Phys. 115, 593-605.

[11] Nachman, A. (1988) 'Reconstructions from boundary measurements' Ann. of Math. 128, 531-576.

[12] Sylvester, J. and Uhlmann, G. (1987) 'A global uniqueness theorem for an inverse boundary value problem' Ann. of Math. 125, 153-169.
[13] Sylvester, J. and Uhlmann, G. (1988) 'Inverse boundary value problems at the boundary—continuous dependence' Comm. Pure Appl. Math. 41, 197-219.
[14] Santosa, F. and Vogelius, M. (1988) 'A backprojection algorithm for electrical impedance tomography' to appear on SIAM J. Appl. Math..

SCATTERING PROBLEMS FOR ACOUSTIC WAVES

G. CAVIGLIA
Dipartimento di Matematica
Università di Genova
V. Alberti 4, 16132 Genova
Italy

A. MORRO
DIBE
Università di Genova
V. Causa 13, 16145 Genova
Italy

ABSTRACT. Inhomogeneous waves in dissipative media are described in terms of complex potentials. Reflection and transmission at a plane interface are examined. Then the scattering of a plane wave by a viscoelastic obstacle immersed in a fluid is investigated by following a Kirchhoff-type approximation. As an application, the scattering by a disk is developed in detail. The expression for the far field, which is a basic tool for the inverse problem, is derived.

1. Introduction

In a typical scattering problem we consider a solid obstacle which is immersed in a different solid material or in a fluid, and a time harmonic acoustic wave that comes from infinity, interacts with the obstacle, and is diffused to infinity. A direct scattering problem consists essentially in the determination of the far field, i.e. of the transmitted field at infinity, provided we know the shape, the size and (possibly) the material properties of the scatterer. Meanwhile the inverse problem aims at an estimate of the characteristic features of the obstacle in terms of the measured scattered wave [1-3].

Such problems are of interest in many areas of scientific research and technology, such as non-destructive testing of materials, sonar exploration, analysis of living tissues, seismology and seismic exploration. In view of these practical issues, in the last years a considerable progress has been made in the analysis of direct and inverse scattering. In this work we aim at a contribution towards considering more realistic models of the scattering by obstacles.

For example it is of interest to examine how the constitutive properties and the curvature of the obstacle affect the field at infinity which originates from the scattering of a plane wave. Specifically we want to take care of dissipation phenomena that occur in the surrounding medium as well as in the obstacle. Accordingly, we assume here that the media are in general dissipative. Incidentally, it seems that only a dissipative model can give a realistic representation of ultrasound propagation in living tissues [4]. As a

R. Spigler (ed.), Applied and Industrial Mathematics, 325–334.

prototype for a lossy medium we consider viscoelastic solids and viscous fluids, whose thermodynamic properties are briefly outlined since they turn out to be crucial in the analysis of the attenuation of plane waves induced by dissipation.

Owing to the dissipative character of the media, time-harmonic waves lead naturally to considering complex wavenumbers, that is inhomogeneous plane waves, for which planes of constant phase and planes of constant amplitude are not parallel. Through a description in terms of complex potentials and an analysis of thermodymamic restrictions, detailed expressions for the wave speed and attenuation are exhibited. The behaviour of inhomogeneous waves at plane interfaces is also examined. As a consequence it has been possible to deal numerically with an inverse problem such as determining the relaxation time that characterizes an exponential-type relaxation function [5].

As a second application, the scattering of a plane wave from a viscoelastic obstacle immersed in a fluid is examined. Following a Kirchhoff-type approximation, the field and its normal derivative at the boundary are assigned by appealing to the behaviour of reflected waves at a plane interface. Then an integral representation for the far-field pattern is determined explicitly and some of its consequences are examined in detail.

2. Material properties of the obstacle and the surrounding medium

A preliminary step in the investigation of scattering problems for time-harmonic waves is the analysis of wave propagation under the influence of dissipation [6]. For definiteness, it is assumed that the obstacle is a linear, isotropic, viscoelastic solid and that it is immersed in a fluid. It is convenient to examine the behaviour of the solid, since that of the fluid will follow as a particular case.

The time-harmonic travelling wave is described through the displacement field $\mathbf{u}(\mathbf{x}, t)$. The material properties are described in the form of Boltzmann's law by letting the Cauchy stress $\mathbf{T}$, at time t, be given by

$$\mathbf{T}(t) = 2\mu_0 \mathbf{E}(t) + \lambda_0 (\mathrm{tr}\,\mathbf{E})(t)\,\mathbf{I} + \int_0^\infty [2\,\mu'(s)\,\mathbf{E}(t-s) + \lambda'(s)(\mathrm{tr}\,\mathbf{E})(t-s)\,\mathbf{I}]ds \qquad (2.1)$$

where $\mathbf{E} = \mathrm{sym}\partial\mathbf{u}/\partial\mathbf{x}$ is the infinitesimal strain tensor; μ_0, and μ' are the initial value and the derivative of the relaxation modulus in shear; the relaxation modulus in dilatation is given by $\lambda + 2\mu/3$, with λ_0 and λ' initial value and derivative of λ. In (2.1) the dependence on the position is understood and not written.

As a consequence of the second law of thermodynamics, $\lambda'(s)$ and $\mu'(s)$ are required to satisfy suitable inequalities coming from the requirement that, when $\mathbf{E}$ is subject to a cyclic history, the energy dissipated along each cycle is non-negative. We find [5,7]

$$\int_0^\infty \mu'(s)\sin(\omega s)ds \le 0, \qquad \int_0^\infty (\frac{2}{3}\mu' + \lambda')(s)\sin(\omega s)ds \le 0, \qquad (2.2)$$

for any positive angular frequency ω. These inequalities are essential in the analysis of wave propagation.

3. Complex potentials and time-harmonic waves

Disregarding body forces, as usual in the modelling of wave propagation, we write the equation of motion as

$$\rho \ddot{\mathbf{u}} = \nabla \cdot \mathbf{T} \tag{3.1}$$

where ρ is the (constant) mass density and a superposed dot denotes the time derivative. In full analogy with the case of purely elastic bodies [8], the displacement $\mathbf{u}$ can be expressed in terms of the *compressional wave scalar potential* ϕ and of the *shear wave vector potential* $\boldsymbol{\psi}$ through

$$\mathbf{u} = \nabla\phi + \nabla \times \boldsymbol{\psi}. \tag{3.2}$$

Since we are dealing with time-harmonic waves the fields under consideration depend on time t through a factor $\exp(-i\omega t)$, ω being the angular frequency. Accordingly we write

$$\mathbf{u}(\mathbf{x}, t; \omega) = \Re[\mathbf{U}(\mathbf{x}; \omega) \exp(-i\omega t)];$$

$$\phi(\mathbf{x}, t; \omega) = \Re[\Phi(\mathbf{x}; \omega) \exp(-i\omega t)];$$

$$\boldsymbol{\psi}(\mathbf{x}, t; \omega) = \Re[\boldsymbol{\Psi}(\mathbf{x}; \omega) \exp(-i\omega t)],$$

where $\Re$ denotes the real part. On comparing with (3.1), (3.2) it follows that

$$\mathbf{U} = \nabla\Phi + \nabla \times \boldsymbol{\Psi}, \tag{3.3}$$

and the equation of motion is equivalent to

$$\Delta\Phi + \frac{\rho\omega^2}{2\hat{\mu} + \hat{\lambda}}\Phi = 0, \tag{3.4}$$

$$\Delta\boldsymbol{\Psi} + \frac{\rho\omega^2}{\hat{\mu}}\boldsymbol{\Psi} = 0 \tag{3.5}$$

where the quantities $\hat{\lambda}$ and $\hat{\mu}$ are defined by

$$\hat{\lambda} = \lambda_0 + \int_0^\infty \lambda'(s)\exp(i\omega s)ds,$$

$$\hat{\mu} = \mu_0 + \int_0^\infty \mu'(s)\exp(i\omega s)ds.$$

Thus the potentials are determined as solutions to the equations (3.4) and (3.5), which have the form of Helmholtz equations with complex coefficients. Hence Φ and Ψ are complex-valued. In the more familiar elastic case the (real) coefficients of the Helmholtz equation are related to the phase speed [8]; in the present circumstances the complex coefficients will be shown to be responsible for the wave speed and the corresponding attenuation. The latter is a mathematical consequence of the thermodynamic inequalities, as it is to be expected. In fact it follows from (2.2) that

$$\Im\left(\frac{\rho\omega^2}{2\hat{\mu} + \hat{\lambda}}\right) \geq 0, \qquad \Im\left(\frac{\rho\omega^2}{\hat{\mu}}\right) \geq 0, \tag{3.6}$$

where $\Im$ denotes the imaginary part, and it will be shown that these inequalities give (ω-dependent) attenuation of plane waves.

The previous analysis applies to viscoelastic bodies but it has been assumed that the medium surrounding the obstacle is a fluid, and this requires propagation properties of time-harmonic waves within the fluid itself. However the previous results can be simply adapted to deal with this medium since the fluid (either viscous or inviscid) can be regarded as a particular case of the viscoelastic model, at least as far as wave propagation is concerned. The same holds for elastic solids. For example, an elastic solid corresponds to

$$\mu'(s) = 0, \ \lambda'(s) = 0, \ s \in (0, \infty).$$

For inviscid fluids in the linear approximation we find [6]

$$\Delta\Phi + \frac{\rho_0\omega^2}{\rho_0 p_\rho(\rho_0)}\Phi = 0, \tag{3.7}$$

and $\Psi = 0$. Here ρ_0 is the density and $p = p(\rho_0)$ is the pressure.

It is worth remarking that although the coefficients of Φ and Ψ in the Helmholtz equation are real when we consider elastic solids and inviscid fluids, Φ and Ψ are allowed to be complex valued. This is explicitly required when dealing with the reflected waves that arise at an interface with either a viscoelastic body or a viscous fluid.

4. Inhomogeneous waves

Suppose we search for solutions of the complex Helmholtz equation in the form of plane waves. Then we find that

$$\Phi = A\exp(i\mathbf{k}_L \cdot \mathbf{x}) \tag{4.1}$$

is a solution to (3.4) if

$$\mathbf{k}_L \cdot \mathbf{k}_L = \frac{\rho\omega^2}{\hat{\lambda} + 2\hat{\mu}}, \qquad \mathbf{U} = i\mathbf{k}_L\Phi. \tag{4.2}$$

Here $\mathbf{k}_L \cdot \mathbf{k}_L$ is given by a complex number and hence the wave vector $\mathbf{k}$ need be complex. Although this does not reproduce the familiar situation of plane waves, it appears that the complex displacement vector $\mathbf{U}$ is parallel to the wave vector $\mathbf{k}_L$; for this reason we find it convenient to refer to solutions with potential of the form (4.1) as *inhomogeneous longitudinal waves.*The word inhomogeneous is a reminder of the complex character of the waves involved.

Similar considerations lead to the definition of *inhomogeneous transverse waves* as solutions to (3.5) generated by a vector potential of the form

$$\Psi = \mathbf{q}\exp(i\mathbf{k}_T \cdot \mathbf{x}); \tag{4.3}$$

here the (complex) wave vector $\mathbf{k}_T$ satisfies

$$\mathbf{k}_T \cdot \mathbf{k}_T = \frac{\rho\omega^2}{\hat{\mu}}, \qquad \mathbf{U} = i\mathbf{k}_T \times \Psi \tag{4.4}$$

To investigate inhomogeneous waves and their meaning, and to what extent they are determined by the material features of the given body, we consider first the phase factor $\exp(i\mathbf{k}\cdot\mathbf{x})$ that enters the expressions (4.2) and (4.4) of $\mathbf{U}$. Then we represent the wave vector through its real and imaginary parts $\mathbf{k}_1$ and $\mathbf{k}_2$, and we introduce unit vectors $\mathbf{n}_1$ and $\mathbf{n}_2$ such that

$$\mathbf{k} = \mathbf{k}_1 + i\mathbf{k}_2 = k_1\mathbf{n}_1 + ik_2\mathbf{n}_2.$$

Since

$$\exp(i\mathbf{k}\cdot\mathbf{x}) = \exp(-\mathbf{k}_2\cdot\mathbf{x})\exp(i\mathbf{k}_1\cdot\mathbf{x})$$

it follows that the condition

$$\mathbf{k}_1\cdot\mathbf{x} = c_1$$

yields planes of constant phase while

$$\mathbf{k}_2\cdot\mathbf{x} = c_2$$

yields planes of constant amplitude. Notice also that $\mathbf{k}_1$ is related to the velocity of propagation of the phase.

The material properties are insufficient for the determination of k_1 and k_2. On setting

$$\mathbf{k}\cdot\mathbf{k} = a + ib$$

we have

$$k_1^2 - k_2^2 = a, \qquad 2k_1k_2\mathbf{n}_1\cdot\mathbf{n}_2 = b \geq 0, \tag{4.5}$$

here the last inequality follows from the thermodynamic relations (3.6); in particular it will be shown to be responsible for the attenuation of the wave while propagating. Equations (4.5) can be solved for k_1 and k_2, provided $\mathbf{n}_1\cdot\mathbf{n}_2$ is given, since a and b are related to the material properties of the viscoelastic body and the angular frequency through (4.2) or (4.4). On the contrary, $\mathbf{n}_1\cdot\mathbf{n}_2$ is a purely geometric quantity but the indeterminacy is removed once one knows how the wave originates. For example this constant may be fixed by the initial conditions, whereas, at an interface, it is determined by standard continuity requirements. Finally, we observe that $b = 0$ for elastic solids and inviscid fluids, in which case inhomogeneous waves are allowed, but $\mathbf{k}_1$ and $\mathbf{k}_2$ are necessarily orthogonal.

To complete our discussion on inhomogeneous waves we observe that the displacement $\mathbf{u}$ is given by an elliptically-polarized wave [6, 9], even in the case of an inhomogeneous longitudinal wave. The geometric features of the polarization ellipse depend on the point $\mathbf{x}$. For example, the length of the axes decreases exponentially in the direction of $\mathbf{k}_1$ in that we have

$$\mathbf{u}(\mathbf{x} + 2\pi n\mathbf{k}_1/k_1^2, t) = \exp(-\pi nb/k_1^2)\mathbf{u}(\mathbf{x}, t)$$

for any positive integer n. On recalling that $b \geq 0$, we find an effective attenuation along the direction of propagation of the phase whenever $b \neq 0$, that is whenever $\mathbf{k}_1$ is not orthogonal to $\mathbf{k}_2$.

5. Behaviour at a plane interface

In this section we consider a plane interface S separating two half-spaces filled by an inviscid fluid and a viscoelastic solid. Our plan is to illustrate what happens when a homogeneous wave propagating within the fluid strikes the interface S. The results will constitute the basis for our analysis of scattering phenomena.

In analogy with the behaviour of interfaces between fluids and elastic bodies, it is assumed that at S three waves originate: a reflected (longitudinal) and two transmitted (longitudinal and transverse). These fields are constrained to satisfy standard junction conditions, namely continuity of the normal component of the displacement $\mathbf{u}$ and of the traction $\mathbf{Tn}$ where $\mathbf{n}$ denotes the normal to the plane.

To describe the results of our analysis it is convenient to introduce a Cartesian coordinate system with the z axis perpendicular to the interface and pointing towards the fluid. We denote systematically by $'$ any quantities pertaining to the solid medium ($z \leq 0$), while it is understood that no $'$ refers to the fluid ($z \geq 0$). It has been assumed that the incident wave propagates in the upper medium. Letting the wavenumber vector $\mathbf{k}^i$ of the incident (homogeneous) wave and the normal $\mathbf{n}$ be coplanar, we define the x axis as the intersection between the plane $(\mathbf{k}^i, \mathbf{n})$ and the interface, in the direction induced by $\mathbf{k}^i$.

The conditions of continuity of displacement and traction across the interface require continuity of the phase, which in turn means that the x components of the incident, reflected (longitudinal), and transmitted (longitudinal and transverse) wavenumber vectors are conserved; namely, with obvious meaning of symbols, we have

$$k^i_x = k_{Lx} = k'_{Lx} = k'_{Tx} =: \xi + i\nu. \tag{5.1}$$

According to our conventions we have $\xi \geq 0$. Equation (5.1) is a mathematical formulation of Snell's law in the case of dissipative bodies. To find the z components of the various wavenumber vectors we observe that they are given, up to a sign, by the principal values of the square roots

$$\sqrt{\mathbf{k}_{L,T} \cdot \mathbf{k}_{L,T} - (\xi + i\nu)^2} =: \varsigma_{L,T} + i\sigma_{L,T};$$

here the value of the scalar product is taken from (4.2) or (4.4), depending on whether we are considering a longitudinal or a transverse wave. The sign is chosen as $+$ or $-$ according as the wave we are considering is travelling in the direction of positive or negative z.

To avoid unnecessary complications we confine our attention to vertically polarized waves for which $u_y = 0$. Then we observe that the wave with vertical polarization involves only the y component of the vector potential, besides the scalar potential. Thus we find $U_y = 0$, $\Psi_x = 0$, $\Psi_z = 0$, and we let Ψ stand for Ψ_y.

The previous considerations hold for inhomogeneous waves at the plane boundary between two different media, independently of their specific nature. The material properties are involved in the determination of the complex amplitudes of the waves arising at the interface, through continuity of displacement and traction. In particular, denoting by Φ_-, Φ_+, Φ'_-, Ψ'_- the amplitude of the incoming, reflected (longitudinal) and transmitted (longitudinal and transverse) waves, we find [6]

$$\Phi_+/\Phi_- = \left\{4m\xi^2\varsigma_L[(\varsigma'_L + i\sigma'_L)(\varsigma'_T + i\sigma'_T) + q'^2] - (\varsigma'_L + i\sigma'_L)(k'_T)^4\right\}/D =: \mathcal{R}, \tag{5.2}$$

$$\Phi'_-/\Phi_- = -[4(\xi + i\nu)(\varsigma_L + i\sigma_L)q'(k'_T)^2]/D, \tag{5.3}$$

$$\Psi'_-/\Phi_- = \frac{\varsigma'_L + i\sigma'_L}{q'}(\Phi'_-/\Phi_-); \tag{5.4}$$

where

$$D = 4m(\xi + i\nu)^2(\varsigma_L + i\sigma_L)[(\varsigma'_L + i\sigma'_L)(\varsigma'_T + i\sigma'_T) + q'^2] + (\varsigma'_L + i\sigma'_L)(k'_T)^4, \tag{5.5}$$

and

$$q' = \frac{1}{\xi + i\nu}\left[(\xi + i\nu)^2 - \frac{1}{2}(k'_T)^2\right].$$

Based on the behaviour of inhomogeneous waves at a fluid-viscoleastic solid interface we have considered the following problems: possible existence and classification of surface waves [10]; determination of the *relaxation coefficient* in the case of an unknown relaxation function of exponential type, through an analysis of the dependence of the reflection coefficient (5.2) on ω [5]; scattering of a plane wave from a viscoelastic solid immersed in a fluid [11]. The second case yields a rather typical example of inverse problem, whose solutions has been given only numerically. The third problem is briefly discussed in the next section.

6. Scattering by an obstacle

Consider a viscoelastic obstacle surrounded by an inviscid fluid and let a time harmonic wave come from infinity and interact with the solid body. Our aim is to find an analytic expression for for the transmitted field at infinity, i.e. for the far field. The scattered field satisfies Helmholtz equation (3.7) within the fluid, but in general is not represented by a plane wave. Starting with Green's identity and using the Sommerfeld radiation condition we obtain an integral representation for the scattered field as

$$-4\pi\phi^s(\mathbf{x}_0) = \int_{\partial\Omega}\left[\frac{1}{r}\exp(ikr)\frac{\partial\phi^s}{\partial n} - \phi^s\frac{\partial}{\partial n}\left(\frac{1}{r}\exp(ikr)\right)\right] da \tag{6.1}$$

where $\partial\Omega$ is the boundary of the region Ω occupied by the viscoelastic solid; $\mathbf{r} = \mathbf{x} - \mathbf{x}_0$, $\mathbf{x} \in \partial\Omega$, is the position of the points on $\partial\Omega$ relative to $\mathbf{x}_0$; $\mathbf{n}$ is the outward unit normal.

The representation (6.1) is given an operative meaning when it is supplemented with the values assumed at $\partial\Omega$ by the uknown scattered field. In general this may be achieved by solving a suitable integral equation defined over the boundary $\partial\Omega$ [1]. However it seems that such a solution can only be found numerically, which prevents any qualitative discussion of the results. To find a way out of these difficulties it is usual to introduce simplifying assumptions. If the obstacle is regarded as sound-soft then the total pressure is taken to vanish on $\partial\Omega$, that is $\phi = 0$ on $\partial\Omega$ (Dirichlet boundary condition). If the obstacle is regarded as sound-hard then the normal component of the total velocity is taken to vanish on $\partial\Omega$, that is $\partial\phi/\partial n = 0$ on $\partial\Omega$ (Neumann boundary condition). Finally, if the normal velocity and the excess pressure are taken to be proportional on $\partial\Omega$ then we write $\partial\phi/\partial n + \eta\phi = 0$ on $\partial\Omega$, (mixed boundary condition) η being a complex-valued constant.

From the physical viewpoint, none of these boundary conditions can be considered as satisfactory. Since the scattered field is the result of the interaction of an incident wave with an obstacle, it is natural to think that the boundary conditions should consist of the continuity of the normal velocity and of the traction, as implied by the balance equations. So any integral formula involving only one of the quantities ϕ, $\partial\phi/\partial n$, and $\partial\phi/\partial n + \eta\phi$ on the boundary seems inadequate to model acoustic wave scattering.

One way of assigning the data at the boundary is through the *Kirchhoff approximation* [12, 13]. Really, under the Kirchhoff approximation various approximation techniques are meant, namely Kirchhoff's one and subsequent refinements. Usually a Kirchhoff-type technique is applied directly in the high-frequency limit or in the far-field approximation. Here we adopt a Kirchhoff-type approximation by following systematically the idea that any point of the reflecting surface behaves as though the surface were locally flat. Then the wave incident at a point is reflected according to Snell's law relative to the tangent plane. Moreover, when the obstacle surface is finite, the scattered field is assumed to be everywhere zero on the closing surface [11]. This is made formal in the following way.

Denoting by $\partial\Omega^-$ the illuminated region, we have

$$\partial\Omega^- = \{\mathbf{x} \in \partial\Omega,\ \mathbf{n}\cdot\mathbf{k}^i < 0\}.$$

Then the wave incident at a point of $\partial\Omega^-$ is reflected according to Snell's law relative to the tangent plane, that is the amplitude of the scattered field ϕ^s is related to that of the incident one through the reflection coefficient $\mathcal{R}$ given by (5.4). The normal derivative $\partial\phi^s/\partial n$ is determined through the condition $\partial\phi^s/\partial n = -i(\mathbf{k}^i\cdot\mathbf{n})\phi^s$ and the previous results. Thus the integral representation for the scattered field in the present Kirchhoff-type approximation reads

$$-4\pi\phi^s(\mathbf{x}_0) = \int_{\partial\Omega^-} \frac{\exp(ikr)}{r}\left[-ik\cos(\mathbf{n},\mathbf{n}^i) + (\frac{1}{r} - ik)\cos(\mathbf{n},\mathbf{r})\right]\mathcal{R}\exp(i\mathbf{k}^i\cdot\mathbf{x})\,da; \quad (6.2)$$

We are interested in determining the far field F, that is the field at large distances with respect to the obstacle. On approximating (6.2) to first order in $|\mathbf{x}|$, with $|\mathbf{x}| \ll |\mathbf{x}_0|$, we find

$$F = \frac{ik}{4\pi}\int_{\partial\Omega^-} \exp\left[ik\left(\mathbf{n}^i - \frac{\mathbf{x}_0}{r_0}\right)\cdot\mathbf{x}\right]\left(\mathbf{n}^i - \frac{\mathbf{x}_0}{r_0}\right)\cdot\mathbf{n}\,\mathcal{R}\,da. \quad (6.3)$$

As an application of the previous result we determine the far field generated by an oblique incident wave hitting a thin circular plate. Assuming cgs units, the region $\partial\Omega^-$ is identified with a disk of radius $l = 50$ cm., whereas the contributions from the lateral surface are disregarded. The solid material is to be copper and the surrounding fluid water. The speed of sound in water, say c, is given by $c = \sqrt{p_\rho} = 1.5\ 10^5$, whence it follows $k_L = \omega/c$. In the following we assume $\omega = 10^5$ and we get $k_L = 0.6667$. As to copper, an analysis of creep tests and the passage from Kelvin and Maxwell models to the stress functional yields the following estimate for annealed copper [14]:

$$\mu_0 = 44.44\,10^{10}, \qquad \lambda_0 = 103.70\,10^{10},$$

$$\dot\mu(\tau) = -\mu_0[1.567\,10^{-6}\exp(-0.1616\,10^{-5}\tau) + 1.565\,10^{-2}\exp(-0.5157\,\tau)],$$

and

$$\dot{\lambda}(\tau) = -\lambda_0[1.567\,10^{-6}\exp(-0.1616\,10^{-5}\tau) + 1.565\,10^{-2}\exp(-0.5157\,\tau)]$$

The density is taken as $\rho = 8.9$.

Introduce spherical coordinates with colatitude ϑ, longitude φ, origin at the center of the disk, and z axis parallel to the normal.

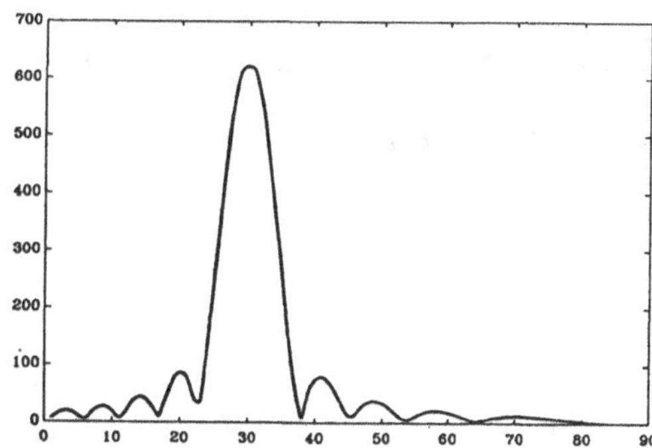

An incident (homogeneous) wave is considered with $\theta^i = 30°$. The figure shows the corresponding behaviour of $|F|$ against the observation angle ϑ while $\varphi = 0$. The modulus of the far field experiences a sharp maximum when the observation point is in the specular direction with respect to the incident wave, as it was expected.

Acknowledgement

The research leading to this work has been partially supported by the Italian CNR through the National Group for Mathematical Physics.

References

[1] Colton, D. and Kress, R. (1983) *Integral Equation Methods in Scattering Theory*, John Wiley, New York.

[2] Kleinman, R. E. and Roach, G. F. (1974) *Boundary integral equations for the three-dimensional Helmholtz equation*, SIAM Review **16**, 214-236.

[3] Colton, D. (1984) *The inverse scattering problem for time-harmonic acoustic waves*, Siam Rev. **26**, 323-350.

[4] Charlier, J. P. and Crowet, F. (1986) *Wave equations in linear viscoelastic materials*, J. Acoust. Soc. Am. **79**, 895-900.

[5] Caviglia, G., Morro, A., and Pagani, E. (1989) *Reflection and refraction at elastic-viscoelastic interfaces*, N. Cim. **12 C**, 399-413.

[6] Caviglia, G., Morro, A., and Pagani, E. *Inhomogeneous waves in viscoelastic media*, Wave Motion, in print.

[7] Fabrizio, M. and Morro, A. (1988) *Viscoelastic relaxation functions compatible with thermodynamics*, J. Elasticity **19**, 63-75.

[8] Achenbach J. D. (1975) *Wave propagation in elastic solids*, North-Holland, Amsterdam.

[9] Buchen, P. W. (1971) *Plane waves in linear viscoelastic media*, Geophys. J. R. Astr. Soc. **23**, 531-545.

[10] Caviglia, G. and Morro, A. *Surface waves at a fluid-viscoelastic solid interface*, Europ. J. Mech. A, in print.

[11] Caviglia, G. and Morro, A. *Curvature and dissipation effects on the scattering of acoustic waves* , submitted for publication.

[12] Ogilvy, J. A. (1987) *Wave scattering from rough surfaces*, Rep. Prog. Phys. **50**, 1553-1608.

[13] Bleistein, N. (1984) *Mathematical Methods for Wave Phenomena*, Academic Press, New York.

[14] Caviglia, G. and Morro, A, *On the modelling of dissipative solids*, Meccanica, in print.

THE USE OF SINGLE-SCATTER MODELS IN MEDICAL RADIATION APPLICATIONS

W. L. DUNN, A. M. YACOUT, AND F. O'FOGHLUDHA
Quantum Research Services, Inc.
Beta Building, Suite 340
2222 E. Chapel Hill-Nelson Hwy.
Durham, NC 27713
USA

ABSTRACT. We consider here the steady-state Boltzmann radiation transport equation, and appropriate boundary conditions, that apply to the irradiation of a uniform plane-parallel medium by a directed beam that is incident locally (at a point) on the surface of the medium. We construct single-scatter solutions for the exit angular and total fluxes, and show that these solutions can be applied to typical problems in medical radiation physics. Finally, we give some results of the application of a single-scatter model to a photon beam-modifier problem relevant to radiation therapy. The presentation stresses the application as well as the mathematical solution.

1. Introduction

The Boltzmann equation, with appropriate boundary and initial conditions, is used to model the transport of radiation through matter; various forms of the equation apply, depending on the radiation type and energy and on the geometry considered. For practical problems in medical radiation physics, such as X-ray imaging and radiation therapy, it is extremely difficult to solve the transport equation rigorously, and various approximate or numerical schemes are commonly employed.

One of the most common approximations is to consider the zeroth-order or uncollided solution, i.e., the solution that results from the assumption that radiation scattering is equivalent to removal. In this case, the so-called inscattering term drops out of the transport equation and a simple exponential solution results. This solution is the basic model of computed tomography (CT) algorithms [1], which are routinely used in medical image reconstruction, and of point-kernel transport models, which are used, for instance, in the buildup factor formulation for the design of shields for both radioisotope and accelerator therapy units.

We consider in this work the construction of first-order, or single-scatter, solutions of the radiation transport equation, and the application of these solutions to problems typical of medical radiation physics. The solutions arose in previous attempts [2,3] to construct semianalytic solutions to the classic searchlight problem of radiative transfer. These attempts led to the development of basic solutions for the angular flux [2] and for the total flux (zeroth angular moment) and current (first angular moment) [3]. For the considered searchlight problem, the single-scatter angular-flux solution is singular, a fact that was first recognized by Siewert [4].

R. Spigler (ed.), Applied and Industrial Mathematics, 335–342.

2. Analysis

We consider a plane-parallel medium of finite thickness, z_0, and infinite transverse extent. We use cylindrical coordinates (ρ,α,z) to designate position, and we let $\Omega = (\mu,\phi)$ represent direction of propagation, where $\mu = \cos\theta$ and (θ,ϕ) are the solid angle variables in spherical coordinates. We consider a monoenergetic beam incident in direction $\Omega_0 = (\mu_0,\phi_0)$ at a point on the z=0 surface of the medium. Without loss of generality, we can place the origin of the coordinate system at the point of radiation incidence, and write the appropriate transport equation and boundary conditions as

$$\left[\mu\frac{\partial}{\partial z} + \omega \cdot \frac{\partial}{\partial r} + \Sigma_t\right]\Psi = \int_0^\infty \int_{4\pi} \Sigma_s\left(E',\Omega'\rightarrow\Omega\right)\Psi'\; d\Omega'\; dE', \tag{1}$$

subject, for $0\leq\mu\leq1$ and $0\leq\phi\leq2\pi$, to

$$\Psi(\rho,\alpha,0,E,\mu,\phi) = \frac{1}{2\pi\rho}\,\delta(\rho)\,\delta(\mu-\mu_0)\,\delta\left(\phi-\phi_0\right)\delta(E-E_0) \tag{2a}$$

and

$$\Psi(\rho,\alpha,z_0,E,-\mu,\phi) = 0\,. \tag{2b}$$

In Eq. (1), Ψ is angular flux, $r = (\rho,\alpha)$ specifies position in the z=0 plane, ω is a two-dimensional vector of the transverse components of Ω, E_0 is the incident radiation energy, Σ_t is the total macroscopic cross section, Σ_s is the differential scattering cross section, and the form Ψ' denotes $\Psi(\rho,\alpha,z,E',\mu',\phi')$. We note that the uncollided solution results from letting $\Sigma_s = 0$, so that the right hand side of Eq. (1) vanishes.

Most imaging and shielding applications require the transmitted flux, and so we seek solutions for the flux on the far (unirradiated) side of the medium; however, our analysis applies, with only minor modification, to the reflected flux as well. We first decompose the desired solution into uncollided, single-scatter, and multiple-scatter terms, *viz.*,

$$\Psi(\rho,\alpha,z_0,E,\mu,\phi) = \Psi_0 + \Psi_1 + \Psi_n. \tag{3}$$

The solution Ψ_0 is clearly singular, but so is the solution Ψ_1. For instance, for the case of normal incidence (μ_0=1) and isotropic scattering ($\Sigma_s(E,\Omega' \rightarrow \Omega)=\Sigma_s(E)/(4\pi)$), we have shown [2] that

$$\Psi_0(\rho,\alpha,z_0,E,\mu,\phi) = \frac{\delta(\rho)\,\delta(\mu-1)\,\delta\left(\phi-\phi_0\right)\delta(E-E_0)\,e^{-\Sigma_{t0}z_0}}{2\pi\rho} \tag{4}$$

and

$$\Psi_1(\rho,\alpha,z_0,E,\mu,\phi) = \frac{\Sigma_{s0}\ \delta(\alpha-\phi)\ e^{-\Sigma_{t0}z_0}\, e^{-\Sigma_t\rho(1-\mu)/\sqrt{1-\mu^2}}}{4\pi\rho\sqrt{1-\mu^2}}\ ,\ 0<\rho<z_0\sqrt{1-\mu^2} \tag{5a}$$

$$\Psi_1(\rho,\alpha,z_0,E,\mu,\phi) = 0\ ,\ \text{otherwise}\ , \tag{5b}$$

where Σ_{s0} and Σ_{t0} are evaluated at the incident beam energy E_0 and Σ_t is evaluated at energy E. The singularity results from the fact that, for given r and Ω, a contribution is obtained only for the unique azimuthal scatter angle $\phi=\alpha$. The solutions of Eqs. (4) and (5) were obtained through a two-dimensional Fourier transform analysis of Eqs. (1) and (2). We note that Larsen [5] and Dunn [6] have independently verified the inverse dependence of the single-scatter solution on the factor $[1-\mu^2]^{1/2}$.

For the more general case of arbitrary direction of incidence (see Fig. 1) but still considering isotropic scattering, we can write the general solution, which is based on that of Siewert [4], for $0\le\mu\le1$, as

$$\Psi_1 = \frac{\Sigma_s\,\mu_0\, e^{-\Sigma_{t0}z_0/\mu}}{4\pi\rho^*\, D(-\mu,\phi)}\,\delta\left[\alpha^*-\chi(-\mu,\phi)\right] e^{-\Sigma_t\rho^*(\mu_0-\mu)/D(-\mu,\phi)}\ ,\ 0<\rho^*\le\frac{z_0\, D(-\mu,\phi)}{\mu\,\mu_0} \tag{6a}$$

$$\Psi_1 = 0\ ,\ \text{otherwise}\ , \tag{6b}$$

where

$$\rho^* = \left[\rho^2 + \left(\frac{z_0}{\mu_0}\right)^2\left(1-\mu_0^2\right) - 2\left(\frac{z_0}{\mu_0}\right)\rho\sqrt{1-\mu_0^2}\,\cos(\alpha-\phi_0)\right]^{1/2}, \tag{7a}$$

$$D(\mu,\phi) = \left[\mu^2\left(1-\mu_0^2\right) + \mu_0^2\left(1-\mu^2\right) + 2\mu\mu_0\sqrt{1-\mu^2}\,\sqrt{1-\mu_0^2}\,\cos(\phi-\phi_0)\right]^{1/2}, \tag{7b}$$

$$\cos\chi(\mu,\phi) = \frac{\left[\mu\sqrt{1-\mu_0^2}\,\cos\phi_0 + \mu_0\sqrt{1-\mu^2}\,\cos\phi\right]}{D(\mu,\phi)}\ , \tag{7c}$$

$$\sin\chi(\mu,\phi) = \frac{\left[\mu\sqrt{1-\mu_0^2}\,\sin\phi_0 + \mu_0\sqrt{1-\mu^2}\,\sin\phi\right]}{D(\mu,\phi)}\ , \tag{7d}$$

$$\cos \alpha^* = \frac{\left[\rho \cos \alpha + \frac{z_0}{\mu_0}\sqrt{1-\mu^2} \cos \phi_0\right]}{\rho^*}, \tag{7e}$$

and

$$\sin \alpha^* = \frac{\left[\rho \sin \alpha + \frac{z_0}{\mu_0}\sqrt{1-\mu^2} \sin \phi_0\right]}{\rho^*}. \tag{7f}$$

We can also construct a solution for the contribution to the total flux, Φ_1, due to single scatters in the medium. We treat the line segment along the intersection of the incident ray and the medium as a line source of exponentially decreasing source strength and use the standard statistical estimator of Schaeffer [7] to construct, for the general case of anisotropic scattering,

$$\Phi_1(\rho,\alpha,z_0,E) = \int_0^{z_0} \frac{e^{-\Sigma_{t0}z/\mu_0}\, e^{-\Sigma_t R}}{2\pi R^2}\, \Sigma_s\left(E_0,\mu_0 \rightarrow \mu\right) dz, \tag{8}$$

in which $\mu = (z_0-z)/R$, and

$$R = \left[(z_0-z)^2 + \rho^2 + z^2\frac{1-\mu_0^2}{\mu_0^2} - 2\rho\frac{z\sqrt{1-\mu_0^2}}{\mu_0}\cos(\alpha-\phi_0)\right]^{1/2}. \tag{9}$$

We have shown, for the case of isotropic scattering and normal incidence, that Eq.(8) gives results that are numerically the same as those given by integrating Eq.(6) over solid angle.

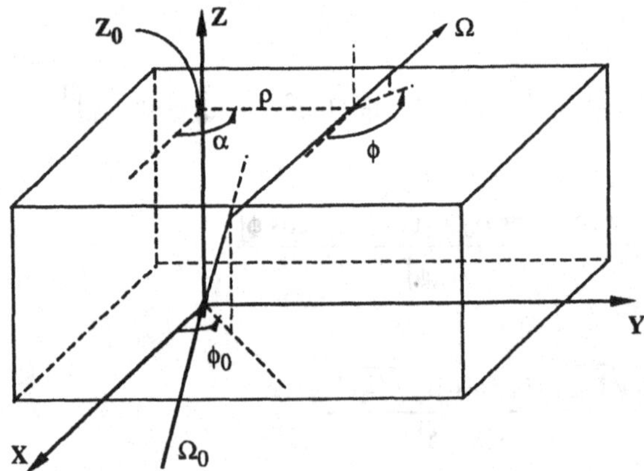

Figure 1. The single-scatter geometry.

3. Medical Applications

2.1 MEDICAL IMAGING

Standard CT models use the zeroth-order thin-beam attenuation model for radiation intensity, I, *viz.*,

$$I(S) = I_0 \exp\left[- \int_0^S \Sigma_t(E,s')\, ds' \right], \tag{10}$$

where s' and S are path length and total thickness, respectively, along the direction of propaga-ion and I_0 is the incident beam intensity. The simplicity of the model makes it mathematically convenient for use in image reconstruction algorithms, which may require the simultaneous solution of hundreds or thousands of equations, but neglecting scattering and spectral effects inevitably leads to inaccuracies or "artifacts" in the reconstructed images, which require post-processing by empirical correction factors [e.g.,1] or filters. The use of models that incorporate single-scatter effects might substantially improve the quality of reconstructed CT images.

2.2 PHOTON BEAM-MODIFIER DESIGN

A principal objective in radiation therapy is to obtain a uniform dose over a certain region while minimizing dose elsewhere. Photon-beam modifiers may be introduced to flatten the intensity distribution at the patient surface and to compensate for tissue deficits. We briefly investigate here the problem of photon-beam modifier design by application of a Monte Carlo technique that incorporates a single-scatter model. This solution is an extension of one reported earlier [8].

For a monoenergetic, collimated photon source projected onto a slab phantom through a modifier, as shown in Fig. 2, we seek the modifier dimensions required to produce a specified dose distribution over a region R in the phantom. In therapy applications, the modifier is usually thin (in terms of the radiation mean free path) and relatively far from the patient or phantom. Thus, the assumptions that multiple scatters in the modifier can be neglected and that the dose distribution in the phantom is insensitive to the precise depth of interactions in the modifier are valid, and use of a single-scatter model for the modifier is appropriate.

The dose distribution, D, in the phantom can be expressed in the general form

$$D(\mathbf{r},z) = \int_{\Delta\Omega} K[\mathbf{r},z,t(\Omega)]\, I_0(\Omega)\, d\Omega , \tag{11}$$

where $\Delta\Omega$ is the solid angle interval defined by the collimator, $I_0(\Omega)$ is the source intensity in direction Ω, $t(\Omega)$ is the modifier thickness along the ray emanating from the source in direction Ω, and K is a transformation function that accounts for photon transport (and is thus a Green's function of the transport equation). We seek to specify D and obtain $t(\Omega)$, and thus Eq. (11) defines an inverse problem for $t(\Omega)$ as a Fredholm Integral Equation of the First Kind.

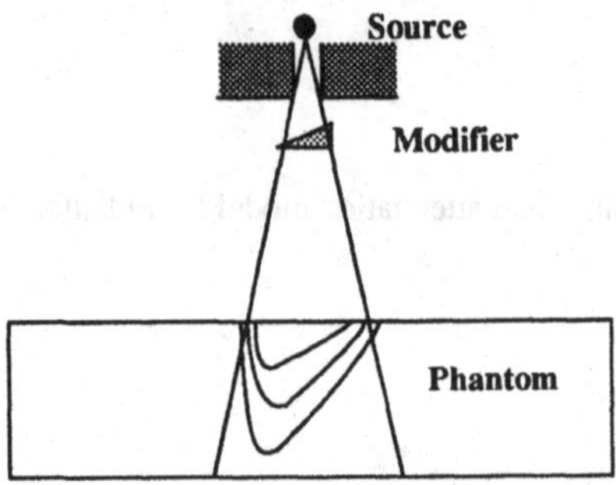

Figure 2. Schematic showing isodose curves in a phantom for a wedge-shaped modifier.

We first discretize the original beam into m solid angle elements, $\delta\Omega$, such that

$$\Delta\Omega = \sum_{j=1}^{m} \delta\Omega_j, \tag{12}$$

and consider the modifier to be composed of segments corresponding to the intersections of the $\delta\Omega$ beams with the modifier, the jth segment having thickness t_j. We then divide the region R into $M \geq m$ regions and define a dose vector $\mathbf{D} = \{D_k; k=1,2,...,M\}$, where D_k is the dose to the kth region. Thus, we seek a solution, $t = \{t_j; j=1,2,...,m\}$ that will produce a specified dose distribution, represented by $\mathbf{D}^*$.

To proceed, we form a vector α of transmission probabilities, the jth one being

$$\alpha_j = e^{-\Sigma_{t0} t_j}, \tag{13}$$

where Σ_{t0} is the total cross section of the modifier at the source energy. We can then construct a Monte Carlo simulation model for Eq. (11) in terms of the transmission probabilities. In constructing the simulation model, we first sample a particular source direction, Ω, within one of the $\delta\Omega_j$. Since we do not know the thickness of the modifier, we do not know the transmission probability and so we cannot sample to determine if the photon interacts in the modifier. Instead, we formally apply the Inverse Monte Carlo logic [8] and sample using an assumed transmission probability, call it α^*, and weight each history by the ratio of true to assumed probability density functions. This introduces the unknown transmission probabilities into the Monte Carlo scores, and a simulation model of the form

$$\mathbf{D} = \mathbf{A}\,\alpha + \frac{\Sigma_{s0}}{\Sigma_{t0}}\mathbf{B}\,\beta \tag{14}$$

results, where $\beta = \{(1-\alpha_j); j=1,2,...,m\}$. Here, $\mathbf{A}$ and $\mathbf{B}$ are simulation matrices such that the kj^{th} elements represent, respectively, the contributions to the dose at location k in the phantom due to photons that passed through element j of the modifier without undergoing an interaction (A_{kj}) and having suffered a single scatter (B_{kj}).

Rearranging Eq. (14) and substituting the desired **D*** for **D**, we obtain

$$\mathbf{D^*} = \mathbf{S} + (\mathbf{A} - \frac{\Sigma_{s0}}{\Sigma_{t0}}\mathbf{B})\,\alpha\ , \tag{15}$$

where **S** is a vector whose k^th^ element is simply

$$S_k = \frac{\Sigma_{s0}}{\Sigma_{t0}} \sum_{j=1}^{m} B_{kj}\ . \tag{16}$$

If M≥m, the unknown modifier thicknesses can be determined by solving Eq. (15) for α and then solving Eq. (13) for the t_j, i.e.,

$$t_j = -\frac{\ln \alpha_j}{\Sigma_{t0}}\ ,\ j=1,2,\ldots,m\ . \tag{17}$$

Thus, the design problem of Eq. (11) has been reduced to the algebraic form of Eq. (15), through the use of a single-scatter model for the modifier. We note that if M=m, Eq. (15) can be solved by matrix inversion, whereas if M>m, a numerical procedure such as least squares can be employed.

As an experimental test of the validity of the method, isodose distributions were measured in a water tank exposed to photon beams delivered by a kilocurie Co^{60} teletherapy source. The beam, which was essentially flat over the entrance field as shown by "in-air" traverses, was modified by inserting a wedge-shaped modifier. Given two sets of measured isodose curves, one for each of two modifiers, we constructed a solution in the form of Eq. (15) after making use of the approximation, which is applicable for a wedge,

$$t_j = t_1 + (j-1)\,\Delta\ ;\ j=1,2,\ldots,m\ , \tag{18}$$

where t_1 is an unknown constant and Δ is a constant increment that is related to the wedge slope. The relative dose values at a set of M=60 points were determined from the dose profiles provided, and the results indicated below were obtained through a least-squares solution of Eqs. (15) and (17).

Nominal Wedge Descriptor	Actual Wedge Angle	Calculated Wedge Angle
30°	7.2°	7.1°
60°	16.1°	16.9°

In generating these results, the modifier was divided into m=30 intervals, the cross sections were averaged over the two Co^{60} photon energies, and the constant α* was taken as 0.2. The fact that the results are better for the thinner wedge is consistent with the restriction to single scatters.

4. Conclusions

We have developed certain single-scatter solutions to the Boltzmann equation for a monodirectional beam incident on a uniform plane-parallel medium. These first-order solutions, which incorporate singularities inherent in the physics of the scattering process, can be used in various radiation physics applications in medicine. We have given an example of the use of single-scatter models in the construction of an approximate solution to an inverse problem basic to the design of modifiers for radiation therapy.

The solutions we have given are fairly general, but further analysis to develop more complete single-scatter solutions is warranted. Specifically, it should be possible to construct a general solution for the once-scattered contribution (to either the angular flux or the total flux) at any arbitrary internal or external spatial point, for a ray emanating in any direction from any point. Such a solution could provide very powerful first-order solutions to problems in imaging, shielding, and therapy, and could possibly form the basis for extensions of the back-projection estimator [6] for numerical calculations using the Monte Carlo method.

Because of the symmetry inherent in constructing the total solution to a radiation transport problem as the sum of contributions from transmission and scattering (both single and multiple), the numerical implementation of single-scatter solutions in practical situations would appear to lend itself to parallel processing. Since the form of the model is similar for each ray from a distributed source, or for each direction of scatter from a single scattering point, the users of single-scatter models may be able to capitalize on evolving parallel computer architectures.

5. References

1. Stonestrom, J.P., Alvarez, R.E., and Macovski, A. (1981) 'A Framework for Spectral Artifact Corrections in X-Ray CT', IEEE Trans. Biomed. Engr. BNE-28(2), 128-141.
2. Dunn, W.L. and Siewert, C.E. (1985) 'The Searchlight Problem in Radiation Transport: Some Analytical and Computational Results', Z. angew. Math. Phys. 36(4), 581-595.
3. Siewert, C.E., and Dunn, W.L. (1989) 'The Searchlight Problem in Radiative Transfer', J. Quant. Spectrosc. Radiat. Transfer 41(6), 467-481.
4. Siewert, C.E. (1985) 'On the Singular Components of the Solution to the Searchlight Problem in Radiative Transfer', J. Quant. Spectrosc. Radiat. Transfer 33(6), 551-554.
5. Larsen, E.W. (1988) 'Explicit Solution of Three-Dimensional Inverse Transport Problems', Trans. Am. Nuc. Soc. 56, 259-260.
6. Dunn, W.L. (1985) 'The Back-Projection Angular Flux Estimator', J. Comput. Phys. 61(3), 391-402.
7. Schaeffer, N.M. (1973) Reactor Shielding for Nuclear Engineers, TID 25951, U.S. Atomic Energy Commission.
8. Dunn, W.L., Boffi, V.C., and O'Foghludha, F. (1987) 'Applications of the Inverse Monte Carlo Method in Photon Beam Physics', Nuc. Instr. Meth. Phys. Res. A255, 147-151.

Mathematical methods for the SP well-logging

Li Tatsien Tan Yongji
Pen Yuejun Li Hailong
Department of Mathematics
Fudan University
200433 Shanghai, China

Abstract. In this paper, the mathematical model and method for spontaneous potential (SP) well-logging are described with some numerical examples.

§1. Introduction

In petroleum exploitation one often uses various methods of well-logging, among which the spontaneous potential (SP) well-logging is one of the most common and important techniques. Since positive ions and negative ions have different diffusion speed in a solution and the grains of mud-stone often adsorb positive ions, there is a steady potential difference called spontaneous potential difference on any interface of different layers. These potential differences cause a spontaneous potential field in the earth. After a well has been drilled, one puts a log-tool with a measuring electrode into the well and then measures the SP on the electrode. Raising the tool along the well-bore one gets the corresponding SP curve as shown in Fig. 1. The SP on the electrode varies with the change of the rock layer and it shows the osmotic layer clearly. As a usual convention, suppose that the layer is symmetric about the well axis and the central plane, we consider the problem on the domain

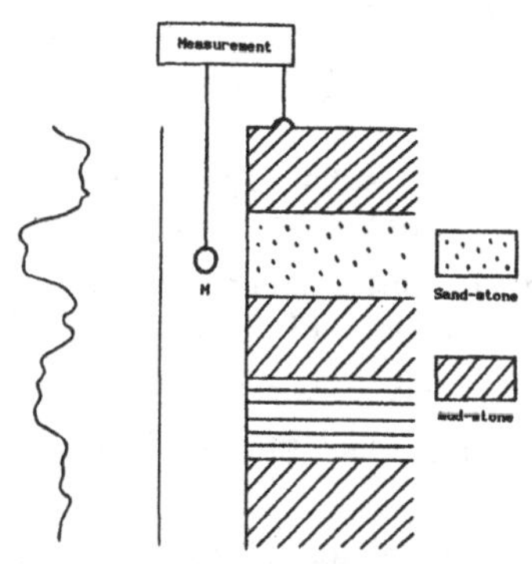

Fig. 1

$\{0 \le r = \sqrt{x^2+y^2} \le R,\ 0 \le z \le Z\}$ in the (r,z) plane, where R and Z are suitable large numbers. Suppose furthermore that the resistivity

R. Spigler (ed.), Applied and Industrial Mathematics, 343–349.

of the earth is piecewise constant (see Fig.2):

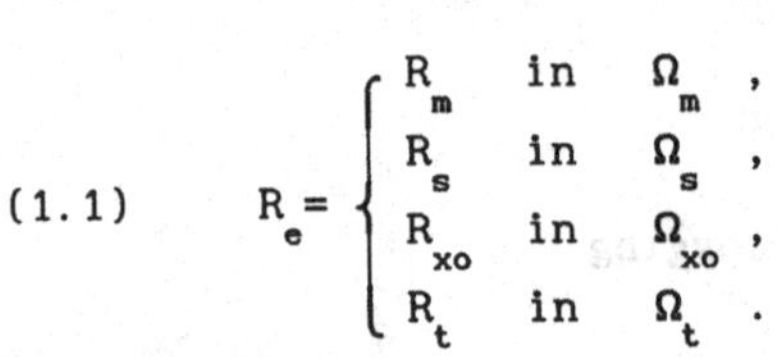

$$(1.1)\qquad R_e = \begin{cases} R_m & \text{in } \Omega_m , \\ R_s & \text{in } \Omega_s , \\ R_{xo} & \text{in } \Omega_{xo} , \\ R_t & \text{in } \Omega_t . \end{cases}$$

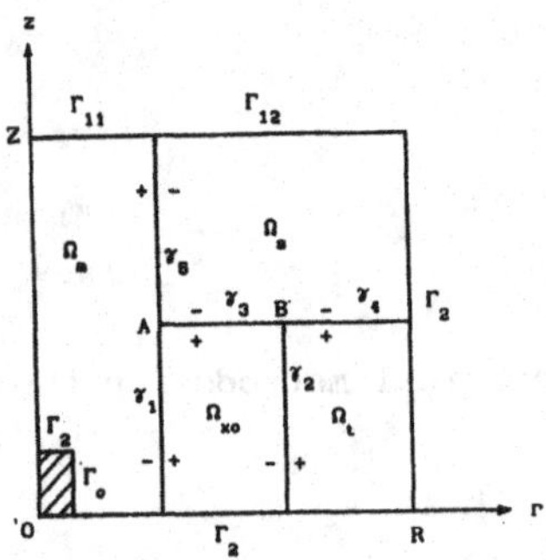

Fig.2

In Figure 2, the shaded part is the area occupied by the log-tool; Ω_m is the well-bore filled by mud; Ω_s is the enclosing rock; Ω_{xo} and Ω_t are two parts of the objective layer, which is the main measuring object. Since the objective layer is usually composed of porous sand-stone, the mud filtrate penetrates into the porous region and changes the resistivity in the domain Ω_{xo} which is then called invasion zone. If the geometric structure of the layer, the resistivity in each subdomain and the SP difference on each interface are all known, as a direct problem the SP $u(r,z)$ satisfies the following quasiharmonic equation in each submain

$$(1.2)\qquad \frac{\partial}{\partial r}\left[\frac{r}{Re}\frac{\partial u}{\partial r}\right] + \frac{\partial}{\partial z}\left[\frac{r}{Re}\frac{\partial u}{\partial z}\right] = 0 .$$

On each segment $\gamma_i (i=1,\dots,5)$ of the interface between layers of different resistivity, we have the conditions

$$(1.3)\qquad \begin{cases} u_+ - u_- = E_i \\ \left[\frac{1}{R_e}\frac{\partial u}{\partial n}\right]_+ = \left[\frac{1}{R_e}\frac{\partial u}{\partial n}\right]_- \end{cases} \qquad \text{on } \gamma_i (i=1,\dots,5) ,$$

where the subscripts + and - stand for the values on both sides of γ_i which have been prescribed in Figure 2 for fixing the idea, and the unit normal vector $\vec{n}$ takes the same direction on both sides of γ_i. (1.3) means that the electric current is continuous on γ_i, while the potential has a jump (the SP difference) on γ_i. As usual the SP differences $E_i (i=1,\dots,5)$ are suppose to be constnts.

On the surface Γ_o of the measuring electrode we have the following nonlocal boundary condition [1]

$$(1.4)\qquad \begin{cases} u = U \text{ (unknown constant) on } \Gamma_o , \\ \displaystyle\int_{\Gamma_o} \frac{1}{Re}\frac{\partial u}{\partial n}\, ds = 0 , \end{cases}$$

where $\vec{n}$ is the unit outward normal vector to Γ_o . On the well axis, the line of symmetry (the bottom boundary in Fig.2), the insulation surface of the log-tool and the distant boundary, the normal derivative of the SP vanishes. Denote these parts of the boundary by Γ_2 , we have

$$(1.5)\qquad \frac{\partial u}{\partial n} = 0 \qquad \text{on } \Gamma_2 .$$

Moreover, on the surface of the earth $\Gamma_1 = \Gamma_{11} \cup \Gamma_{12}$, we have

$$(1.6)\qquad u = E_5 \quad \text{on } \Gamma_{11} ; \qquad u = 0 \quad \text{on } \Gamma_{12} .$$

Here, condition (1.6) is compatible with the jump condition in (1.3) on γ_5 .

As a direct problem, the goal of the SP well-logging is to solve the boundary value problem (1.1)-(1.6).

§2. Reduction of the problem

First, the electrode can be simplified into a point since its size is relatively small and there is no electric current emitted from it. The domain in which the boundary value problem will be solved then reduces into a rectangular $\Omega = \{0\le r\le R, 0\le z\le Z\}$ shown in Fig.3. Next, we make a transformation

$$(2.1)\qquad \bar{u} = u - u_o ,$$

where

$$(2.2)\qquad u_o = \begin{cases} 0 & \text{in } \Omega_s , \\ E_5 & \text{in } \Omega_m , \\ E_1+E_5 & \text{in } \Omega_{xo} , \\ E_1+E_2+E_5 & \text{in } \Omega_t . \end{cases}$$

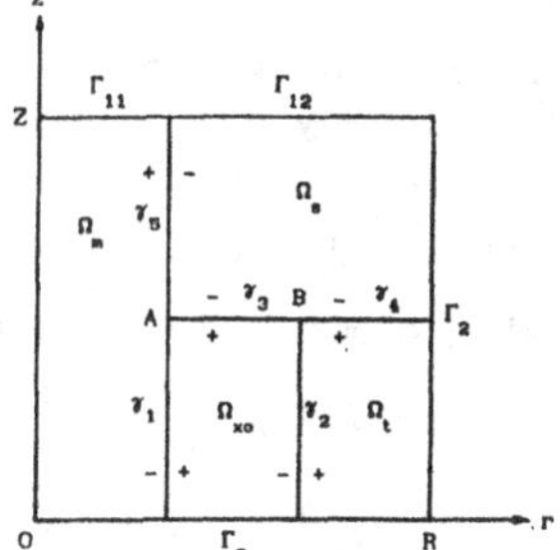

Fig.3

It is easy to see that $\bar{u}$ will be the solution to the following new boundary value problem:

$$(2.3)\qquad \frac{\partial}{\partial r}\left[\frac{r}{Re}\frac{\partial \bar{u}}{\partial r}\right] + \frac{\partial}{\partial z}\left[\frac{r}{Re}\frac{\partial \bar{u}}{\partial z}\right] = 0 \qquad \text{in each subdomain,}$$

$$(2.4)\qquad \bar{u} = 0 \qquad \text{on} \qquad \Gamma_1 = \Gamma_{11} \cup \Gamma_{12} ,$$

(2.5) $$\frac{\partial \bar{u}}{\partial n} = 0 \qquad \text{on} \qquad \Gamma_2 ,$$

(2.6) $$\begin{cases} \bar{u}_+ - \bar{u}_- = \bar{E}_i \\ \left(\frac{1}{R_e} \frac{\partial \bar{u}}{\partial n} \right)_+ = \left(\frac{1}{R_e} \frac{\partial \bar{u}}{\partial n} \right)_- \end{cases} \qquad \text{on } \gamma_i (i=1,\ldots,5) ,$$

where $\bar{E}_1 = \bar{E}_2 = \bar{E}_5 = 0$, $\bar{E}_3 = E_3 - E_1 - E_5$ and $\bar{E}_4 = E_4 - E_1 - E_2 - E_5$.

Evidently, the new problem is much simpler than the old one, since $\bar{u}$ satisfies a unified vanishing condition on boundary $\Gamma_1 = \Gamma_{11} \cup \Gamma_{12}$ and the discontinuity of $\bar{u}$ no longer happens on those vertical interfaces. Therefore the process of numerical solution becomes much simpler. But the most important advantage of the transformation is that it improves the accuracy of the numerical solution essentially, as we shall see in §4.

If the algebraic sums of the SP differences around the joint points A and B of the interfaces are just equal to zero, that is

(2.7) $$\begin{cases} \Delta_A \triangleq E_3 - E_1 - E_5 = 0 , \\ \Delta_B \triangleq E_4 - E_2 - E_3 = 0 , \end{cases}$$

we have $\bar{E}_3 = \Delta_A = 0$, $\bar{E}_4 = \Delta_A + \Delta_B = 0$, then $\bar{u} \equiv 0$. Hence, u_o given by (2.2) is the piecewise constant solution to the oringinal problem (1.1)-(1.6). In this case, the SP diferences are called to be compatible.

§3. Existence and uniqueness

In this section we shall consider the existence and uniqueness for SP boundary value problem. We shall mainly investigate the general compatible case in which the jumps on the intefaces are Lipschitz continuous functions $\bar{E}_i(s)(i=1,\cdots,5)$ and satisfy the following compatible condition:

(3.1) $$\begin{cases} \Delta_A \triangleq \bar{E}_3(A) - \bar{E}_1(A) - \bar{E}_5(A) = 0 , \\ \Delta_B \triangleq \bar{E}_4(B) - \bar{E}_2(B) - \bar{E}_3(B) = 0 . \end{cases}$$

We have

Lemma 3.1: the set

(3.2) $$\bar{U}_1 = \{\bar{u} | \bar{u} \in H^1 \text{ on each subdomain, } \bar{u} = 0 \text{ on } \Gamma_1 , \quad \bar{u}_+ - \bar{u}_- = \bar{E}_i(S) \text{ on } \gamma_i (i=1,\cdots,5)\}$$

is nonempty, if and only if (3.1)holds.
Set

$$(3.3) \qquad U_1 = \{v \mid v \in H^1 \text{ on each subdomain, } v = 0 \text{ on } \Gamma_1, \; v_+ = v_- \text{ on } \gamma_i (i=1,\cdots,5)\} = \{v \mid v \in H^1(\Omega), \; v = 0 \text{ on } \Gamma_1\} .$$

U_1 is a Hilbert space. $\bar{u} \in \bar{U}_1$ is a weak solution to problem (2.3)-(2.6), if

$$(3.4) \qquad \sum \iint \frac{r}{R_e} \left[\frac{\partial \bar{u}}{\partial r}\frac{\partial v}{\partial r} + \frac{\partial \bar{u}}{\partial z}\frac{\partial v}{\partial z} \right] drdz = 0 \;, \quad \forall \; v \in U_1 \; .$$

Since $\bar{U}_1 \neq \phi$, for any given element $\bar{u}_o \in \bar{U}_1$, setting

$$(3.5) \qquad u = \bar{u} - \bar{u}_o \in U_1 \;,$$

It follows from (3.4) that

$$(3.6) \qquad B(u,v) = L(v), \qquad \forall \; v \in U_1 \;,$$

Where

$$(3.7) \qquad B(u,v) = \iint_\Omega \frac{r}{R_e} \left[\frac{\partial u}{\partial r}\frac{\partial v}{\partial r} + \frac{\partial u}{\partial z}\frac{\partial v}{\partial z} \right] drdz \;,$$

$$(3.8) \qquad L(v) = - \sum \iint \frac{r}{R_e} \left[\frac{\partial \bar{u}_o}{\partial r}\frac{\partial v}{\partial r} + \frac{\partial \bar{u}_o}{\partial z}\frac{\partial v}{\partial z} \right] drdz \; .$$

Hence, by Lax-Milgram's theorem [2], it is easy to see that there exists a unique $u \in U_1$ such that (3.6) holds. Therefore, the original problem admits a unique weak solution $\bar{u}=\bar{u}_o+u\in\bar{U}_1$ and the standard finite element method can be applied to get an approximate solution [3].

Now we briefly discuss the case when the compatible condition fails. Lemma 3.1 shows that it is impossible to get a piece-wise H^1 weak solution for this case. So we try to set up a broader function class. For simplicity, we consider the situation of consant jumps instead of the jumps with Lipschitz continuity.
We have

Lemma 3.2: For any fixed p with $1<p<2$, the set

$$(3.9) \qquad \bar{U}_1 = \{\bar{u} \mid \bar{u} \in W^{1,p} \text{ in each subdomain, } \bar{u} = 0 \text{ on } \Gamma_1, \; \bar{u}_+ - \bar{u}_- = \bar{E}_i \text{ on } \gamma_i (i=1,\cdots,5)\}$$

is always nonempty.

From lemma 3.2, it is reasonable to expect an existence of the

$W^{1,p}$ weak solution for imcompatible SP problem. But for discontinuous resistivity, it has not been proved yet.

§4. Numerical results

As we have shown in §3, the piecewise H^1 weak solution exists so that the standard finite element method can be applied to solve the SP prlblem numerically only if the SP differences are compatible. Nevertheless in practice we often have to face the data for which the compatible condition fails, i.e.

$$(4.1)\qquad \begin{cases} \Delta_A = E_3 - E_1 - E_5 \neq 0 , \\ \Delta_B = E_4 - E_2 - E_3 \neq 0 . \end{cases}$$

To overcome this difficulty, we modify the SP differences E_i into $E_i(s)$ such that $E_i(s)(i=1, \cdots,5)$ are Lipschitz continuous functions composed of affine functions in suitable neighbourhoods of points A and B and original E_i elsewhere. By choosing the sizes of these neighbourhoods carefully, we can get satisfactory approximate solutions. For examples, taking two groups of data for SP differences (See Fig.5 and Fig.6 respectively) and taking R=1000m, z=800m, we solve problem (2.1)-(2.6) numerically by use of the technique described above and get the approximate SP on the well-axis respectively for a homogeneous layer model with

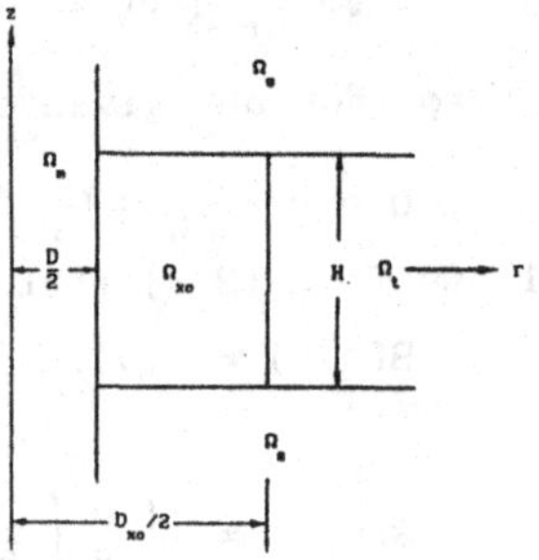

Fig.4

$$(4.2)\qquad R_s = R_m = R_{xo} = R_t = 1$$

and well radius $D_o/2$=125mm, invasion depth $D_{xo}/2$=625mm, the thickness of objective layer H=4m=4000mm (see Fig.4). For homogeneous layer with infinite radius, there is an explicit formula for the SP on the well-axis as follows [4]:

$$(4.3)\qquad u(o,z) = E_5 + \frac{1}{2}\Delta_A \left[\frac{2z+H}{\sqrt{D_o^2+(2z+H)^2}} - \frac{2z-H}{\sqrt{D_o^2+(2z-H)^2}} \right] + \frac{1}{2}\Delta_B \left[\frac{2z+H}{\sqrt{D_{xo}^2+(2z+H)^2}} - \frac{2z-H}{\sqrt{D_{xo}^2+(2z-H)^2}} \right] .$$

With the same data of SP diffrences, the exact SP curves can be computed by means of formula (4.3).

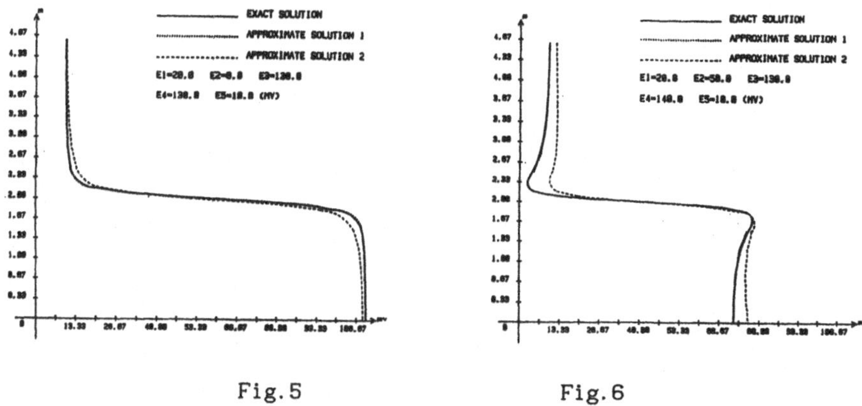

Fig.5 Fig.6

From Fig.5 and Fig.6, it can be found that the approximate SP curve (approximate solution 1 in the figures) almost coincides with the exact SP curve (exact solution in the figures) for both given SP differnces. Indeed, the maximal error between the approximate SP and the exact SP on the well-axis is only 1.27mv, and the relative error is less than 1.3%. As a comparison, another numerical solutions for the same cases have also been computed by solving problem directly without using transformation (2.1). Even though the similar modification technique to make SP differences compatible has been applied, the results, shown in Fig.5 and Fig.6 as approximate solution 2, are much worse. The biggest error attains about 18.7mv.

References

[1] Li Tatsien (1989), "A class of nonlocal boundary value problem for partial differential equations and its applications in numerical analysis", J.of Comp.& Appl.Math., 28,pp.49-62.
[2] Bears,L. et al (1964), "Partial differential equations", Wiley, New York.
[3] Li Tatsien et al. (1980), "Applications of finite element method to electric well-logs", Petroleum Industry Press, Beijing.
[4] Pen Yuejun (1988), A mathematical method in electric logs by the spontaneous potential, Comm. on Appl. Math. and Comput., 2, pp.35-43.

EFFECTIVE COMPUTATION OF THE SYMMETRIC LENS

C. DENSON HILL
Department of Mathematics
SUNY at Stony Brook
Stony Brook, NY 11794 USA

PERRY SUSSKIND
Department of Mathematics
Connecticut College
New London, CT 06320 USA

VINCENT GIAMBALVO
Department of Mathematics
University of Connecticut
Storrs, CT 06268 USA

ABSTRACT. We present an effective numerical scheme to compute, to any desired degree of accuracy, the shape of a symmetric lens of unit radius with given index of refraction and prescribed foci on the axis of symmetry. When the right focus is fixed at infinity there is a minimal focal length for the left focus. The scheme converges when the left focus is not too near the minimal focal distance. Interesting phenomena occur when the left focus is very close to the minimal focal length.

1. INTRODUCTION

Let an index of refraction $n_o > 1$ and a choice of on-axis foci A and B be given. For illustration in Figure 1 we take B at infinity. The problem is to find the symmetric lens having unit radius, so that all rays from A which impinge upon the lens should focus exactly at B.

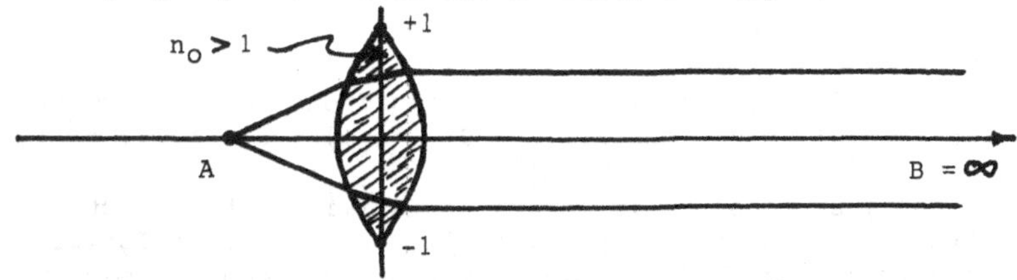

Figure 1. A symmetric lens with right focus B at infinity. The index of refraction outside the lens is equal to 1.

R. Spigler (ed.), Applied and Industrial Mathematics, 351–357.

It is assumed that the lens is rotationally symmetric about the x-axis. The term "symmetric lens" here refers to the fact than the lens not only has S^1 symmetry, but also a Z_2 symmetry with respect to reflection in the y-axis. We assume our lens has a "tip" at +1 and -1, and that at least near the tip the rays inside the lens behave in the manner shown in Figure 1. It is a different problem to consider rays which cross one another inside the lens.

We are concerned here with a problem in pure geometrical optics, so we use only Snell's law at the interfaces where a ray enters or exits from the lens. There is no spherical or thin lens approximation.

It is well known that for a configuration as in Figure 2, Snell's law implies that there is a conservation law (Fermat's principle) which says that the "optical path length" is constant.

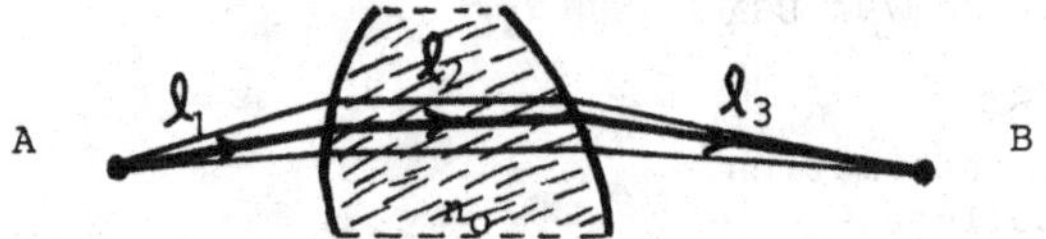

Figure 2. The optical path length $\ell_1 + n_0 \ell_2 + \ell_3$ = const.

It is actually this conservation law that is of practical interest in the design of a lens in our sense, since it means that in a wave train, wave fronts or pulses that are emitted in phase from A also arrive in phase at B.

2. FUNCTIONAL DIFFERENTIAL EQUATIONS

The requirement that Snell's law be satisfied on both sides of the lens, together with the Z_2 symmetry leads to a system of rather messy ordinary functional differential equations for three unknown functions $x(t)$, $x^*(t)$, $t^*(t)$ where $-1 \leq t \leq 0$. They have the form

$$\frac{dx}{dt} = f(t,x,x^*,t^*)$$

$$\frac{dx^*}{dt} = g(t,x,x^*,t^*)\frac{dt^*}{dt}$$

$$x^*(t) = -x(t^*(t))$$

x
(0,a) A
a > 0
(t,x)
-1
t*
t
t
(t*,x*)
$a \leq |b|$
b < 0
(0,b) B

Figure 3. The functional differential equations for the symmetric lens have the initial conditions $x(-1) = 0$, $x^*(-1) = 0$, $t^*(-1) = -1$.

It does not seem possible in general to eliminate the functional dependence from these equations by any simple trick or transformation.

Such functional differential equations may exhibit perhaps unexpected behavior, as can be seen from the example of the initial value problem $\frac{dy}{dt} = y(y(t))$ with $y(0) = 0$. It has at least 3 solutions: One is $y(t) = 0$ and there are two other complex ones of the form $y(t) = Ct^p$ for an appropriate choice of C and p.

For finite foci A & B the functions f and g appearing in Figure 3 are given explicitly by

$$f = \frac{[t^2+(x-a)^2]^{1/2} n_o(t^*-t) - [(t^*-t)^2+(x^*-x)^2]^{1/2} t}{[(t^*-t)^2+(x^*-x)^2]^{1/2}(x-a) - [t^2+(x-a)^2]^{1/2} n_o(x^*-x)},$$

$$g = -\frac{[(t^*)^2+(x^*-b)^2]^{1/2} n_o(t^*-t) + [(t^*-t)^2+(x^*-x)^2]^{1/2} t^*}{[(t^*)^2+(x^*-b)^2]^{1/2} n_o(x^*-x) + [(t^*-t)^2+(x^*-x)^2]^{1/2}(x^*-b)}.$$

When the right focus B is at infinity, the formula for f remains the same, but the one for g is replaced by

$$g = -\frac{n_o(t^*-t)}{n_o(x^*-x) + [(t^*-t)^2+(x^*-x)^2]^{1/2}} \qquad (b = -\infty).$$

3. RELATION TO OTHER WORK

First and foremost one should mention Sternberg [1],[2],[3] who has greatly inspired our interest in this kind of problem, and who, in a different formalism, has used functional differential equations that are equivalent to the ones in Figure 3. Friedman and McLeod [4] have considered a different but related problem for the symmetric lens. They formulate their functional differential equations somewhat differently, and they start at the base instead of having a tip, by prescribing the thickness of the lens where it intersects the axis. By means of the Schauder fixed point theorem in a complex Banach space, Friedman and McLeod prove a local existence theorem yielding a local real analytic solution that can be continued until the incoming ray from A is tangent to the surface of the lens. There is no a priori prediction of just when this breakdown will occur. Thus their result is entirely analogous to the classical Picard existence theorem for a nonlinear system of ordinary differential equations. Actually they treat directly the case of two finite foci, and pass to the case where one focus is at infinity by some limiting procedure. By its very nature their approach does not seem to be utilizable for numerical calculation, as it is a nonconstructable approach. Rogers [5] has treated the same problem as Friedman and McLeod by a different method which also does not appear to be well suited to numerical calculation. For related work see Ockendon and Howison [6] and especially the Ph.D. Thesis of van-Brunt [7] where an extensive bibliography can be found.

4. THE PRISM CONDITION AT THE TIP

It seems to be a more difficult problem to begin at the tip of the lens and to prescribe that the lens have unit radius (the problem for a lens of any given radius can be reduced to unit radius by rescaling). But it is this problem that is fundamental for the understanding of a stepped lens, as shown in Figure 4.

For such a lens the quantity dx/dt(-1) can be determined a priori, once the foci A & B are prescribed, by the prism condition at the tip:

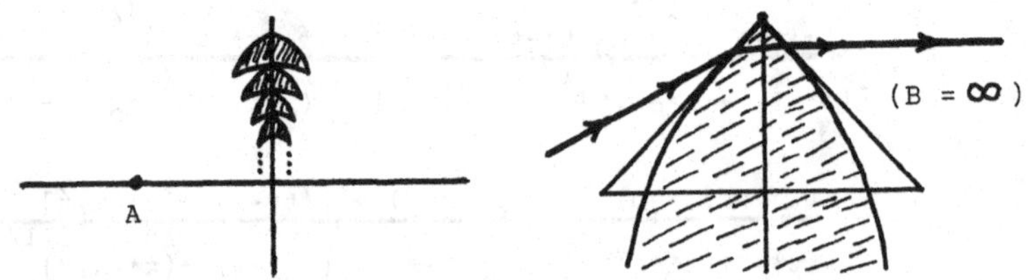

Figure 4. A stepped lens and a prism tangent to the tip.

The opening angle of the prism is determined by requiring that rays infinitesimally near the tip should focus correctly.

For such a lens there is a critical quantity a_0 (a_0 = dx/dt(-1)) determined by the condition that the incoming ray from A should be tangent to the prism at the tip. We shall call a_0 the minimal focal length, just to give it a name.

5. FAILURE OF STANDARD ITERATIVE METHODS

One may regard t as a time parameter in the equations of Figure 3, and adapt some explicit or implicit iterative finite-difference scheme, such as the Euler, Heun or Runge-Kutta methods to approximately solve them. The authors have not been able to make such schemes work. Typical results are depicted in Figure 5.

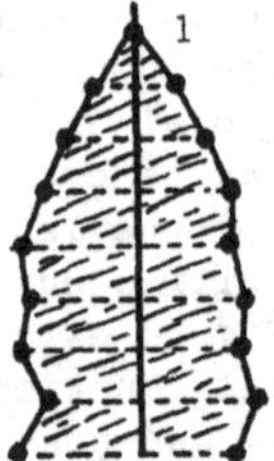

Figure 5. Typical results for large step size are on the left, and for a smaller step size on the right.

As the step size tends to zero, initial errors which are very small appear to grow exponentially, as numerical instability sets in, and the scheme diverges. The conservation law, that the optical path length should be constant, is badly violated.

6. A SIMPLE FIRST ORDER METHOD

This method can be thought of as being analogous to the Euler method for a first order system of ODE's, in the sense that a linear prolongation is made at each time step; the resulting approximate lens is piecewise-linear. But instead of having, at each time step, the new position determined by a fixed recursive formula based on the functional ordinary equations, the new position is determined from the basic geometry of the lens. The method can be more clearly understood not by writing down any specific formulas, but rather by looking at the picture on the left in Figure 5:

The lens is constructed out of slabs of height h (h will play the role of the mesh size), with a triangle at the tip, and with all other slabs being symmetric trapezoids: imagine a perfect version of the picture on the left in Figure 5. The symmetric triangular tip is determined uniquely by the requirement that the ray from A wich enters at the lower left vertex of the triangle should exit and focus exactly at B. The first trapeziodal slab is determined uniquely by requiring that the ray from A which enters the lower left vertex of the trapezoid should exit somewhere above on the right, and focus exactly at B. Now the ray may exit either from the triangular tip or from the side of the first slab. We check each case, working from the tip down, until a solution is found. Each case involves solving some nonlinear equation by, say, Newton's method. (Actually we often find it more convient to use the secant method.) At any rate the main point seems to be that the new position is found using essentially a Newton's method type of nonlinear iteration, rather than by using a fixed scheme obtained by replacing derivatives by difference quotients directly in the functional differential equations of Figure 3. This process is continued: the kth slab is determined by checking all the possible exit conditions from the tip down through the k-1 trapezoidal slabs above the one being constructed. The Z_2 symmetry is built into the procedure.

When A is taken sufficientlv far to the left ($a \gg a_0$) this simple scheme works quite effectively, and seems to be of practical value. But the path length error grows rather significantly as one proceeds down the lens. It is not clear just for which values of a this scheme converges, in some weak sense, and just for which values of a it behaves more like an asymptotic series.

But if A is taken too near the minimal focal length ($a \approx a_0$) then things definitely can go wrong: typically, after traveling almost all the way down to the base, the scheme crashes and there is a discontinuous jump in the path length error from A to B. Often a concave or a convex bubble begins to form around the time where the optical path length error takes a jump. Thus the behavior of this first order method for the case where A is too close to the minimal focal length, as $h \to 0$, is not clear.

Figure 6. The First Order Scheme crashes for A near minimal focal length.

Decreasing the mesh size h improves accuracy near the tip, but does not make the glitch or bubble go away near the bottom of the lens. A closer inspection of the situation suggests that there is a type of behavior as indicated in Figure 7.

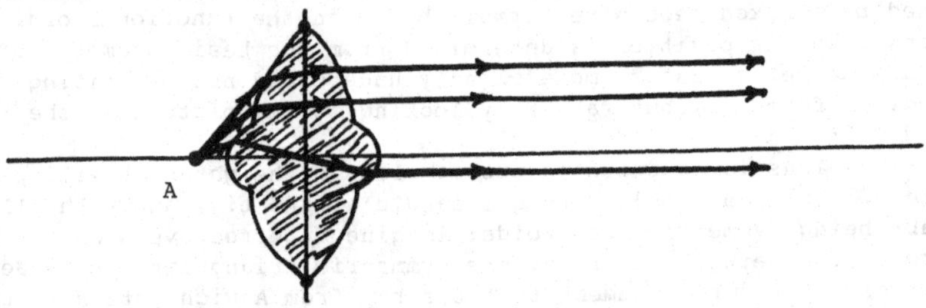

Figure 7. The solution has jumped onto another lens. The two lens have different optical path lengths.

7. A SECOND ORDER METHOD

The simple method described above can be improved significantly: Instead of constructing approximating lenses with piecewise linear sides, we use quadratic splines. This gives approximating lenses of class C^1, and the use of quadratic splines rather that cubic splines guarantees that convexity can be maintained during the course of the calculation. Once again we make symmetric slabs of height h, but use a triangular tip and successive trapezoids having quadratic sides as in Figure 8.

Figure 8. The use of quadratically curved sides.

The new tip is uniquely determined by requiring that the prism condition is satisfied at the vertex and that the ray which enters at the lower left vertex should exit somewhere and focus at B. Each successive trapezoidal slab is then determined by requiring that the ray from A which enters at the lower left vertex should exit somewhere above on the right, together with the requirement that the quadratic splines should fit together in a C^1 manner. Just as for the simpler method given above, a search from the tip down must be done to find the correct exit point for each ray. This involves solving some nonlinear equations by, say, the secant method k times to build the kth slab. Thus as $h \rightarrow 0$, the amount of computation required grows extremely rapidly.

But fortunately this method gives much greater accuracy that the simple first order method. Even for moderate mesh size h, the authors have obtained extremely nice results in which the path length error appears to stay within the limits imposed by the finite arithmetic of digital computation. At least this is the case for A sufficiently far from the minimal focal length a_0. For such A the second order method converges to a C^1 solution which, by Friedman & McLeod must actually be a real analytic curve.

But when A is very close to a_0 the situation is not so clear. The second order method sometimes does not crash as soon as the first order method; but on the other hand, even with it certain kinds of bubbles may eventually appear before reaching the bottom of the lens. There seems to be strong evidence that in this case the boundary value problem does not have a C^1 solution. For further work on these questions, we refer the reader to the forthcoming publications of the authors.

8. ACKNOWLEDGEMENT

The authors would like to acknowledge the support they obtained while working on this problem from the U.S. Navy-ASEE Summer Faculty Research Program at the Naval Underwater Systems Center in New London, CT.

9. REFERENCES

1. Sternberg, R. (1955) 'Successive Approximation and Expansion Methods in the Numerical Design of Microwave Dielectric Lenses', J. Math. and Phys. 34, 209-235.
2. Sternberg, R. (1963) 'On a Multiple Boundary Value Problem in Generalized Aplanatic Lens Design and its Solution in Series', Proc. Phys. Soc. 81, Part 5, No. 523, 902-924.
3. Sternberg, R. (1977) 'Optimising and Extremizing Nonlinear Boundary Value Problems in Lenticular Antennas in Oceanography, Medicine and Communications: Some Solutions and Some Questions', in Lakshmikanthan (eds.), Nonlinear Systems & Applications, Academic Press.
4. Friedman, A. and McLeod, J. (1987) 'Optimal Design of an Optical Lens', Arch. Rat. Mech. An. 99, 147-164.
5. Rogers, J. (1988) 'Existence, Uniqueness, and Construction of the Solution of a System of Ordinary Functional Differential Equations, with Application to the Design of Perfectly Focusing Symmetric Lenses', IMA J. Appl. Math. 41, 105-134.
6. Ockendon, J. and Howison, S. (1986) 'The Antenna Design Problem', RPI Report.
7. van-Brunt, B. (1989) 'Functional Differential Equations and Lens Design in Geometrical Optics', D. Phil. Thesis, St. Catherine's College, Oxford.

MATHEMATICAL MODELLING OF STRUCTURAL INDUSTRIAL PROBLEMS: METHODOLOGIES AND ALGORITHMS

L. BRUSA
CISE Tecnologie Innovative S.p.A.
P.O. Box 12081, I
20134 Milan
Italy

ABSTRACT. The paper describes CISE's activity for the development of application software for the solution of industrial problems. After some comments on the work organization, the main features of some structural computer programs are presented. As a consequence of the increasing need for computing power to investigate complex large scale problems, studies on parallel computing were undertaken. The work in this field is described and some of the obtained results are presented.

1. INTRODUCTION

CISE has been engaged for many years in the development of production codes for the solution of structural and thermal-fluid dynamic industrial problems.
At the beginning of this activity special attention was paid to the work organization, since the production of good quality software could not be considered as an extention of do-it-yourself programming. In fact the design of computer codes requires strict cooperation between engineers and numerical analysts/programmers, to obtain proper matching of physicals models and numerical methods and to produce well-documented and modular software.
This cooperation is not always easy but at present the development of application codes is performed by a multidisciplinary team who can cover the different aspects of the problems.
Special attention is paid to the modular implementation of numerical algorithms in order to obtain a library of high-level mathematical subroutines which can be used in different codes. At present the CISE library is mainly oriented to finite element applications.
One of the main advantages of this work organization is the possibility of reducing the cost of application software

R. Spigler (ed.), Applied and Industrial Mathematics, 359–367.

development by assembling the modules of a well-tested library to obtain new finite element programs or to extend the capabilities of the existing ones.
The fundamental requirements of industrial application codes are not only modularity but also flexibility, reliability and, last but not least, computational efficiency.
Simulation of complex phenomena, such as three-dimensional non-linear structural or fluid-dynamic problems, is heavily limited by the use of traditional computers and in many cases realistic results can be obtained by using parallel supercomputers only.
The computing power of these machines, however, can be fully delivered only if the software is designed to exploit their architectural features.
Practical experience shows that many present application codes, designed in a scalar computing environment, only partially fulfil this requirement.
As a consequence, the simple migration of traditional codes to supercomputers does not often produce the performance improvement necessary for large scale scientific computing.
For these reasons researches on parallel computing were started five years ago to develop parallel algorithms and programming methodologies for production of a new generation of application codes.
As an example of the type of problems examined, the paper describes the main features of some structural computer programs developed at CISE.
The codes concern the dynamic analysis of fluid-structure interaction problems and the simulation of highly non-linear thermal-mechanical phenomena.
The description includes a brief outline of the physical model and a more detailed presentation of the algorithms used to obtain efficient numerical simulations. Some results obtained by using parallel algorithms are also presented.

2. NUMERICAL SIMULATION OF STRUCTURAL PROBLEMS

2.1. Dynamic analysis of fluid-structure coupled systems

The dynamic behaviour of structures can be significantly affected by the presence of a fluid. The effect of the fluid is of primary importance for instance in the seismic analysis of many fluid-structure coupled systems, such as reactor internals immersed in liquid coolant or concrete dams coupled with the reservoir. These problems may be tackled using mathematical formulations based on the following assumptions:

- the structure has a linear elastic behaviour;
- the fluid is compressible and its motion is restricted to small displacements and low velocities, so that the fluid

pressure field is described by the wave equation.
By applying the finite element method the discretized equilibrium equations of the fluid structure coupled system are:

$$K\delta + C\dot{\delta} + M\ddot{\delta} = H^T p + f_1 \qquad \delta(t_o) = 0 \; ; \; \dot{\delta}(t_o) = 0 \; ,$$
$$Ap + B\ddot{p} = -\varrho H\ddot{\delta} + f_2 \qquad p(t_o) = 0 \; ; \; \dot{p}(t_o) = 0 \; . \qquad (1)$$

where δ and p are the vectors of the nodal structural displacements and fluid pressures; K, C, M are stiffness, damping and mass matrices of the structure; $H^T p$ stands for the load on the structure deriving from the fluid pressure field; A and B are the volumetric and inertial fluid matrices; $\varrho H\ddot{\delta}$ is the load on the fluid deriving from the structure; f_1 and f_2 are additional loads.
Eqs. (1) can be conveniently solved by the modal superposition method using the undamped modes of the coupled fluid structure system, obtained by solving the eigenvalue problem:

$$K\zeta - \omega^2 M\zeta - H^T\eta = 0 \; ,$$
$$A\eta - \omega^2 B\eta - \varrho\omega^2 H\zeta = 0 \; . \qquad (2)$$

where ζ and η are the displacement and fluid pressure modal shapes corresponding to frequency ω .
The mathematical model previously described has been implemented in the computer code INDIA[1,2].
Special attention was paid to the definition of techniques to improve or to control the accuracy of the modal superposition method and to the choice of the algorithm for the solution of the eigenvalue problem (2), which represents the most time consuming computational task.
Modal analysis provides an approximate solution of eqs. (1), because only a limited number of modes is considered. The modes taken into account are in general those relevant to frequencies below a cut-off value above which the excitation frequency spectrum is supposed to have a negligible value. This approximation is usually satisfactory for displacement computations. Yet, the error may be substantial for stresses near the constraints because their values can be significantly influenced by the high frequency modes.
In this case the "missing mode correction"[3] can be applied which is aimed at evaluating the contribution of the neglected modes without the need for their computation.
In the case of seismic excitation in which all the support nodes are subjected to the same acceleration, the code computes "modal masses" which can give significant information on the accuracy of the modal superposition

method. In fact the sum of the modal masses relevant to all modes is approximately equal to the sum of the masses of the structure and of the fluid. The code computes the sum of the modal masses relevant to the modes used for the response computation and compares this value with the mass of the fluid-structure system.
The computation of a limited number of modes relevant to the lower frequencies of problem (2) is performed by means of the simultaneous iteration method.
By this technique the sparse structure of the matrices in eq. (2) can be fully exploited so that the computational effort is comparable to that required by a separate vibration analysis of structure and fluid.
When a portion of fluid is completely surrounded by a structure or if a free surface in a gravity field is considered, the first frequency of problem (2) is zero and the standard simultaneous iteration method cannot be applied because a singular matrix should be inverted.
In these cases the Householder-Givens method is generally used but its computational cost is prohibitive for large matrices. To overcome this drawback and to retain the efficiency of the simultaneous iteration method, the algorithm was modified to allow the computation of a limited number of modes relevant to the lower non zero frequencies[4].

2.2. Analysis of highly non-linear thermal-mechanical problems

Simulation of highly non-linear thermal-mechanical phenomena is the basis for the study of many technological processes such as welding or quenching.
The interest for these analyses mainly lies in the computation of residual stresses whose presence in the material can cause many drawbacks such as brittle fractures, stress corrosion fractures, reduction of the fatigue strength.
Although from a theoretical point of view the thermal and mechanical problems are coupled, the real interaction between the two phenomena is very weak.
Therefore the two problems can be assumed to be decoupled and the analysis can be performed by sequentially using two different computer codes: the results of the thermal code being the input data for the structural one.
The thermal problem is highly non-linear owing to the presence of state and phase changes of the material. As a consequence special purpose computer programs are required, the usual non-linear thermal and structural codes being inadequate. To solve these types of problems ATEN[5] and ASFOR[6] codes has been developed.
ATEN-2D and ATEN-3D programs perform the computation of

thermal transients for two-dimensional and three-dimensional structures respectively.
The finite element method is used for spatial discretization, while two methods are available for time discretization, namely the one step implicit α-method, suggested by Hughes,[7] and a three-level unconditionally stable explicit scheme described in[8]. The efficiency of these two-methods is in many cases comparable but the choice of the initial values necessary for starting the explicit method may be critical for the stability of this algorithm.
The application of the α method requires the solution of the following non linear system at each time step:

$$[\frac{1}{\Delta t} B^* + A^*] U^{n+1} = [\frac{1}{\Delta t} B^* -(1-\alpha) A^*]U^n + F^* \qquad (3)$$

where U^n is the vector of nodal temperatures computed at $t = t_n$; A^*, B^*, F^* are the thermal conductivity matrix, the heat capacity matrix and the heat generation vector respectively, computed at the temperature $U^* = \alpha U^{n+1} + (1-\alpha)U^n$.
Parameter α must verify the condition $0.5 \le \alpha \le 1$ to satisfy the unconditional stability requirement.
Two iterative techniques can be used to solve eq. (3), based on the fixed point iteration and on a residual minimization respectively.
The structural equilibrium equations are solved by ASFOR-2D and ASFOR-3D programs. The 2-D version in available while the development of the three-dimensional code is now in progress.
The inertial effects being negligible, the finite element discretized equations at time $t = t_n$ assume the form:

$$K^n \delta^n = F^n + R^n; \; R^n = \int_V B^T C_E^n (\epsilon_p^n + \epsilon_{th}^n)dV \qquad (4)$$

where V is the volume of the structure; K is the elastic stiffness matrix; δ is the nodal displacement vector; F is the vector of the nodal external loads; B is the total displacement transformation matrix; C_E is the elasticity matrix; ϵ_p and ϵ_{th} are the plastic and thermal strain vectors, respectively.
The behaviour of the material during the change of state is simulated by means of a cut-off surface which limits the values of the tension stresses. The thermal strain vector includes the effects of the phase changes.

Eq. (4) is solved by the following iterative scheme:

$$K^n \delta_{(i)}{}^n = F^n + R_{(i-1)}{}^n \qquad (5)$$

Computation of vector R implies the evaluation of the plastic strain which can be obtained by solving an ordinary differential equation. Owing to the strong non linearities of the problem, this equation must be solved by using an implicit scheme which in turn requires a further iteration cycle.
ATEN and ASFOR codes use the same module to perform the solution of linear systems during the iterative processes.
The algorithm is designed to sequentially solve several linear system derived from the application of the finite element method on a domain in which not only the physical properties of the elements may change but also the geometry may be modified by adding and eliminating groups of elements. The numerical method is based on a frontal technique with substructuring.
The finite element discretization is performed on a virtual domain which includes all the elements present during the transient.
The transient is subdivided into intervals and for each interval the user must specify the group of elements (substructures) which are present and the possible changes in the boundary conditions.
The features of the algorithm help the user in simulating particular processes, such as material cutting or multipass welding, which otherwise would require the redefinition of the finite element discretization and of the node numbering.

3. PARALLEL COMPUTING ACTIVITY

The studies on parallel computing were started five years ago and are concerned with vector computers and shared-memory multiprocessors with vector architecture.
The activity was planned having in mind the necessity both of fully exploiting the potentiality of supercomputers, by developing "machine-oriented" software, and of preserving the existing codes as much as possible, owing to the huge investments done for their production.
In order to fulfil these two opposite requirements, a revision of CISE library routines, aimed at the solution of large scale mathematical problems, was undertaken to produce equivalent modules based on vector and parallel algorithms.
By replacing the modules based on sequential algorithms with the "machine-oriented" ones, the existing application

software may be only partially revised. At the same time its performance can be substantially increased because the mathematical kernel of the codes in general requires the major part of the global analysis running time.
An example of the results obtained is shown in Table 1 which reports the running times for the computation of the first seven vibration characteristics of a large structure. The computations were performed on CRAYX-MP/48 using INDIA and INDIAVET codes.
INDIA code, previously described, uses scalar algorithms and was installed on CRAY X-MP/48 without any modification except for the use of compiler directives to favour automatic vectorization when necessary. INDIAVET differs from INDIA program only for the use of a sparse linear equation solver designed for vector computers[9].

TABLE 1. Performances of INDIA and INDIAVET codes.

Number of D.O.F.	Max. front width of the stiffness and mass matrices	CPU time (s)	
		INDIA code	INDIAVET code
20004	740	9762	685

Present activity not only deals with the study of vector and parallel algorithms but with a deeper optimization of the global structure of the codes. This seems to be important especially for parallel computers because proper data structure and I/O management may reduce the interruption of concurrent computations with subsequent increase of performance.
Table 2 presents preliminary results obtained with ATEN-3D program on the parallel computer ALLIANT FX/80 with four processors. The optimization of the code is in progress and at present concerns the use of a sparse linear equation solver, designed for shared-memory multiprocessors[10], and a partial revision of the code structure.

TABLE 2. Performances of scalar and parallel versions of ATEN-3D program.

Number of D.O.F.	Max. front width of matrices	CPU time (s)	
		Scalar code	Parallel code
2459	137	1357	114

4. CONCLUSIONS

The paper has presented some results on the application software developed at CISE. The decision of undertaking a software development activity has been taken mainly to consolidate the cultural background on mathematical modelling and computational simulation. The main adavantages are: possibility of easily extending the capabilities of the existing codes to cover new application fields; better understanding of the major features of the commercial computer programs; greater ease in their use.

5. REFERENCES

1. Bon, E. et al.(1986) 'INDIA: Computer Code for Dynamic Analysis of Three-Dimensional Fluid-Structure Coupled Systems', CISE Report 3123.

2. Brusa, L. et al. (1988) 'Fluid-Structure interaction: Experimental and Numerical Analyses', in D. Ouzar and C.A. Brebbia (eds), 'Computer Methods and Water Resources: Computer aided Engineering in Water Resources', Springer Verlag, pp.447-458.

3. Powell, G.H. (1979) 'Missing Mass Correction in Model Analysis of Piping Systems', 5th SMIRT Conference, Paper K 10/3.

4. Brusa, L. et al. (1982) 'Vibration Analysis of Coupled Fluid-Structure Systems: a Convenient Computational Approach', Nuclear Eng. and Design 70, pp. 101-106.

5. Bon, E. et al. (1986) 'Numerical Solution of Thermal Processes with State and Phase Change: the Computer Code ATEN-2D' in Taylor C. et al. (eds), 'Non Linear Problems', Pineridge Press, pp. 951-963.

6. Belà, T. et al. (1989) 'Residual Stress Evaluation: the RES Computing System', 10th SMIRT Conference, pp. 251-256.

7. Hughes, T.J.R. (1977) 'Unconditionally Stable Algorithms for Non Linear Heat Conduction', Compt. Meth. Appl. Mech. Eng. 10, pp. 135-139.

8. Lees, M. (1966) 'A Linear Three-Level Difference Scheme for Quasi-Linear Parabolic Equations' Math. Comp. 20, pp. 516-522.

9. Brusa, L. and Riccio, F. (1989) 'A Frontal Technique for Vector Computers' Int. J. Num. Meth. Eng. 28, pp. 1635-1644.

10. Brusa, L. and Riccio, F. (1988) 'Substructure Technique for Parallel Solution of Linear Systems in Finite Element Analyses' in Evans D.J. and Sutti C. (eds), 'Parallel Computing: Methods, Algoritms and Applications', pp. 127-141.

AUTHOR INDEX

Subject Index